P9-EJU-721

MONSTER OF GOD

MONSTER OF GOD

The Man-Eating Predator in
the Jungles of History and the Mind

DAVID QUAMMEN

W. W. NORTON & COMPANY

NEW YORK * LONDON

The Publisher and author make grateful acknowledgment for permission to reprint the following:

From *The Epic of Gilgamesh*, translated by Andrew George (Allen Lane, the Penguin Press, 1999).
Copyright in this translation © Andrew George 1999. Used by permission of Penguin Press.

From *Myths from Mesopotamia: Creation, the Flood, Gilgamesh, and Others* by Stephanie Dalley.
Oxford World's Classics, Oxford University Press, 1998. Copyright © Stephanie Dalley 1998.
Used by permission of Oxford University Press.

From *Beowulf*, translated by Seamus Heaney. Copyright © 2000 by Seamus Heaney.
Used by permission of W. W. Norton & Company.
For U.K. rights please contact Faber and Faber Limited.

From "Being Prey," published in *Terra Nova*, Vol. 1, No. 3. Copyright © Val Plumwood.
Used by permission of Val Plumwood.

For information about permission to reproduce selections from this book, write to Permissions,
W. W. Norton & Company, Inc., 500 Fifth Avenue, New York, NY 10110

Manufacturing by The Haddon Craftsmen, Inc.
Book design by Lovedog Studio
Production manager: Andrew Marasia

LIBRARY OF CONGRESS CATALOGING-IN-PUBLICATION DATA
Quammen, David, 1948–
Monster of God : the man-eating predator in the jungles of history and the mind /
by David Quammen
p. cm.
Includes bibliographical references (p.   ) and index.
ISBN 0-393-05140-4
1. Dangerous animals. 2. Dangerous animals—Psychological aspects.
I. Title.
QL100.Q36 2003
591.6'5—dc21   2003007812

W. W. Norton & Company, Inc., 500 Fifth Avenue, New York N.Y. 10110
www.wwnorton.com

W. W. Norton & Company Ltd., Castle House, 75/76 Wells Street, London W1T 3QT

1 2 3 4 5 6 7 8 9 0

To
the good Dr. Byers and Heather
and to
E. Jean

AUTHOR'S NOTE

The idea for this book took hold of me during a visit to the Gir forest, in the state of Gujarat, India. The lions of Gir were its starting point. Because my observations of those lions were made over the shoulder of Ravi Chellam, of the Wildlife Institute of India, and were informed by his years of field experience, his published work, and his unpublished data and insights, I owe him a special debt of gratitude, gladly acknowledged here.

The many other scientists and nonscientists to whom I'm also indebted are mentioned in my acknowledgments, at the back.

CONTENTS

THE FOOD CHAIN
OF POWER AND GLORY

1

Great and terrible flesh-eating beasts have always shared landscape with humans. They were part of the ecological matrix within which *Homo sapiens* evolved. They were part of the psychological context in which our sense of identity as a species arose. They were part of the spiritual systems that we invented for coping. The teeth of big predators, their claws, their ferocity and their hunger, were grim realities that could be eluded but not forgotten. Every once in a while, a monstrous carnivore emerged like doom from a forest or a river to kill someone and feed on the body. It was a familiar sort of disaster—like auto fatalities today—that must have seemed freshly, shockingly gruesome each time, despite the familiarity. And it conveyed a certain message. Among the earliest forms of human self-awareness was the awareness of being meat.

Nowadays the term "man-eater" may seem retrograde. Some people who care about large carnivorous animals would like to be rid of it altogether. A case can be made that it's sexist: *man* eater. A case can be made that it's misleading and sensationalistic, that it tends to reinforce an excessively fearful attitude toward those species of which some individuals occasionally kill and eat a human. The first objection—about sexism in the use of "man"—is a semantic argument that I'll leave to

semanticians. The second objection—about sensationalism and fear—
is the one that's relevant here.

There's some validity to that discomfort with the term. The shock
value of the man-eater notion has been more than sufficiently
exploited. The bookshelves of my office, crammed with the literature
of predation, harbor some unabashedly lurid books with titles such
as *The Jaws of Death*, *Crocodile Attack*, *Man Is the Prey*, and simply
Attacked! The cover photo on that last one shows the gape of a grizzly
bear, lips pulled back in a snarl (or maybe it's a yawn, or what biologists call a *flehmen* grimace, which involves smelling) to display the big
canine teeth and the mottled pink-and-gray tongue. This is a closeup
so intimate you can almost look down the bear's throat and imagine
yourself in its stomach, along with the yampah roots and the huckle-
berries and the whitebark pine nuts. Three other books in my motley
collection carry the titles *Maneaters*, *Man Eaters*, and *Man-Eater*, each
with another snarly-faced cover image and one bearing the subtitle
*True Tales of Animals Stalking, Mauling, Killing, and Eating Human
Prey*. Also here at hand I've got *The Man-Eaters of Tsavo*, a classic of
the genre by Lieutenant Colonel J. H. Patterson, who as a construction
supervisor on the Uganda Railway in 1898 shot the two marauding
lions of his title. Adorning the cover of my Patterson paperback is—
can you guess?—the face of a snarling lion. The purpose of all these
menacing images, in accompanying the word "man-eater," is to market zoological melodrama. Less charitably, we might even call it predator pornography. Such melodrama, such toothy porn, gives a skewed
impression of the fraught, ancient relationship between large carnivores and the ubiquitous primate that, in moments of reckless desperation, they sometimes turn upon as prey.

Despite those objections, I'm unwilling to see "man-eater" erased
from our dictionaries altogether. Inflammatory or not, sexist or not,
the word serves a purpose within the English language. There's just no
precise and gender-neutral alternative that says the same thing with

the same degree of terse, atavistic punch. It deserves preservation because it labels and commemorates an elemental experience—the experience in which, on rare occasions, members of our species are relegated to the status of edible meat. It's a reminder of where we have stood, for tens of thousands of years, on the food chain of power and glory. That is, not always and indisputably at the top.

Who are these man-eaters? Defined broadly, the group would include a number of small-bodied, gregarious species of flesh-eating predators as well as some large-bodied, solitary species. It would include hyenas and jackals and wolves and feral dogs and piranhas, and probably some other species of mammals and fish that travel in packs or schools and sometimes gang up on a human. But those aren't the man-eaters that concern us here. What I'm asking you to contemplate are the psychological, mythic, and spiritual dimensions (as well as the ecological implications) of a particular sort of relationship: the predator-prey showdown between one dangerous, flesh-eating animal and one human victim. That relationship, I believe, has played a crucial role in shaping the way we humans construe our place in the natural world.

There's no collective scientific name, no formal category, for the animals I'm talking about. Lacking a better label, I'll call them *alpha predators*. They belong to a select but diverse group that transcends zoological boundaries to encompass some mammals, some fish, and some reptiles. In purely scientific terms, the grouping is artificial; it has no taxonomic or ecological basis. Its reality is psychological, as registered in the human mind. It includes the tiger (*Panthera tigris*), the brown bear (*Ursus arctos*), the great white shark (*Carcharodon carcharias*), the Nile crocodile (*Crocodylus niloticus*), the saltwater crocodile (*Crocodylus porosus*), the lion (*Panthera leo*), the leopard (*Panthera pardus*), the Ganges shark (*Glyphis gangericus*), the polar bear (*Ursus maritimus*), and the Komodo dragon (*Varanus komodoensis*), as well as a few other species. The cougar (*Puma concolor*) seems to be reemerging as a candidate. The African python (*Python sebae*), the reticulated python (*Python reticulata*), the anaconda (*Eunectes murinus*), and the jaguar (*Panthera onca*) might also qualify, plus several additional species of crocodilians and sharks. But that's about all. Big cats, some cartilaginous fish, a few reptiles, a couple of bears—it's a short, formidable list. What sets them apart from all other creatures, and places them in commonality with one another, is that each of these species has members big enough, fierce enough, voracious and indiscriminate enough to—occasionally—kill and eat a human. The danger they present to people is different from all other zoological dangers.

Elephants commit fatal tramplings every year, both in Africa and in Asia, but they don't feed on the victims. Bison and rhinos can be as lethal as runaway trucks, but they aren't carnivorous. Hippopotamuses, notwithstanding their vegetarianism, are dangerous to rural people who live and work along certain rivers. Hyenas will attack humans, but hyenas are social hunters, not solitary predators. Likewise, wolves have been known to attack humans in India and elsewhere, but they generally do so in packs, not as individuals. Cobras, mambas, and other poisonous snakes cause many human fatalities every year; scorpions and spiders rack up a few. And malarial mosquitoes could be considered the deadliest form of wildlife on the planet. But all those death-dealing animals fall outside the category I'm framing here. They're not maneaters. They're not alpha predators.

The alpha predators, and the responses they evoke, have transcended the physical dimension of sheer mortal struggle, finding their way also into mythology, art, epic literature, and religion. In Egypt there was Sekhmet, the lion-shaped goddess, famously bloodthirsty in her association with war, plague, and death. The sphinxes were lion-bodied creatures with human heads, sometimes also wings, whose disposition was more ambiguous than Sekhmet's. Not just in Egypt but throughout the ancient Middle East, lions were the preeminent predator and the first template for predatory icons, as reflected too in

Judeo-Christian scripture. They're mentioned at least 130 times (by the count of one patient scholar) in the Bible. A lion invoked in the Book of Job serves as a reminder, like so much in that richly glum tale, of disasters awaiting the prideful. "The roar of the lion, the voice of the fierce lion, and the teeth of the young lions are broken," reads Job 4:10. "The strong lion perishes for lack of prey, and the whelps of the lioness are scattered." The lions faced by Daniel, when he's sealed into their den, perform as arbiters of righteousness by declining to eat him; later, they obligingly devour the scheming satraps who put Daniel into bad light. In the first Book of Samuel, an obscure young buck named David, recommending himself to Saul as the man to take on Goliath, brags that in his shepherding days he often killed lions attacking his flock. He'll handle the big Philistine lummox, David promises, just like he handled those cats. And from Psalm 7 comes this:

> Oh Lord my God, in you I take refuge;
> Save me from all my pursuers, and deliver me
> Or like a lion they will tear me apart.

Were these, and the other 126 or so biblical lions, purely imaginary beasts? Were they phantasms concocted from distantly rumored archetypes? No, they were real lions cast in a pageant of holy parable. They were theological correlatives of local fauna.

In India there was Narasimha, the lion-headed god-man, venerated as the fourth avatar of Vishnu. In northern Australia, along the east coast of a big Aboriginal reserve known as Arnhem Land, the Yolngu people anciently recognized—and still embrace—intricate totemic connections with familiar native animals, one of which is the saltwater crocodile, respectfully addressed as Bäru. The Inuit of Greenland and northern Canada have their legends of the polar bear, including one about the she-bear who ate a pregnant woman but lovingly raised the unborn boychild she'd torn from the woman's womb. The Maasai of eastern Africa have their tradition of the formalized lion hunt, *ala-maiyo*, during which warriors prove their courage and win glory, with the first man to spear the lion claiming its mane and tail as trophies. On the island of Komodo, in the Indonesian archipelago, early humans buried their dead in shallow graves (digging deeper was presumably impossible on that rocky, volcanic landscape) and then topped the graves with cairns of large stones, evidently to discourage *Varanus komodoensis* from scavenging bodies. Among the Ainu of Japan's Hokkaido island the brown bear, *higuma*, was venerated as God of the Mountain; yet the Ainu practiced a rite in which a bear cub was hand-reared and then killed, "sent home," at the age of two or three.

Shark worship was the equivalent on certain Pacific islands, at least until Christian missionaries arrived to disapprove of it. Solomon Islanders erected stone altars and, according to one account, made human sacrifices to a shark god named *takw manacca*. Fijians performed a shark-kissing ceremony twice a year, partly to buy safety for their swimming areas.

In a mountainous region of west-central Sumatra, the Kerinci people have sacralized their view of *Panthera tigris* with a distinction between two forms of big cat, the physical tiger (*harimau biasa*) and the spirit tiger (*harimau roh*). The former is feared, the latter revered as an ancestral guardian and judge. When faced with dire trouble, a Kerinci might invoke *harimau roh* and become suffused with and emboldened by its tigerish energy. In the eastern Congo, there was a parallel notion about shape-shifting between humans and leopards, which invited abuse in the form of leopard-men, known as Anioto, who sometimes used clawlike weapons in order to blame their murders on real leopards. And among the Udege folk of southeastern Russia, whose traditional culture revolves around hunting and trapping, the sovereign beast of the forest is Amba, their name for what we call the Siberian tiger. Amba is sometimes considered a benign watcher and guardian, sometimes resented as a competitor for prey, but rarely feared as a direct threat. Leave her to her business, the Udege seem to believe, and Amba will leave you to yours.

Amba the tiger and Bäru the crocodile will both receive fuller attention in this book, along with the brown bear, an impetuous omnivore that inspires complicated apprehension not just on Hokkaido but across the northern reaches of three continents. So will a little-known subspecies of lion, *Panthera leo persica*, which survives today in only a single woodland enclave in western India. These four cases define the geographical itinerary that I've traveled in the course of my research: from the Gir forest (with its lions) in the Indian state of Gujarat, to the Arnhem Land Reserve (with its crocodiles) in northern Australia, to the Carpathian Mountains of post-Communist Romania (with their surprising abundance of brown bears), to the snowy Sikhote-Alin range (the last stronghold of Siberian tigers) in the Russian Far East. India, Australia, Romania, Russia—it's an errant and far-flung circuit, but big predators are where you find them. Although each of those situations is peculiar, and seemingly marginal to the world's larger concerns (even the larger concern of big predators where they're more famously known), each is also, in its own ways, emblematic and telling. Landscapes have the power to teach, if you query them carefully. And remote landscapes teach the rarest, quietest lessons.

My itinerary through the mythic and literary sources has been equally spotty and roundabout. I have gone back to *Beowulf* for a fresh look at the man-eater Grendel; have considered a few memorable monsters from ancient Babylonian poetry (Humbaba in *The Epic of Gilgamesh*, Tiamat in *Enuma Elish*); have turned to *The Saga of the Volsungs* from medieval Iceland for its portrait of a wormy dragon; and have swung forward into the future (at least as Hollywood imagines it) to connect those rough beasts with the extraterrestrial predator faced by Sigourney Weaver's character in the *Alien* movies. Does a flick like *Alien Resurrection* constitute literature? No, but it certainly partakes of the process whereby mythic perceptions and anxieties are reinforced.

Beowulf too, in its own day, was a form of popular entertainment.

Scripture is another matter. Scriptural monsters tend to have didactic purposes, not just lurid narrative roles. And certain instances combine the didactic with the lurid especially well. Stepping past the lions mentioned in Job 4 and elsewhere—which, though fearsome enough, are ordinary animals of no outsize scale or unearthly menace—I come to the real bugaboo of all the monsters portrayed in the Bible, the archetype of alpha predators: Leviathan.

2

Leviathan appears several times in the Old Testament and the Apocrypha, but nowhere more vividly than in chapter 41 of Job. Unlike the lions of Daniel and David (or the wolf and the leopard passingly mentioned in Isaiah 11, which fraternize gently with lambs and kids), Leviathan is bigger than life and, especially in Job 41, preternaturally dreadful: a long-toothed monster with armored skin, fire-blasting mouth, smoking nostrils, heart firm as stone, and a dim, smoldering gaze—or, more poetically, eyes "like the eyelids of the morning." According to one theory, this figure may have derived from an earlier Phoenician monster called Lotan, a huge, seven-headed dragon representing primordial chaos, conquered by the deity Baal. Within the Hebrew scriptures, it seems more firmly subordinate to divine power. Yahweh is Almighty, Leviathan is mighty, and then comes everyone else. The passage from Job 41 is a portrait of God's handmaiden predator, a creature that exists to remind humans—poor Job himself, and the rest of us—that we stand no higher than third on the food chain of power and glory.

Don't confuse this original Leviathan with a whale. In later times and looser usage, the word took on that meaning, but the biblical Leviathan was something weirder and scarier. It was a chimerical thing, part crocodile, part dragon, conjured up for spiritual purposes from materials of psychological and zoological reality. Isaiah 27, for instance, promises a judgment day when "the Lord with his cruel and great and strong sword will punish Leviathan the fleeing serpent, Leviathan the twisting serpent, and he will kill the dragon that is in the sea." That's a modern translation, from the New Oxford Annotated Bible; the King James Version, slightly less limpid but to me preferable for its broody resonance, calls Leviathan "the crooked serpent." In Psalm 74 God is praised and thanked for having "crushed the heads of Leviathan" and having given its multiheaded carcass as food to the people of the wilderness, a bit of emergency provisioning that must have seemed even less savory than manna. How many heads did Leviathan have? Probably seven, like Lotan, though in the definitive verses of Job 41 the beast seems reduced to just one. Despite that minor comedown, it's here that Leviathan gets its full due:

> Who can open the doors of his face? His teeth are terrible round
>   about.
> His scales are his pride, shut up together as with a close seal . . .
> Out of his mouth go burning lamps, and sparks of fire leap out.
> Out of his nostrils goeth smoke, as out of a seething pot or caldron.
> His breath kindleth coals, and a flame goeth out of his mouth.

The Lord is lecturing Job on the awful majesty of this animal. The Lord's purpose, at least initially, is to deepen Job's humility and reverence by reminding him that the world contains certain beings beside which man is ineffectual.

> Canst thou draw out leviathan with an hook? or his tongue with a
>   cord which thou lettest down?
> Canst thou put an hook into his nose? or bore his jaw through with
>   a thorn?

The answer, as Job well knows, is no. But the Lord pushes his point, in a tone of teasing irony:

> Will he make many supplications unto thee? will he speak soft
>   words unto thee?
> Will he make a covenant with thee? wilt thou take him for a ser-
>   vant for ever?
> Wilt thou play with him as with a bird?

Not hardly. No, the Lord says, Job as a sensible man will always bear in mind Leviathan's savagery, strength, daunting appearance, and invincibility, and will accordingly steer clear. Then comes the Lord's take-home message: "None is so fierce that dare stir him up: who then is able to stand before me?"

This much is consonant with the story's larger theme—Job afflicted, Job mortified, Job ineradicably pious. But the Lord gets a little carried away, rambling on into a celebration of Leviathan for its own sake. The sheer dastardly robustness of the creature seems momentarily to have sidetracked the Lord's (or at least the scriptural author's) imagination, much the way Satan stole off with Milton's in *Paradise Lost*. A good villain, as we know, is more interesting than an impeccable hero. Chapter 41 ends:

> The arrow cannot make him flee: slingstones are turned with him
>   into stubble.
> Darts are counted as stubble; he laugheth at the shaking of a spear.
> Sharp stones are under him: he spreadeth sharp pointed things
>   upon the mire.
> He maketh the deep to boil like a pot; he maketh the sea like a pot
>   of ointment.
> He maketh a path to shine after him: one would think the deep to
>   be hoary.
> Upon earth there is not his like, who is made without fear.
> He beholdeth all high things: he is a king over all the children of
>   pride.

Those children of pride bring God back onto point. The point is that Leviathan, though awesome and dreadful, owes its existence to the Lord himself. *I made this monstrous beast. No one is brave enough or reckless enough to tangle with it. Who then can stand before me?* The role of Leviathan, here in Job 41 and elsewhere in the Bible, was to keep people humble.

Meanwhile, real animals with big teeth and long claws were accomplishing the same thing. For as long as *Homo sapiens* has been sapient—for much longer if you count the evolutionary wisdom stored in our genes—alpha predators have kept us acutely aware of our membership within the natural world. They've done it by reminding us that to them we're just another flavor of meat.

Leviathan is one culture's sanctified exemplum of the alpha predator. Amba is another, Bäru another. Similar beliefs and traditions could be cited for jaguars, Nile crocodiles, cougars, reticulated pythons, and every other species of large predators that have shared landscape in uneasy proximity with human beings, sometimes treating them as prey. Clearly, the mythic dimension of alpha predators, as reflected and amplified in those beliefs and traditions, has played an important role in shaping the way our species construes its own place in creation.

Have lions, tigers, and bears made the dark forest scary? They have indeed, and in some ways it has been a good thing. Have crocodiles and sharks committed ugly, horrific acts of homicide and anthropophagy? Yes, and by doing so they've offered us a certain perspective. While we humans may be the most reflective members of the natural world, we're not (in my view, anyway) its divinely appointed proprietors. Nor are we the culmination of evolution, except in the sense that there has never been another species so bizarrely ingenious that it could create both iambic pentameter and plutonium. Throughout the course of the human story, one reminder of our earthly status has been that, at some times, in some landscapes, we have served as an intermediate link in the food chain. I'm not talking now about that scriptural food chain of power and glory, which the Lord so firmly impressed upon Job. I mean the literal food chain—who eats whom.

Those times and those landscapes are disappearing. Alpha predators face special troubles in the struggle for collective survival, because they live at low population densities (spaced apart by their own hunger and ferocity), require a high energy input per individual (especially the mammals among them, less so the reptiles and sharks), and need a large area of habitat to sustain a viable population. Many of them have vanished within the last couple of centuries—the Barbary lion, the Atlas bear, the Javan tiger, the California grizzly—and many other populations, subspecies, and whole species are in jeopardy. Because of their charisma—their handsome scariness and their thrill value— they'll probably long remain popular as zoo attractions. But it won't be the same. When they're lost from the wild, they're lost in the deepest sense. Though samples of their DNA may still exist, twitching innocuously in cages or test tubes, their survival as functional members of intact ecosystems is another matter.

Six billion humans currently weigh upon this planet. According to the most authoritative projection now available (from the United Nations Population Division), five billion more may be added within 150 years. With every additional child comes additional pressure on the productivity of landscape, turning forests into crop fields and rivers into gutters. Under pressure of this kind, alpha predators face elimination. Already they're being marginalized, diminished in number, deprived of habitat, leached of genetic vigor, constricted within insufficient refugia, extirpated here, extirpated there. One aspect of that trend is that they're becoming disconnected from *Homo sapiens* and we're becoming disconnected from them. Throughout our history as a species—tens of millennia, hundreds of millennia, going on two million years—we have tolerated the dangerous, problematic presence of big predators, finding roles for them within our emotional universe. But now our own numerousness, our puissance, and our solipsism have brought us to a point where tolerance is unnecessary and danger of that sort is unacceptable. The foreseeable outcome is that in the year 2150, when human population peaks at around eleven billion, alpha predators will have ceased to exist—except behind chain-link fencing, high-strength glass, and steel bars. After that time, as memory recedes and the zoo populations become ever more genetically attenuated, ever more conveniently docile, ever more distantly derivative from the real thing, people will find it hard to conceive that those animals were once proud, dangerous, unpredictable, widespread, and kingly, prowling free among the same forests, rivers, estuaries, and oceans used by humanity. Adults, except a few recalcitrant souls, will take their absence for granted. Children will be startled and excited to learn, if anyone tells them, that once there were lions at large in the very world.

ONCE THERE
WERE LIONS

3

The teak trees are bare of leaves and powdered with tawny dust. The acacias and banyans are dusted too. The whole forest is flocked like broccoli tempura, parched with heat and drought, thirsty for monsoon. It's late May, which in this part of the world is late summer. As dawn light begins seeping in, the tropical darkness fades to a palette of umber, dun, tan, and other dull variants of brown. Even the peacocks, asleep in their treetop roosts, tails folded, offer no flash of color. Night doesn't belong to them, not in the wild. Peacocks do business by visual display, whereas night stalkers rely on seeing before being seen, on acute hearing, and on smell—the smell of water, the smell of fear, the smell of blood.

Through this sepia landscape walks a sepia animal, ghostly, along the side of a forest road. It's an improbable creature—a lion, in a country famed for tigers. If someone were watching, it might seem translucent and incorporeal, holographic, projected here into western India by laser gimmickry. But no one is watching, and the animal is real. It has substance and heft. It's a native denizen of a place known as Gir, the last natural refuge of *Panthera leo persica*, the Asiatic lion.

The road dust, which is fine like ground coriander and inches deep, takes the marks of the lion's big, four-toed paws. Pug marks of such

size generally indicate an adult male, so picture him that way: a large
tom with a sparse mane. Maybe the mane is blackish, setting off his
face against all the brown. Imagine a loose, confident gait. His belly
hangs low, his shoulders rise and sink like pistons, each step is placed
soundlessly. If a lion strides through a forest and no one is there to see
him, is he still kingly? The pug marks say yes.

The very word "lion" carries kingly force. Richard I of England was
renowned as lionhearted, after all, not bearish. Fashionable people get
lionized, not tigerized. Early kings in India sat on a throne called *sin-
hasana*, the seat of the lion, when they dispensed justice. In America, it's
a flattering metaphor as applied to our cougar, *Puma concolor*, the
"mountain lion," which inhabits lowlands as well as mountains and
isn't a lion. (A cougar of either sex does roughly resemble a lioness, but
is actually a different sort of cat, more closely related to the cheetah.)
Europe once contained lions, and not just in Roman circuses. They're
gone with the Pleistocene but remain reverentially pictured on the
walls of Chauvet Cave, a recently discovered Paleolithic art site in
southeastern France. Chauvet is spectacular and perplexing on several
grounds, not the least being that it contains gorgeous images of lions,
more of them than are known from all other European art caves combined. The big cats portrayed at Chauvet represent a species commonly
called the cave lion and formally labeled, by various experts disagreeing
over fine points in the fossil evidence, with several different scientific
binomials, including *Panthera spelaea* and *Panthera atrox*. (I'll come back
to that little taxonomic question, and to the larger meaning of Chauvet
itself, near the end of this book.) But none of the experts denies that,
whether placed within *spelaea* or *atrox* or considered a subspecies of
Panthera leo itself, the cave lion was indeed a true lion, not a metaphor-
ical one, and very similar to lions as we know them today. Nor can it be
doubted, after one look at the images of Chauvet, that the artists who
painted them by torchlight, more than thirty thousand years ago, recognized a monarchical grandeur in those dangerous cats.

Of course most people think of lions as an African phenomenon, embodying the eat-or-be-eaten severity of the East African savanna, the bush country, or the southern African veldt in its noble quintessence. The Pleistocene lions of Europe are a matter of distant memory, ancient art, and paleontology. The lion in India is something else: an outlier, a relict, a modern actuality representing the wildly improbable notion that alpha predators might have a future on planet Earth.

This ghostly male, on the dusty shoulder of a dirt road, is a survivor descended from pioneers. His ancestors arrived here millennia ago by dispersal eastward from Asia Minor. Their ultimate source, and the source of all other lions in the strict sense of the word, was an aboriginal population of *Panthera leo* in Africa, first known from Tanzanian fossils dating back 3.5 million years. Within recent centuries, as humanity transformed the landscape with axes and firearms and plows, lions disappeared almost everywhere throughout what had been their Asian range. They survived only at Gir, a small remnant of forest on the Kathiawar Peninsula of western India, in the state of Gujarat, not far from the Arabian Sea. Most of that remnant is now statutorily recognized as the Gir Wildlife Sanctuary and National Park, under the management of the Gujarat Forest Department. The park, a strictly protected core of 25,900 hectares (roughly 100 square miles), is nested within the larger and less restricted sanctuary. Together they constitute the protected area—or PA, as it's labeled for short—encompassing 141,200 hectares (550 square miles) of rolling volcanic hills, dry forest, scrub, and scattered patches of savanna, an island of ecological richness surrounded by a sea of human dominion. The island is roughly as large as Oahu. Most of the land surrounding it has long since been deforested, cropped down by goats and hungry cattle, striated by plows, kiln-fired with the forces of tropical sun and drought upon naked soil, and picked clean by needy people. Gujarat contains forty-five million humans in an area the size of Nebraska, and amid them, within the Gir forest, about 325 lions.

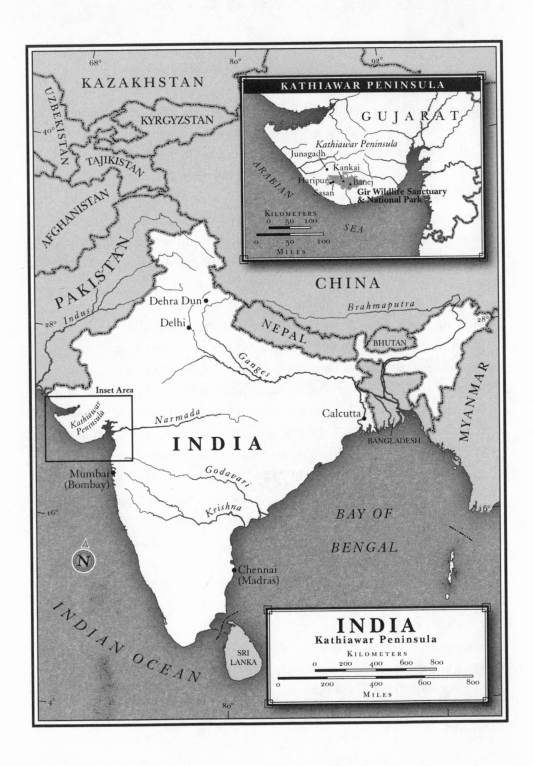

KATHIAWAR PENINSULA

GUJARAT

Kathiawar Peninsula

Junagadh

Kankai

Haripur

Sasan

Banej

**Gir Wildlife Sanctuary
& National Park**

ARABIAN

SEA

KILOMETERS
0   50   100

0   50   100
MILES

KAZAKHSTAN

UZBEKISTAN

KYRGYZSTAN

TAJIKISTAN

AFGHANISTAN

PAKISTAN

Indus

CHINA

Dehra Dun

Delhi

NEPAL

Ganges

Brahmaputra

BHUTAN

Inset Area

*Kathiawar
Peninsula*

Narmada

INDIA

Calcutta

BANGLADESH

MYANMAR

Mumbai
(Bombay)

Godavari

Krishna

BAY OF

BENGAL

N

Chennai
(Madras)

INDIAN OCEAN

SRI
LANKA

INDIA
Kathiawar Peninsula

KILOMETERS
0   200   400   600   800

0   200   400   600   800
MILES

The lions of Gir aren't constrained by any perimeter fence; they are merely marooned on their island by the ocean of humanity. They can leave at will. But when they do—as happens regularly, owing to reproductive increase and habitat constraints within the PA—they commit themselves to a life of scrounging in a far less hospitable landscape.

Roughly three dozen lions, a tenth of the total population, exist in satellite subpopulations outside the PA. They find cover among small patches of natural vegetation, agricultural fields, and tree plantations to the south, near the seacoast. Lacking wild prey there, they feed on cows, buffalo, dogs, and any other sizable domestic animals they find unattended. In rare cases, having made themselves unwelcome, lions are poisoned or electrocuted by irate villagers. Or a lion, dashing recklessly at the wrong moment, may be hit by a car, a truck, or a train. It's a precarious life, survivable for the short term but for the long term untenable. Sometimes in the course of surviving, or before dying, these satellite lions kill. During the last major outbreak of lion-human conflicts—it was the aftermath of a bad drought, in the late 1980s—most of the lion attacks against people occurred in a peripheral zone outside the PA. To cross the sanctuary boundary and take up residence in the agricultural countryside is, for a lion, an act of desperation that carries mortal risks.

This strolling male, who left his footprints in road dust, isn't so reckless as that—at least not today. He's headed north and east, away from the PA perimeter, back into the security of its forest. For him the dirt track is just one segment of a convenient path leading toward localized prospects of what he wants and needs: food, water, mating, and comfy repose. Having traveled along the roadside under cover of darkness, at first light he turns off and melts into the woods.

I arrive, with all my questions and notions, about half an hour later. By then he's gone. What remains are his pug marks, impressive but transitory. The dust is so light that a person could almost erase them with a sneeze.

4

Once there were lions across southwestern Asia. There were lions in Syria. There were lions in Mesopotamia, along the bottomlands of the Tigris and Euphrates rivers. There were lions in the part of the Persian empire that today is coastal Iran. There were lions in Baluchistan, an area now known as southern Pakistan. There were lions all the way into India—from the Sind borderlands in the west to as far east as Palamau, within several hundred miles of Calcutta. Although they don't seem to have crossed below the Narmada River, which cuts the subcontinent at about midpoint, they spread themselves well across India's northern half. There were lions, quite a few of them, in the outskirts of Delhi.

These aren't fossil records of some paleontological lionlike precursor. In saying "lions here, lions there," I'm referring to historical records from humans who saw, lived among, killed, and in some cases were killed by *Panthera leo,* the species you picture when you hear the word "lion." *Panthera leo* as it is commonly known, with its flowing mane, its noble nose, its nocturnal roars on a thorn-tree savanna, nowadays seems inseparable from Africa. But the fact is, it *was* separable. Laboratory measurements of genetic divergence suggest as much as two hundred thousand years' worth of separation between the Gir lions and their African relatives—a significant degree of genotypic difference between them, though not quite enough to indicate distinct species. Hence the subspecies designation: *Panthera leo persica.* The founding generations of that subspecies included restless lions who colonized eastward as far as the Ganges valley. For centuries, for millennia, they seem to have made a nice living there, feeding on deer, antelope, and boar.

There were lions, as I've mentioned, in Palestine and Egypt, where they played charismatic roles in iconography and scripture. There were lions in Turkey, Macedonia, and Greece. Herodotus reported that in 480 B.C., when Xerxes of Persia marched his army through the Greek peninsula, lions "came down from their haunts at night" to attack his baggage train. That report mentions a curious fact, one that puzzled Herodotus himself: The lions ignored Xerxes' men, ignored too his horses and other beasts of burden, and preyed exclusively on his camels.

The particular area where Xerxes had trouble, between the rivers Nestus and Achelous, at that time teemed with lions, according to Herodotus. Elsewhere in Greece, the writer claimed, there weren't any. None whatsoever? Well, it's hard to prove a negative, and Herodotus was after all the father of history (or of lies, depending on your viewpoint) but certainly not of biogeography. Anyway, another source reports that by A. D. 100, if not earlier, lions were entirely absent from Greece. Civilization rose, lions fell.

That negative correlation prevailed throughout these regions. With the growth of towns and cultures and kingdoms, the spread of agriculture and iron weaponry, lion populations began to disappear. Crop production was one problem, shrinking lions' habitat as fast as axes and plows could move. Pastoralism was another problem, putting livestock in competition with native prey animals and thereby tempting or forcing lions into conflict with the humans who tended the herds (such as young David, before Goliath and kingship). Hunting was another factor driving down lion populations, one that grew more severe as weapons improved. Emerging traditions of heroism, vainglory, and royal ostentation exacerbated that factor.

In Egypt, where lion hunting was a prerogative of the pharaohs, a decorated chest from the trove of Tutankhamen portrays young Tut in action, with seven lions already bagged and his arrow about to strike an eighth. An engraved scarab totem from the reign of Amenophis III tells of that ruler's having killed 102 lions. Among the Assyrian kings also, lion killing passed as a voucher of majesty. In an inscription from the seventh century B.C., one monarch boasted: "I am Ashurbanipal, King of the World, King of Assyria. For my pleasure, and with the help of the God Ashur and the Goddess Ishtar, Mistress of the Battle, I pierced a wild lion with my spear." Several centuries earlier, an alabaster relief from a palace wall in the city of Calah depicted Ashurnazirpal II as an archer, in a chariot, aiming an arrow toward a wounded, screaming lion. The lion is handsome and lifelike as well as ferocious, carved with knowing anatomical appreciation. In a related inscription, Ashurnazirpal gloats at having killed this many elephants, that many wild bulls, and adds that "with a stout heart I caught 15 strong lions in mountains and forests." Whatever he means about having "caught" lions, we can be sure that it wasn't to take DNA samples and release them unharmed. Earlier still, an Assyrian ruler named Tiglath-Pileser hit a high note of self-congratulation: "On the order of Ninurta, my patron, I have, with a brave heart, killed 120 lions in a terrible fight, facing the animals on foot. I have also killed 800 lions from my war chariot." Although the trend through time seems to have been downward—from Tiglath-Pileser's 920 lions, to Ashurnazirpal's 15, to Ashurbanipal's one—we can only wonder whether that represents a decline in the proficiency of Assyrian royal hunters, a relaxation of the need to exaggerate, or a decreasing supply of lions.

During the late Roman republic and the early empire—by which time Italy had no lions of its own—captive animals were brought from Syria, Mesopotamia, and Africa to play their doomed roles in garish entertainments. According to Pliny the Elder, a minor official named Quintus Scaevola was the first to stage a fight between lions at Rome, and others soon outdid him. Lucius Cornelius Sulla, a cold-hearted but capable man who eventually ruled as dictator, offered the people a blood show involving "100 maned lions." Maned lions—that is, mature males—seem to have been the premium attraction. Pompey the Great topped Sulla with a series of spectacles putting six hundred lions (if you trust Pliny's account), or at least five hundred (according to Plutarch's short biography), into the Circus. Pliny noted that, of the six hundred, 315 were maned. Plutarch was more specific about what happened: "Five hundred lions were killed in the course of these shows, but the main event, a truly terrifying sight, was a battle between elephants." Maybe the numerical discrepancy implies that one hundred individuals from a six-hundred-lion pool *weren't* killed but were spared—at least long enough to be available for later shows, such as the one Julius Caesar promoted, with four hundred lions. Then there was Augustus, who as "ruler of the world" as it was then known, or known to him, presented celebrations of his own augustness featuring a thousand gladiators and thirty-five hundred animals, including more hundreds of lions. Rome in those years, for *Panthera leo*, was what biologists today would call a population sink.

It couldn't last, and it didn't. A scholar of animal-human relations named Juliet Clutton-Brock reports that, despite a Roman ban on gladiatorial contests instituted in A.D. 325, "the 'amusements' with wild animals continued, even though lions were becoming scarce." She adds that a later decree, promulgated in A.D. 414, guaranteed everyone "the right to kill lions" as necessary for personal safety, but specified that systematic hunting and commercial trading of the animals required a license. What she doesn't explain is, who issued the licenses? By that time Rome itself had been sacked by Alaric and the empire was collapsing inward, a reversal that presumably offered some respite to wild populations of lions in the provinces, since Alaric's Visigoths, as well as the Huns and the Vandals and the rest of the newcomer hordes, all had other forms of violent adventure with which to amuse themselves. This sort of inverse relationship between centralized, imperial power and the health of alpha-predator populations on the colonized landscapes— lions in Roman Syria, tigers in British India, crocodiles in British Australia—is a larger matter to which I'll return. For now it's enough to note that *Panthera leo* survived, famously in Africa, more obscurely and marginally in southwestern Asia.

By the twelfth century the lion was extinct or nearly extinct in Palestine, though in Syria and farther eastward it hung on much longer, and the written records seem to skip forward into the nineteenth century. A Colonel Chesney, making a trip down the Euphrates in 1830, saw a lion on the riverbank in what is now northwestern Iraq. Twenty years later Henry Layard, in a book describing his exploration of the ruins of Nineveh, the ancient Assyrian capital, wrote: "The lion is frequently met with on the banks of the Tigris below Bagdad, rarely above." On an earlier trip, the local tribesmen with whom Layard traveled had lit fires around camp to repel lions, and his raftsman had refused to pull into shore during night floats, "for fear of marauders and thieves and also he averred lions, which are occasionally, but very rarely, found so far north on the banks of the Tigris." Outlaws and lions lumped together—here's another theme that will recur.

As settlements and agriculture increasingly claimed the landscape, lions lost caste as well as habitat; they were transformed from kings of the forest into ecological brigands. In 1880 an English scientific journal carried a report, passed along from a sheikh, that "five years ago a lion appeared near Biledjik and after destroying many horses was done to death." If the sheikh's "Biledjik" was the "Birecik" of modern maps, a town on the upper Euphrates, that reckless animal may have been the last lion in Turkey, probably driven to killing horses because its natural prey had been usurped, its habitat had been turned to firewood and crop fields, or its teeth were old and broken. As last evidence of the Syrian lion, another commentator reported five years later that "a few years ago the carcass of one was brought into Damascus."

Eastward, in Mesopotamia and beyond, the species lingered. As late as 1910, Sir Percy Sykes wrote in *The Field* that "lions still exist along the banks of the rivers in Arabistan, but in very small numbers." What he called Arabistan we would call southwestern Iran, where the Karun River drains into the Persian Gulf. "I once saw a dead one floating down the Karun being eaten by sharks," Sykes added. But the real problems were on land, and the real sharks were humanity.

On the upper Karun River, winding out of the Kuhha-ye Zagros mountains, lions seem to have endured a bit longer. A report from the time of World War I states that Indian soldiers spotted a lioness with cubs near Ahwaz, and another sighting was claimed in 1929 by American engineers at work on a railroad line near Dizful, along a tributary of the Karun. But both these reports have come down through the lion literature in vague, thirdhand form. They don't carry the authority of an eyewitness account delivered in a reputable journal, or even in a signed memoir like Henry Layard's. That's generally how it is with extinctions, which by their nature are seldom witnessed and hard to record. Although the *Journal of the Bombay Natural History Society* in 1944 reported another Iranian lion sighting, again near Dizful, by a contingent of Indian surveyors, the reality is that no one knows, no one *can* know, when *Panthera leo* ceased to exist in Iran or Iraq or any of those other lost homelands. Chances are that the last big cat in the Zagros mountains didn't vent its desperation on horses and then die violently in someone's corral. Chances are that it passed away quietly, alone and unnoticed. Once there were lions, in this place or that, and then there weren't.

With the lions of India it will be different. The lions of India are a story in themselves—a big, sweeping, gaudy, sad, hopeful story that has come to focus in a single small place. The Asiatic lion, having disappeared everywhere else, is now synonymous with those animals in and around the Gir Wildlife Sanctuary and National Park. When that lineage dies out, if it does, there will be no uncertainty about when or where, although people may still ask themselves why.

5

Of the six billion humans already scratching and nibbling at Earth's landscape, one billion live in India, yet by some miracle there are still lions at Gir. A male of the population has just walked down this road. How long ago? Is he gone, or has he lingered nearby? Squatting curiously over his pug marks in the dust, I defer to the expert judgment of Dr. Ravi Chellam, who defers to a fellow named Mohammad Juma.

Dark-eyed and articulate, urbane and well-educated and as sharp as a meat knife, Ravi Chellam is a biologist with the Wildlife Institute of India, a governmental research institution based in the hill town of Dehra Dun, north of New Delhi. Mohammad Juma is a tracker, a thin man in a brimmed cap and an olive uniform, laconic in speech but eloquently perceptive. They are field partners from the old days, when Ravi did his doctoral research here at Gir. That was more than a decade ago. Ravi had come as a young graduate student, a high-caste city lad raised and educated in Madras. He stayed for most of four years, adapting himself to the hard, lean regimen of an impecunious field biologist, acquiring research skills, survival skills, and social skills applicable to rural Gujarat, and completing the most thorough study of the Gir lions within recent decades. Nowadays he's widely recognized as an authority—probably the world's leading authority—on the ecology and population status of *Panthera leo persica*. But he still relies on the local knowledge and keen attunement of trackers such as Mohammad, who've spent their lives in and around the Gir forest.

Mohammad and Ravi inspect the pug marks. They exchange a few words in Gujarati. The tracks are fresh, Ravi tells me. Within the last half-hour, he says. How can he know that? Concentrating, he declines to elaborate. Since Muhammad speaks no English and I no Gujarati, I can only guess. Newly trodden road dust has a certain angle of repose, maybe, which can't hold through many hours of changing temperature and moisture? Anyway, it's now 6:35 A.M., so by Mohammad's estimate the lion must have passed here in the tranquil moments just before dawn. Headed where? Presumably toward water and a shady place in which to snooze after a night's hunting. The only water to be found, at this end of the dry season, would be a tepid pool lingering in one of five little rivers that drain south from the Gir hills. The riverbeds are deep notches overhung with gallery forest that remains green even when Gir's upland woods and savannas go brown and bare. A sandy patch in the nearest stretch of river bottom might be a good, cool refuge for this night-prowling male, but that's just speculation. Our own plans for the morning don't involve tracking him to his day bed.

Mohammad walks the road's shoulder, noting other tracks. Porcupine, jackal, peafowl, wild boar. The large, cloppy hoofprints of a camel. Evidently there has been plenty of faunal traffic in recent hours. Peacocks and peahens customarily descend from their roosts at daylight, and a cock has now begun advertising himself near the road-side—strutting, skrawking, fanning his tail. The common peafowl, *Pavo cristatus*, is a native bird here, not an exotic decoration, and its very presence, its ability to survive and reproduce among lions, leopards, ravenous pigs, and snoopy langurs, is like a magic reminder of primordial plenitude as painted by Henri Rousseau. It's also a reminder that the male's ludicrous, extravagant tail is a product of sexual selection in the wild, not of demented tinkering by human breeders. Camel tracks, on the other hand, reflect human intervention; the dromedary is domesticated and exotic. A few are kept as cargo animals by an indigenous pastoral people, known as Maldharis, who still live in the forest. Ravi Chellam's own research revealed the fact, among many other points, that camels (like peafowl, though no doubt for different reasons) rank low on the list of preferred lion prey. Whatever seemed so enticing about those baggage camels of Xerxes to the lions of central Greece doesn't appear to hold sway at Gir.

The Maldharis themselves are a crucial element hereabouts—crucial both to the Gir ecosystem and to the political considerations that influence its management. The word *maldhari*, Ravi has told me, simply denotes "people who take care of cattle or livestock." *Mal* means livestock, *dhari* means guardian. An anthropologist named Harald Tambs-Lyche, in his study of traditional society on the Kathiawar Peninsula, gives a translation in the same vein, "milkmen," and adds that Maldharis are "regarded with the mixture of suspicion and exotic appreciation that Europeans reserve for Gypsies." When capitalized, as it usually is, the term seems to suggest an ethnic or tribal name, but in fact the Maldharis of Kathiawar belong to several different groups, each of which is considered a "pastoral caste," according to Tambs-Lyche. So the word itself can be better understood as a generic, indicating an occupational specialization embedded in a strong matrix of cultural heritage, roughly similar to a medieval guild. Then again, nothing is really similar to the Indian caste system, which to an outsider seems not just unplumbably complex but also cruel and weird.

Tradition dictates an eccentric, old-fashioned uniformity in the way Maldharis live, work, and dress. Their origins are clouded in an unwritten past that seems to have included nomadic pastoralism and, in some cases, long-distance migrations, which Tambs-Lyche suggests are still going on: "Migrating with flocks of a hundred animals or more, several families together, they stay out of villages. During the dry season they graze their animals on the stubble of the fields: in summer they retire into desert or semi-desert areas. They specialize in camels or bovines, in both cases keeping a few camels for transporting their tents and other gear." But the Maldharis of Gir are no longer nomadic. Given the constraints of the crowded Kathiawar landscape, it's hard to imagine how they could be. No one seems to know just where they came from, though necessarily it was somewhere north of the peninsula, possibly the Sind area along what's now the Pakistan border. No one knows how long they've been resident in the Gir forest, though guesses at the timing of their migration into Kathiawar run back to a thousand years ago. An ethnologist from Berlin with a deep interest in the subject, Sigrid Westphal-Hellbusch, has suggested that at least one of the Maldhari groups moved south from the Sind around the year 1000, to escape the Muslim rulers up there, and arrived in Kathiawar no later than 1400.

Nowadays at Gir the Maldharis occupy semi-permanent forest camps. Each camp, known as a *ness*, is protected by a perimeter fence of piled thorn branches, like a hasty emergency barricade. The ness encompasses roughly an acre of ground and therefore demands a fair bit of thorn fencing, but acacias are abundant in this scrubby forest. Inside the compound stand a half-dozen or so mud-and-timber huts, one hut to a family. Each hut fronts to its own stock pens and its own gate through the thorn fence, the gates framed by timber gateposts. Between the gateposts swing iron doors that can be closed at night against lions and leopards. Essentially the Maldharis are living inside a hive of corrals. Until recent decades they would abandon a ness site every five or ten years and shift to another, putting themselves closer to fresh grazing and forage, but nowadays their choice of sites is more restricted by forest regulations. The sanctuary portion of the Gir protected area contains about sixty nesses, widely scattered and fairly inconspicuous. The inner portion of the PA also contained nesses once, but not since the mid-1970s, when it received its national-park status and the resident Maldharis were required to move, under a well-meant but ultimately unsatisfactory resettlement scheme. The remaining Maldharis raise cattle and domestic buffalo for dairy products, sharing their landscape—and often, unwillingly, their livestock—with the lions.

Life, as they choose to live it, is hard and elemental. It's harder still for their animals. A milk cow is no challenge to a hungry lion but, defended by a herder with the light tool known as a *kuwadi* that is commonly carried, the cow has at least a chance of escape. Buffalo are slightly more capable of lowering their horns and taking a stand on their own, which may be why they predominate in Maldhari herds. Occasionally, though not often, lions kill a camel.

Insofar as it's possible to make a categorical statement about the private lives of an extraordinary, little-studied group of people, here's one: Maldharis don't own guns. Their battles with the lions approximate hand-to-hand combat, not safari hunting or varmint eradication. Their chief weapon, the kuwadi, is barely more than a short-handled hoe. Their relationship to *Panthera leo persica* is wary but intimate. Rarely does a herder get hurt. Part of being a Maldhari, at least at Gir, is coping routinely with lions through the use of caution, bluff, and an occasional kuwadi-thunk on the skull.

Just up the road from the pug marks, we encounter a group of Maldhari men, smartly turned out in their blousy white cotton shirts (called *kadia*), their white peg-leg trousers (*chorni*), and their high-rising white knit caps, odd headgear that would seem more appropriate for Norwegian ski troops. Having seen this traditional uniform on other Maldharis, I'm puzzled by all the crisp whiteness—yes, no doubt it reflects sunlight and so mitigates the afternoon heat, but isn't it damned hard to keep clean? A ness doesn't typically contain even a well, let alone such amenities as electricity or running water, not to mention—not to dream of—a washing machine. But Maldharis maintain their adherence to old ways with fastidious dignity, rather like the Amish of Pennsylvania or the Hutterites of Montana, and despite their life close to the soil, the white outfits somehow stay white. On a festive occasion, the knit cap may be replaced by a carefully wrapped white turban. Today is no festive occasion, merely a workaday morning, and the white-capped men stand with their large cans of milk, waiting for a refrigerator truck that will carry their product to market in a village outside the forest.

This business of milk selling is new, another incursion of modernity, Ravi tells me. Until recently it was impossible to move fresh milk out of the forest, and the main product from the nesses was ghee, a form of clarified butter, preservable at warm temperatures for use as cooking fat. Traditional Maldharis still sell ghee in large cubish cans; they make yogurt and more ghee for their own consumption and feed the butter-milk back to their livestock. But fresh milk yields more cash income. Meanwhile, the increased ease of travel in and out of the forest, by refrigerator trucks and other traffic, can be expected to bring irre-versible change not just to the Maldhari lifestyle but also to the lions.

We're headed north across the protected area this morning, in a lit-tle Jeep-like vehicle, on a looping survey of Gir and its surroundings. Just up the road we pass a ness, tranquilly sealed behind its thorn fence. We meet a few dozen buffalo parading slowly toward us under the gentle nudging of their Maldhari herders. We meet an orange bus carrying passengers southward on a shortcut toward the village of Sasan, our own point of departure, just outside the southwest entrance. We meet several old white Ambassador cars, several three-wheeled taxis, a few motorcycles. It's a public thoroughfare, not just a road serv-ing the PA, and though the hilly forest of teak, acacia, and various other scrubby species of tree and brush offers good cover for lions, good foraging for ungulates, the sense of refuge is fragile. The bound-ary of the protected area is porous, and "protected" is a relative, not absolute, term. From outside the Gir perimeter, India's billion people are pressing in. You can almost taste their collective hunger in the air. You can almost smell their determination to survive and procreate. You can almost hear their resolute striving, like the faint crackle of jaws from a vast column of termites on the march.

By 7:30 A.M. we've crossed the PA and exited through the north gate. Within less than a mile the terrain changes drastically and we find ourselves on a bare, dreary plain of red dirt and heavily grazed stubble. No trees grace the horizon. Hedge fences of euphorbia, prickly pear cactus, and agave stand firm between nothing much and nothing much else, like barbed wire on a deserted prairie. A few small patches of plowed, rocky soil seem to represent the hope of a peanut or lentil crop here or there. The dirt road turns to pavement, along which we pass several villages of red brick houses with red tile roofs. We see a herd of floppy-eared black goats—no surprise, those, since the whole landscape looks goat-eaten. At the edge of one village, a dog tugs at the remains of a bullock, mostly bone and stiff hide, drier than jerky, while gray-necked crows wait their turn and cattle egrets stand by like dispassionate professional mourners. The egrets are in their breeding color, sandy orange across the head and shoulders. They live as commensals, not scavengers, and a dead bullock, incapable of scaring up insects as it grazes, doesn't interest them.

"That's probably a lion kill," Ravi says. Then again, he corrects himself, there's really no way of knowing. So many other factors hereabouts could account for the death of a bullock—drought, starvation, leopards, disease. But lions, I ask, do they come out of the forest this far? Oh, definitely, he says. Farther. A young male with no territory, an old female that's failing and has been displaced, a healthy lion that rambles out in response to hunger or thirst—no telling how far such an animal might go before taking up residence within a skimpy patch of forest or, in some cases, getting into trouble. Ravi mentions the satellite subpopulations to the south and the west, which have found a tense sort of accommodation within the wider landscape.

How should these outlier lions be handled? Well, because they prey on what's available—out here, mainly livestock—and thereby raise resentment against lions generally, they need to be captured if possible, Ravi says. Once trapped, they can be shipped off to any zoo or captive-breeding program that will take them. They should not—repeat, not—be returned to the Gir PA, because the habitat there is already full. If they can't be trapped, then they must be shot. It's an unpleasant necessity, but it's more judicious than letting the situation fester, more humane than starving the lions by total interdiction of cattle killing, and more honest than the sentimental pretense of carting them back where they came from. To pluck them up and return them to the forest, the overcrowded forest, accomplishes nothing. In fact, worse than nothing, since an absence of firm control measures against errant animals, out here in the human zone, only makes the lion still more unpopular among local people. Ravi has expressed this point repeatedly in policy discussions—no doubt in his characteristically limpid, blunt style, which can seem threatening to timorous bureaucrats. For such heretical talk, he says, others have suggested that Ravi himself should be shot.

We drive east on the potholed road, passing more villages, bullock teams, euphorbia fences, crop fields in which people scrape and poke to tease out a few bushels of food. Any lion that would venture across this terrain must be a very desperate animal. The soil looks hard and weary, there's no sign of flowing water, and in some places lava-rock slabs lie naked to the sun. We ride in silence for a few moments, until Ravi says, "A lot of India is like this."

6

The Indian lion is counterintuitive, and not just because it inhabits a tiny remnant of verdant landscape on a generally sere, bedraggled subcontinent. Say the words "Indian lion" to an average Westerner, and that person will imagine you meant to say "Indian *tiger*." No no, you'll need to repeat, the *lion*—there's a lion, too, surviving over there. Even in India itself, *Panthera leo persica* is an obscure, poorly known creature, overshadowed by *Panthera tigris tigris*. Hindu theology still honors the lion as the steed of the goddess Durga in one of her avatars, and painted lion effigies stand guard atop many shrines. Narasimha, the lion-headed god-man, was an incarnation assumed by Vishnu in order to wreak gory justice on a demon-king. For a while the lion was considered India's national animal. But no longer. Tigers, more widely distributed in the country, more inherently Asian, and also sorely endangered (though not so sorely as the Indian lion), now have a higher profile. Tigers are lionized.

In earlier centuries, the lion was relatively abundant in those states later known as Rajasthan, Gujarat, Punjab, Haryana (surrounding Delhi), Uttar Pradesh, and Madhya Pradesh, with a marginal presence as far east as Bihar. That is, it had colonized and maintained itself in hospitable forest and savanna habitats across the northern half of the country. The written records of its presence here or there consist largely of accounts of individual animals being shot—enough lion-shooting reports to go some way toward explaining why the creature suffered nearly total extirpation. As in ancient Assyria, so in medieval and colonial India: Shooting lions was a favorite sport of royalty and lesser princes. With the Indian subcontinent divided into many minor principalities (ruled by this or that rajah, maharajah, thakur, or nawab), each subsumed or at least overshadowed by successive imperial powers (the Mauryas, the Mughals, the Marathas, and finally the British), there were all too many regal hunters. But it doesn't seem to have been fashionable here to leave self-congratulatory bag lists like Ashurnazirpal's. In the early seventeenth century the Mughal emperor Jehangir shot at least one lion; by Jehangir's testimony, his father (the famous Akbar, who consolidated Mughal power) had also been a lion hunter. Jehangir even shared his lion hunting with the first English ambassador to India, Thomas Roe, thereby establishing a precedent. Several centuries later, during those years when the imperial lion roared from London, politically astute princes of semi-independent Indian states would invite British dignitaries to visit and shoot a lion.

So many of the officers and administrators who ran Britain's Raj took to lion hunting, in fact, that well-catered lion hunts became a sort of recreational grease for the gears of diplomacy and business, rather like pricey cigars, prostitutes, or golf. In some cases, lion killing seems also to have served as a kind of cavalry drill. An old sporting magazine mentioned that, back in 1832, officers of the 23rd Bombay Cavalry hunted lions on horseback, and a later cavalry veteran recalled that officers of his regiment shot at least twenty-six lions. In other cases it became almost a continuation of warfare by other means. During the violent uprising of 1857–1858 (which British historians call "the Sepoy Rebellion" or "the Mutiny" and which Indians consider their first war of national independence), Colonel George A. Smith reportedly killed three hundred lions, fifty of those in the Delhi district alone. Smith's rampage sounds like displacement activity more than sport hunting, presumably intended to make some sort of bloody, vehement point. It was a nervous time for everybody.

Full extirpation of lions happened first in eastern India, at the far limit of their range, and in far western India (except for the Kathiawar Peninsula), where the habitat was dry and difficult. In central India, the hunting continued. Colonel Martin of the Central Indian Horse, along with his hunting partner, a deputy commissioner, killed eight lions at a place called Patulghar in 1863. Three years later a hunting party near Kotah, in what's now Rajasthan, claimed nine lions. Sir Montagu Gerard killed a lion near the city of Guna, north of Bhopal, in 1872, and another trophy taken near there the following year is considered the last native lion killed in central India.

In the state of Gujarat they held on. Why? Possibly because Gujarat was more remote, peripheral to the main region of British control, with a landscape severe enough to remain wilder longer. In 1878 a lion was killed by one Colonel Heyland on the racetrack at Deesa, a northern Gujarati town. Heyland's kill, like the lion carcass reduced to shark bait on the Karun River in southern Iran, seems poignantly emblematic. Details aren't available, but I picture that racetrack lion confused and panicked, dashing across a prim, grassy infield on a festive Saturday, crawling under fences or jumping over them, doubling back when cornered, while the brave colonel galloped after it to the roaring amusement of the crowd. The lion's world had changed, and the race was lost.

Four lions killed elsewhere in Gujarat, in 1888, bring the records to an end for all India, with the exception of that final enclave in Kathiawar.

Some authorities on the Indian lion have offered thoughts about why this cat was so vulnerable to hunters—in particular, why it could be eradicated so much faster than the tiger. "The lion is a far more noisy animal than the tiger," wrote L. L. Fenton in 1909, "and for this reason is more easily brought to bag, being so much more in evidence." Forty years later another credible source, M. A. Wynter-Blyth, a British educator who played a big role in monitoring population trends among the Gir lions, wrote: "It is unnecessary to look far for the cause of their extinction in other parts of India because their fearlessness of man and liking for fairly open country has always made them an easy mark for sportsmen and others." The co-authors of a recent book on the subject, M. A. Rashid and Reuben David, hold the same view: "The lion, with his comparatively bold nature and greater tolerance of human presence, adapted himself to the plain and open areas in which there was greater human movement and activity, while the sly and secretive tiger preferred the more remote and less disturbed hilly and wooded areas where he felt more at ease." Those open plains and savanna areas were quickly transformed by India's fast-growing human population, with their livestock, their crops, their settlements. "The lion's bold nature and more social way of life," say Rashid and David, made it "an easier victim of human aggression in comparison with the tiger." True enough, it's hard to imagine sporty British cavalrymen chasing tigers through the moist bamboo jungles of Madhya Pradesh or the swampy mangrove flats of the Sundarbans. And the wetter forests generally favored by tigers in India do seem to have been less susceptible to plowing and planting than the drier lion habitats.

The Kathiawar Peninsula was an exception, a region of rocky soils, flat plains, heavy monsoon rains alternating with scorching dry seasons, modest rivers draining out of a few scattered patches of hill and mountain. The mountain slopes were too steep for tilling, so they retained much of their natural vegetation—mostly dry deciduous forest with openings of scrubby savanna—even when the Kathiawar plains began filling in with crop fields, roads, and towns. The Barda Hills punctuated the westernmost wedge of the peninsula, within a minor princely state called Navanagar. On the east side of Kathiawar in the princedom of Bhavnagar, the Sihor Hills and the Ramdhari Hills were covered with scrub and thorn forest. In south-central Kathiawar, a massif called Girnar rose imposingly just east of Junagadh, an old fortress city. Thirty miles farther to the southeast was a wrinkled zone of ridges, peaks, hidden tablelands, and gulches, all cast up by volcanic oozing and buckling above the surrounding plain, over which spread the forest known as Gir. Collectively these highlands—Barda, Sihor, Ramdhari, Girnar, Gir—represented an archipelago of natural refuges for the last holdouts of Indian lion.

In medieval times, Kathiawar had a tradition of piracy (against trading ships just offshore, crossing the Arabian Sea toward the Middle East), banditry, proud warriors and chieftains of the Rajput caste, bardic poetry, pastoralism, petty princes, and recurrent armed scuffling over land. The name Kathiawar itself derived from the Kathis, a roguish tribe of fighters and cattle thieves from the north. The Kathis had earlier been driven out of Punjab, then out of the Sind; like the Maldharis, but more rambunctiously, they kept moving south, eventually to try their luck on this peninsular frontier. "The Kathis were a brave and warlike race, and acquired great reputation from their plundering forays," wrote a British officer named Wilberforce-Bell in his history of Kathiawar, which appeared in 1916. "Their women are said to have been very beautiful, and the breed of Kathi horses became as well known as the people who fostered it. They were formerly sun worshippers." With their dashing rascality, their sun worship and horsiness, the Kathis set the tone for Kathiawar, but they didn't rule it.

Nobody ruled it, except partially and temporarily. For centuries it was a buckaroo place, relatively underpopulated, politically unstable, a gladiatorial arena for opportunists. Ambitious warlord types from various tribes and families, Muslim as well as Hindu, took hold of fiefdoms, raised armies, and bumped one another around. In the mid-eighteenth century the Babi family established themselves at Junagadh, and each reigning Babi thereafter held the Muslim title *nawab*, which meant deputy, implying a deputized subordination to the Mughal emperor. The first Nawab of Junagadh was Sher Khan Babi, self-declared to that status in 1748. Navanagar and Bhavnagar also became regional powers, with rulers who claimed roughly equivalent Hindu titles— the Jam of Navanagar, the Thakur of Bhavnagar. Besides these, there were many smaller principalities. As the Mughal empire declined and the Maratha confederacy replaced it, Maratha armies made tribute-extorting expeditions into Kathiawar, where they forcibly persuaded the local princes to serve as their franchisees in gathering protectionist payments from the minor chiefs and the populace. With a cut taken at each level, by this nawab or that thakur or the officers and soldiers who did their enforcement, what wealth came up from the sweat of the lower orders didn't go far or satisfy everybody. According to Wilberforce-Bell, "everything was chaos and confusion, and great misery was the lot of all those who were unable to exact a livelihood from others less fortunate even than themselves. Rapine and robbery were rampant throughout the country, and the hand of every man was against his neighbor." A lion killing a milk cow would have made life still harder for the struggling cowherd, but at this point *Panthera leo persica* was not the most troublesome predator in Kathiawar. *Homo sapiens* was worse.

Then came the British, replacing the Marathas as overlords and establishing a new degree of stability, if not liberty. Kathiawar would never be fully absorbed into the British empire but became a sort of tribute-paying protectorate—or rather a cluster of protectorates, directly ruled by the Nawab of Junagadh and those other princelings, with British sufferance, administrative services, and advice. By one account, the last surrender of old-fashioned feudal autonomy occurred in 1822, when the Nawab reached a revenue-sharing settlement with the British for the tax payments they would henceforth collect on his behalf. Among the lands contained within greater Junagadh, his domain, was most of the Gir forest.

Although the British arrival had advantages for the nawabs and thakurs, it offered little to the Kathis, that recalcitrant tribe of fighters and rustlers, who were dispositionally unsuited to settling down, defending borders, truckling to overlords. As Wilberforce-Bell explained it, the Kathis "began to tire of the, to them, strange and peaceful existence" under the British, and so "to appease their hunger for fighting they attacked Bhavnagar territory." Wilberforce-Bell was himself an imperial administrator. He might condescend wryly to the Kathis for what he took to be their habitual bellicosity, but from another perspective they were indigenous guerrillas trying to reclaim ancestral lands.

In the nineteenth century, Kathi bands operated as rebel raiders against the princedoms, sometimes using the Gir forest as a refuge. In 1820, for instance, a gang of Kathis under a leader named Hada Khuman burnt and plundered several Bhavnagar towns, then skittered off to Gir like Butch Cassidy and Sundance heading for Hole-in-the-Wall. Four years later, after another raid on Bhavnagar territory, the Kathis were caught by Bhavnagari soldiers and one leader killed, but the rest escaped again into Gir, where the troops wouldn't follow them. Another Kathi band had the temerity to grab a British hostage, one Captain Grant of the Indian Navy, who while traveling overland from the coast on administrative affairs was protected by only a small escort, his own British aplomb, and a riding crop. Taken prisoner, he

was held in the Gir forest for two and a half months. After some nego-
tiations between the British and a Kathi chief, with the Nawab of
Junagadh as intermediary, Captain Grant was released, to be found
"wandering in a field at night in a state of delirium, covered with vermin, and severely ill with ague and fever caused by exposure and
fatigue." It had been the rainy season, a bad time for roughing it in Gir.

During the latter half of the century, an occasional rebel or outlaw
still used Gir as a place of escape. Lions found refuge there too. All the
insular mountain ranges of Kathiawar supported small populations of
lions, or at least the intermittent presence of a few individuals, until
well into the nineteenth century. But as human density increased, as
crop fields replaced forest in the lowlands and the political scene stabi-
lized, both Kathi guerrilla bands and lions became increasingly rare.
The Gir forest, because it spanned a large area (roughly two thousand
square miles) of continuous habitat, considerably more than any other
single patch, emerged as the last hope for the Indian lion.

For a while it looked as though the lion wouldn't survive even at
Gir. In 1880 a British officer named Watson, reporting on conditions
throughout Junagadh for a geographical survey, offered his guess that
"not more than a dozen" lions remained within the forest. Watson was
an avid hunter and a knowledgeable field man. Although his estimate
shouldn't be taken as precise, it conveys the sense of a dire plunge, and
in his day it alerted a few people among the British and the princely
houses to recognize that the Indian lion was not an infinite source of
sport or nuisance, as they'd let themselves think, but a precious rarity.
In 1893 a more official estimate put the number at thirty-one lions in
all of Junagadh, including both the Gir population and those at Girnar,
the big mountain thirty miles away. Considering the likely margin of
error and the split between the two populations, that wasn't much less
discouraging than Watson's guess. Another estimate, in 1900, came up
with nineteen lions, of which eight were cubs. Cubs have a low survival rate in the best of circumstances, so the more telling number was

the eleven adults. Again, it shouldn't be mistaken for a precise count, but it was eloquently gloomy. Maybe a dozen lions survived, maybe twenty or thirty. One dissenter, a Major Carnegy, stationed in Junagadh from 1903, put the number as high as sixty or seventy. In any case, there was troubling uncertainty, and even if Carnegy was right, the population was perilously low.

Two events occurred then, at the turn of the century, to make the situation more complicated. The first was a matter of social politics, the second a vagary of climate.

The incumbent Viceroy of India, a brilliant and snooty English aristocrat named George Nathaniel Curzon, received an invitation from the Nawab of Junagadh to come lion shooting. By then the Nawab's family had ruled their little princedom for 150 years, during which the Gir forest had gradually transmogrified from an outlaws' haven into a de facto wildlife reserve. As lions disappeared elsewhere, they became much appreciated at Gir for what passed at the time as sport hunting—that is, standing amid a crowd of tony chums, with servants and native *shikaris* (professional hunters) nearby, waiting to shoot a driven lion and then adjourn for drinks. For the Nawab's purposes, this carried value as a hard-to-match form of prestige hospitality. Many minor princes around the empire could lay on some tiger shooting, but only he could offer lions. And although Junagadh didn't encompass quite the whole of the Gir forest, successive nawabs had tended increasingly to view Gir's lions with jealous pride.

Lord Curzon, after accepting the Nawab's invitation, belatedly heard from someone about the lion's population status—that only a dozen remained. Curzon followed through on the Junagadh visit but nixed the lion hunt. What were his motives? It's hard to guess whether he really felt concern for the survival of the subspecies or had merely calculated the adverse publicity and political costs. (I find no mention of his pro-lion gesture in Curzon biographies, not even David Dilks's two-volume *Curzon in India*, though the anecdote is told and retold in

the lion literature.) In any case, Curzon encouraged the Nawab to protect this noble cat from extinction. The Nawab, in agreement, issued a
ban on lion shooting.

But the lions continued to make trouble for themselves, as lions will
do, by attacking livestock in or near their habitat. When they caused
damage in the neighboring state of Jetpur, which claimed a small part
of the Gir forest, tension arose across the border. Baroda, another adjacent princedom, also embraced a bit of Gir but didn't care to embrace
lions. The rulers of both Baroda and Jetpur seem to have felt that if the
Nawab of Junagadh chose to view these beasts as *his* lions, then he
could bloody well pay compensation for what they did to other people's cows and goats.

The second complicating event was a terrible drought that afflicted
the peninsula during 1899–1900, bringing what's remembered as the
Great Kathiawar Famine. For the lions, this must have been—initially, anyway—a feast, not a famine, with thirst-weakened prey concentrated around a few shrinking water holes. The bounty of food
may even have raised the rate of cub survival and thereby boosted the
lion population, exacerbating the difficulties that came next. Before
long, virtually all the wild ungulates were dead. The lions became desperate. Driven out to the forest edges and beyond, they preyed on livestock. A 1901 report noted: "Lions are much bolder and enter villages
during the daytime and fearlessly attack men and cattle. The police
have had to exert themselves to protect human life." Within a short
time lions had killed 352 domestic animals, two women, and a boy.
Further reports over subsequent years continued the tally: two men
killed in the Kodinar district, which was twenty-some miles south of
Gir; in the Ghantwad district, a boy killed and eaten, and hunger-
crazed lions there had ripped the roof off a village hut to get at the
goats inside. In Gir itself, lions killed thirty-one men and mauled eight
others during 1901. After a two-year gap in the records, twenty-nine
fatalities and eleven maulings were cited for 1904, in a report that

described the lions as "very bold." Under this pressure, the Nawab revoked his ban, ending the first experiment in full protection.

How close was the Indian lion to extinction? Opinions differed. At one extreme was that optimistic population estimate—sixty to seventy animals—from Major Carnegy. Then again, Carnegy got himself killed by a lion soon afterward, during one of those big social hunts, which might cast some doubt on his expert judgment. A breezy history of Junagadh and its sovereign family, titled *Ruling Princes of India: Junagadh* and published in 1907, dismissed any concern for the lion's population status: "Lord Curzon, when he abandoned his projected lion shoot in the Gir some years ago, unconsciously helped to spread the misapprehension prevailing." This volume, evidently part of a fulsome series offered by the *Times of India* newspaper, was authored by S. M. Edwards and L. G. Fraser, who put Curzon's decision in a sanguine perspective:

> Apparently in consequence of the outcry in the Press, or because His Excellency was not correctly informed of the real situation, or because careful enquiries were not made in the proper quarter, he declined to shoot. There is every reason to believe that even five years ago his decision was based on misleading intelligence. According to the trackers the inmost recesses of the forest now contain very many lions which lurk for the most part in places which are comparatively inaccessible. Hardly any Europeans know the Gir well.

One European who did know it well was L. L. Fenton, and he held an opposing view. Fenton was a British field officer, like Carnegy, but perhaps more judicious. He wrote in 1909 that despite some marginal protection accorded to the lions, "there cannot be the slightest doubt that they are gradually, but surely, approaching extinction."

At the time of Fenton's prognosis, lions were being shot at the rate of four or five per year within Junagadh's portion of Gir, eight per year in the outlying portion. That pace continued until 1911, when the old

Nawab died and was succeeded by his son. The new Nawab was just a boy, so while he went off to England for schooling, the British assumed administrative control of Junagadh on a regency basis. In 1913, after a two-month tour through Gir, Junagadh's chief forestry officer judged that the lion population might be down to six or eight animals, and that it certainly wasn't above twenty. The British administrator, in response, put another ban on lion shooting. Like the old Nawab's ban, it wasn't destined to be permanent, but it did help the population recover from an extreme low point. The worst years had passed.

"However, a most interesting fact to observe," noted M. A. Wynter-Blyth, "is that it was during these years that the habits of the lions underwent a profound change, for never again are they heard of as a menace to human life." Wynter-Blyth, who later organized some of the first systematic lion-census efforts at Gir, was a careful observer. He was right when he wrote that, in 1950. He's not right anymore.

7

By 1986 the picture had changed, in some ways for the better. Extinction of the subspecies had been averted, barely. Protective regulations were in place. What had started with the old Nawab had been continued by the government of India; lion hunting was forbidden. With their death rate lowered, the lions of Gir had bred themselves back up to a population of roughly 250, and that demographic recovery was a positive development, provisional but cheery.

Their recovery of genetic diversity, on the other hand, was probably minimal in so short a time. Their gene pool presumably remained tiny (given that genetic diversity rebounds far more slowly than sheer population), still reflecting the crisis of rarity when they'd been reduced to just a couple dozen survivors. Although they were now counted in hundreds, that raw number belied the continuing precariousness of the population.

With or without a solid genetic base, they were now recognized as the last wild remnant of *Panthera leo persica*, an endangered subspecies unique to the Gir Wildlife Sanctuary and National Park. A few animals had been moved elsewhere—to a zoo in Junagadh, a translocation attempt (which failed) in the northern Indian state of Uttar Pradesh, other zoos as distant as London and Washington. As for the Gir population itself, an estimate of 250 animals seemed much better than six, eight, or twenty, but it wasn't enough. Even without the lingering genetic consequences of that bottleneck at the turn of the century, the long-term survival prospects of such a modest population weren't good. Isolated within their little patch of forest, surrounded by the enterprising scramble and gasping desperation of modern India, the Gir lions were a textbook case of the jeopardies facing small populations insularized by habitat constriction. Then, in March of 1986, another bad drought began.

This drought, like the one of 1899, created circumstances—wild prey concentrated thirstily at water holes, neglected and dying domestic livestock—that at first favored the lions. For a while, as herbivores suffered, lions benefited. Then the drought broke, in 1988, and a reversal of conditions drove them toward increasingly reckless behavior.

By this time there were lions living as resident subpopulations in those scraps of marginal habitat outside the protected area. Other lions were passing in and out across the PA boundary as opportunity or necessity guided them. During the worst of the drought, the peripheral fields and villages had offered considerable lion food in the form of stressed, abandoned, and dead domestic animals. But when the drought ended, that situation changed. What was good news for livestock and for the folk who tended them—that is, a lifesaving new supply of water and fodder—was bad news for the lions, because far fewer cows and buffalo lay dead to be scavenged, and far fewer were wandering neglected. People could again afford to feed and water their animals and to

protect them within corrals or other enclosures. Lions continued to
range through those outer locales, beyond the PA boundary, in search
of food. Of course this meant trouble. Facing a shortage of vulnerable
prey animals, they again became a menace to human life, invalidating
the blithe observation Wynter-Blyth had made back in 1950. By the end
of 1991, desperate lions had attacked 120 people, killing 20 of them.
Seven of the corpses had been at least partially eaten.

Three years later, a paper appeared in the journal *Conservation
Biology* under the title "Lion-Human Conflict in the Gir Forest,
India." It was co-authored by four researchers, the first of whom was
Vasant K. Saberwal, of the Wildlife Institute of India. Basing their
analysis on interviews with people in fifty-six villages surrounding the
Gir forest and on data from other sources, Saberwal and his colleagues
found a few interesting trends and patterns. They tabulated 193 cases
of lion attack on humans, spread across a thirteen-year period (1978–
1991) that encompassed the drought and its aftermath. Of those
attacks, 28 were fatal. The rate of such conflicts before and during the
drought averaged 7.3 per year; after the drought, the rate jumped to
40.0 per year. Subadult lions (that is, adolescents who have reached
nearly full size and may have recently become independent from their
mothers) of both sexes were more frequently implicated than adults.
Most attacks occurred outside the protected area, committed by some
of the lions that had long since taken harbor out there or by others that
had lately dispersed, driven by exigence, from inside the PA. These
outlier lions had traveled fifteen or twenty miles through the peanut
fields, mango orchards, and villages roundabout, crossing roads, finding cover and food where they could, managing long, stealthy odysseys
before setting up shop and then, in certain cases, not all, getting into
serious trouble. It was especially interesting that, during the post-
drought period, attacks increased dramatically outside, but not inside,
the protected area. Evidently the Maldharis and the lions who co-
occupied the Gir forest had a degree of compatibility—a tense but

steadying mutual tolerance—that didn't exist in the outer world. Another difference between the drought and the post-drought periods was noted dryly by Saberwal and his co-authors: "Following the drought, lions began to feed off corpses of humans they had killed."

Having read this paper and wanting to learn more, I tracked down Vasant Saberwal by telephone. He was in New Haven, Connecticut, under some sort of temporary association with Yale University. When I began peppering him with questions about the ecology and population status of the lions, he explained that field biology wasn't his expertise. He was a sociometrician, a maker of surveys, a poller and analyzer of human attitudes and experiences. I should talk to his third co-author, Saberwal said, the fellow who actually studied the lions at Gir. That was Ravi Chellam.

Saberwal gave me a phone number in Dehra Dun, the old hill-station town north of Delhi where the Wildlife Institute of India is based. Waiting until well after midnight in order to catch working hours on that end, I called. I hollered my way through a switchboard. Finally I got this fellow on the line. He was cordial, responsive to my interest, but brusque, in what I would eventually come to know as the Ravi Chellam way. My God, man, why are you telephoning? he said. Do you have any idea what an international call costs in India? But it's my nickel, I reminded him. Haven't you heard of email? he demanded, and hung up.

8

Not long after that phone contact I was in Dehra Dun myself, giving a talk at the institute and, as a happy bonus, meeting Ravi Chellam for the first of our many conversations. He was in his mid-thirties, slender, with a dark full beard, an incisive but very orderly mind, a set of lucid professional convictions, an impeccable sense of the limits of his own knowledge combined with an adamantine confidence in what he did know, a tendency to answer every question thoroughly, a vigilant precision about facts and details, and an insistent intelligence that sometimes seemed to blur the line between explanation and argument. In other words, he would have been a pain in the ass if he didn't also have a quick, ironic sense of humor. But he did. I liked him immediately. Let's go to Gir together, I said. And so a few weeks later here we are, following this road eastward across a blighted plain just outside the protected area's northern boundary.

On our left, the moonscape sweeps away to a cheerless horizon, interrupted sparsely by euphorbia fences, doleful little villages, and goats. But the road has now angled south again to converge with the PA's perimeter, bringing us back toward the consoling presence of trees. We cross a dry riverbed and catch sight, down among its rocks and gravel, of a female mongoose and her three skittering young, like a family of otters wondering what became of the water. Reentering the PA through a checkpoint, we're surrounded once more by dry deciduous forest, scraggly but diverse. There are big-leafed *Butea* trees, stunted teak, acacia, several species of thorny *Zizyphus*, and a bushy plant called *Carissa opaca,* which tends to grow into bouffant domes with cavelike hollows underneath. The *Carissa* domes, shady and cool, sometimes serve as shelters for lions taking afternoon rest. The leaves and fruit of the same shrub, Ravi tells me, are important as browse for wild ungulates. And the fruit—tangy blue spheroids like serviceberries—is also coveted by humans from surrounding villages, who invade the PA and carry out bucketfuls. These competing attentions paid to *Carissa opaca* are representative of larger conflicts at Gir.

Browse is limited in such a hardscrabble forest, but browsing is crucial to wild herbivores because of the shortage of grass. Grass is meager, especially in the dry season, and heavily grazed by Maldhari cows and buffalo. To bring a little extra browse within reach of their live-stock, Maldharis sometimes cut small trees or lop the leafy limbs off large ones. They also scrape up the cow and buffalo dung, along with a bit of topsoil, and sell it as fertilizer. Villagers from outside the perimeter, besides collecting *Carissa* berries and several other kinds of wild fruit, sneak in and cut trees for firewood, either to meet their own fuel needs or to sell. In earlier times, before the sanctuary and park were declared, outside villagers inflicted still greater impact, driving tens of thousands of their own cows and buffalo into the forest each day to graze. Within just six miles of the PA boundary are more than two hundred villages containing 160,000 people and almost that many head of livestock. Inside the sanctuary are about 2,500 Maldharis, along with their 15,000 cows and buffalo. The Maldhari economy, though still grounded in subsistence activities, is changing inexorably in response to new opportunities and enticements. The Maldharis are becoming ever more connected to the outside world. And the sur-rounding villagers haven't entirely stopped making their own extrac-tions—urgent and understandable, whether legal or not—from the forest. As a result of all this, nutrients and other resources are flowing out of the Gir ecosystem at an unsustainable rate.

It's an incremental, cumulative problem, not a glaring catastrophe, but from a close reading of ecological studies done by Ravi Chellam, his colleagues, and his predecessors over the past thirty years, it emerges clearly: Gir is eroding. Small as the area is, its margin of tolerable loss isn't great. Meanwhile, there sit the planet's last few hundred Asiatic lions, along with some tens of thousands of wild ungulates, their natu-ral prey.

One bit of good news is that the ungulate populations have increased dramatically in three decades, since the designation of Gir's core as a national park and the prohibition (not absolutely enforced, but helpful) on bringing domestic animals into the sanctuary from out-side. Those measures have carried troubling implications for local people—especially for some Maldharis, as I'll explain in due time—but for native herbivores, the effect has been blessed. The most common of the native species is the chital, *Axis axis*, a small spotted deer that now constitutes the lions' main prey. There are also sambar (a large, elk-like deer), chinkara (Indian gazelle), nilgai (a hefty antelope), chowsingha (loosely known as the four-horned antelope), and wild boar. Of this native menagerie, some are relatively rare and some tend to occur only in one part of the ecosystem or another. Chital are likely to turn up anywhere, but their overall abundance has changed within recent decades. Sambar prefer the rich, thick forest of the national park.

Back in the early 1970s, according to work done by two North American researchers in the course of their doctoral studies, the forest harbored only about four thousand chital. Domestic livestock far outnumbered them, and the patterns of lion predation reflected that imbalance. The lions ate what was most available, with livestock accounting for roughly seventy-five percent of their diet. By the time Ravi Chellam did his own dissertation work, in the late 1980s, the chital population had risen to almost fifty thousand, there were far fewer cows and buffalo in the forest, and the lions were mainly eating chital. This shift went some way to mitigate conflict between the Maldharis and lions but didn't eliminate it completely. Ravi's mathematical analysis of the lions' preference for this or that species—not just in absolute numbers, but relative to the availability of each—revealed those big, meaty sambar as the lions' favorite prey. Cattle ranked second. Chital were abundantly available but too small individually to satiate a group of lions—and therefore, though frequently killed, they weren't especially sought after.

Up here along the north boundary, I've noticed a conspicuous absence of chital. In fact, we've seen almost no animals at all. We drive south for an hour, stop for a late breakfast of *roti* (Indian flatbread) in the shade of a gully, and go on. The dearth of wildlife continues until, crossing another dry streambed, we hear a hubbub of raucous caws.

Not far upstream, dark against the sky, crows are circling and diving excitedly. Their whirly pattern stands out like a marker buoy. It could be a kill, Ravi figures.

So we stop. Led by Mohammad, we pick our way carefully upstream through thick brush to the spot. As the crows scatter, we find the carcass of a young sambar, stashed beneath a tree for future delectation by whatever predator killed it. Covered with flies, the dead deer buzzes like an electrical transformer.

Nearby lies the rumen, pulled out intact and then half buried. That's the act of a fastidious carnivore, taking care to avoid fouling good meat with a compote of stomach-soured grass. The carcass itself hasn't yet been ripped apart, although a single detached foreleg lies ten feet away. Is this a lion's kill, I wonder, or a leopard's? Ravi suspects the latter, evidently because a lion wouldn't limit itself to such a dainty first meal. A moment later, sure enough, Mohammad finds leopard tracks in the loose soil of a game trail along the stream bank. The carcass looks several days old, Ravi says. Where that leopard has been in the meantime, and why it hasn't returned to continue feeding, are questions unanswerable from the evidence on the ground. Where it might be at this moment is anybody's guess.

9

The leopard, *Panthera pardus*, is the most versatile and widespread of all the world's big cats. The global distribution of the species spans an amazing range of distance and habitat variety, from the rainforests of western Africa, through the deserts of southwestern Asia, to the rainforests of Indochina and the temperate woods of southeastern Russia.

Its altitudinal range goes from sea level to more than 17,000 feet. In India, the leopard survives in dozens of different parks, sanctuaries, and fringe areas, from Tamil Nadu in the country's southernmost tip to medium-high Himalayan slopes, and from Kathiawar in the west to the eastern highlands of Arunachal Pradesh. It shares habitat with tigers (in some Indian parks) and with lions (in parts of Africa, as in Gir); elsewhere, it presides as the top native predator. The leopard's smaller size, its lesser food needs, its ability to climb trees and stash carcasses up there, its adaptability to dry habitats, its tolerance of ecological edges, marginal conditions, and human proximity—all these attributes make it compatible, not directly competitive, with the two bigger cats. Furtive and enterprising, it tolerates human proximity far better than lions or tigers do, even pushing its luck into human-dominated landscapes, as evidenced by the leopards who venture out of Sanjay Gandhi National Park to feed on stray dogs in the slums of Bombay. Its coexistence with the lions at Gir can't be explained by any great divergence in food preferences. Ravi's study showed that although the leopard is more likely to content itself with a langur or a peafowl than to bring down a buffalo, both cats tend otherwise to target the same prey: chital, sambar, and cattle.

A leopard is neither too furtive nor too small to kill a person. The occurrence of man-eating by *Panthera pardus* is common enough—or, anyway, dramatic enough—to generate many vivid accounts in the hunting and wildlife literature of India. The most famous of those is *The Man-Eating Leopard of Rudraprayag*, a popular book by Jim Corbett, the storied hunter and author better known for his rogue-tiger tales, as collected in *Man-Eaters of Kumaon*. Somewhat drier accounts of leopard-human conflict have appeared in the *Journal of the Bombay Natural History Society*. That society, the BNHS, has been a distinguished body of wildlife observers, biologists, and conservationists for more than a century, and its *Journal* papers tend to be sober reports written for a serious audience. Many of those relate to *P. pardus*, which by an earlier fashion of speech was called "the panther." For instance, one early contributor to the *Journal*, R. G. Burton in 1918, described the paradox of "panther" behavior:

> Their conduct is frequently characterized by extreme boldness and extreme timidity. Though so bold that they have been known to enter a tent and even a house, they will seldom take their prey in the presence of man when they are aware that they can be seen. Thus a herd of goats watched by a small herd boy will probably be unmolested, but stragglers will be seized.

Another account in the *Journal* offered this:

> Like the tiger, the panther sometimes takes to man-eating, and a man-eating panther is even more to be dreaded than a tiger with similar tastes, on account of its greater agility, and also its greater stealthiness and silence. It can stalk and jump, and . . . can climb better than a tiger, and it can also conceal itself in astonishingly meager cover, often displaying uncanny intelligence in this act. A man-eating panther frequently breaks through the frail walls of village huts and carries away children and even adults as they lie asleep.

It's a recurrent theme: Lions and tigers may be big and fearsome, but leopards are stealthy.

Jim Corbett, in the *Rudraprayag* book, reaffirmed that point with the story of a preternaturally sneaky attack by the cat of his title. It occurred at a pilgrim shelter along a road to the high-country shrines. *Prayag* is a Hindi word meaning "confluence," and this leopard-terrorized place, Rudraprayag, sat at the confluence of two Ganges headwater streams—a mountainous, half-wild place where spring meltwater "cascades unconfined and merrily over rocks draped with moss and maidenhair fern" and where the path has been worn by the bare feet of millions of pious Hindu travelers. The victim was an unlucky local woman who stopped there for the night rather than risk walking back in the dark to her village. She had lain down in what seemed the safest part of the shelter—far from the road, and surrounded by fifty pilgrims. As Corbett tells the tale, this amazing leopard padded in among the rows of people and padded back out with her body in its jaws, accidentally clawing one other woman as it stepped past. No one saw it. No one heard it. But the clawed woman awoke, thinking she'd been stung by a scorpion. She yelped, other pilgrims roused, and though they were slightly perplexed by the fact that her "scorpion sting" was a bleeding scratch, they comforted her. Then everyone went back to sleep. In the morning they discovered that the first woman was gone, her bloody sari lying outside in the road.

Corbett heard this account, he says, from the Hindu pundit who ran the shelter. Whether or not we can swallow it as literal fact (personally I can't), it does reveal something about the leopard's reputation in India.

R. G. Burton, besides noting the unsteady combination of boldness and timidity, suspected that leopards turn more readily to man-eating than tigers do, partly because of circumstantial temptations: "Their habits bring them into closer and more frequent contact with human habitations." Careless mothers leave children unattended, Burton wrote, and a panther, skulking around the village outskirts after dark, "even though not a confirmed man-eater, will always be ready to carry off a child if no one is watching." Burton himself helped kill a leopard that had specialized in snatching children from beside their parents when, on hot summer nights, whole families slept outside in the cool air. His account, along with Corbett's from Rudraprayag, suggests that the most deft of these animals could pick through a pile of people like jackstraws.

Another contributor to the BNHS *Journal*, identified as E. Brook Fox, in 1920 reported a man-eating leopard in the Gir forest: "I believe it is a female—probably with cubs. She kills spasmodically; for four

successive years she has killed and eaten one child in each monsoon."
The children in question were Maldharis. The leopards (the lions too)
of Gir, according to Brook Fox, had developed a habit of leaving their
lairs around sunset, headed straight for the nearest Maldhari ness;
from there, "if they find no victim, they move on to the next settlement," and so on, until they luck upon "a straggler." The straggler
might be a cow, a calf, a herder, or, as in the particular cases cited, a
herder's child, caught isolated at one of those penumbrous, twilight
times when livestock are being moved between forest and ness. "The
four little girls . . . were carried off about dusk or dawn when visiting
the edge of their camp," and in each instance, Brook Fox claimed, "the
only trace left was a bare skull."

A bare skull left behind—in four different cases? That stretches
credulity even thinner than do some of Jim Corbett's shapely tales.
Predators generally crack bone and gnaw on it; vultures, crows, beetles,
and ants pick it clean, if given the time. It's hard to imagine that
Maldharis would grant such time in the case of a lost daughter, or that a
leopard itself would be able to chew (peel? suck?) the flesh off a victim's
severed head. Still, apart from melodramatic elaborations, the death of
four little Maldhari girls from leopard attack is entirely plausible.

Although the leopards of Gir haven't been well studied, Ravi suspects that they may constitute the largest single population of *P. pardus*
in India. There are many other patches of leopard habitat in the country, but probably none so sizable and intact. His impression of Gir as
an optimal haven was reinforced some years ago by an encounter on
the same road we're driving today. It was just after the monsoon, with
tall grass lining the road's shoulder on both sides. As Ravi rode along
in a vehicle, one leopard popped into view like a highwayman—then
another leopard, still another, eventually *five*. He recognized them as a
female with four subadult cubs. The subadults jounced and jostled,
wrestling playfully with one another as cubs do, and then continued on
their way, disappearing into grassy thicket on the far side of the road.

"The experience may have lasted three, four minutes," Ravi remembers. "But it says so much." What it says, at least to his attuned mind, is that for now the Gir forest is healthy, rich, and robust. Alpha predators such as leopards face life-threatening challenges constantly, and cub mortality in most ecosystems is high; if a leopard can raise four cubs to adulthood, he figures, she must be living in a hospitable place.

Other accounts of man-eating leopards have appeared, over the past century, in the BNHS *Journal*. J. C. Daniel, formerly curator of the society and the *Journal*'s editor, has collated those accounts for his book *The Leopard in India*. Daniel himself is a respected Indian naturalist, well qualified to give an overview of leopard behavior and leopard history on the subcontinent. "Leopards have lived in close association with man for centuries," he writes, noting that the closeness entails a mixed bargain for *P. pardus* in a human-dominated world. On the positive side, it reflects the leopard's tolerance for degraded habitats, forest edges, and agricultural intrusions into wild landscapes. On the negative side, it goes some way to explain why man-eating leopards have occurred as often and operated as lethally as they have. In the district of Bhagalpur along the lower Ganges, leopards reportedly killed 350 people between 1959 and 1962. The man-eating leopard that Corbett hunted at Rudraprayag in the late 1920s had taken at least 125 human victims. Even in recent times the problem is serious. From 1982 to 1989, by one tally, leopards killed 170 people in India. Just within greater Bombay, in and around Sanjay Gandhi National Park, there were fourteen fatal leopard attacks during the ten years ending in 1996. Those and other records support Daniel's assessment: "Among the larger mammals, the leopard comes second to the elephant in causing human fatalities."

But of course there's that crucial difference: Elephants don't eat their victims, they merely trample and gore them in outbursts of confusion and rage. Daniel notes too that, whereas elephant-caused fatalities are usually accidental, "fatalities caused by leopards are in the majority of cases the result of deliberate attacks." The incidence of attacks tends to rise in response to environmental stresses, such as drought, famine, or the constriction of habitat. That can lead to a cycle of grim consequences. "The leopard once it takes to man-cating becomes a scourge and is often a deadly menace to children," Daniel warns, adding tersely, "and is therefore eliminated." He means that the particular man-eating individual is eliminated. It's also true—not just with leopards but with every species of alpha predator—that man-eating is the most fatal of indiscretions, in that it often provokes retaliatory eradication. Man-eating invites demonizing, and demonizing a species is the first step toward total war. Yet the man-eating habit at least among leopards, as Daniel reminds us, usually arises after extermination of their natural prey. So extermination follows extermination, as inexorably as civilization follows the plow.

Panthera pardus is an extraordinary animal, even among the large cats. It's not just ferocious and sneaky, it's also tenacious. J. C. Daniel captures its character in an apt phrase when he writes that the leopard, "the perfect predator for this day and age," will probably survive when other big carnivores have disappeared. But for now, at Gir, the perfect predator remains an ecological understudy to the magisterial but imperfect lion.

10

The lion is a godly animal as well as a kingly one. This becomes clear when Ravi and I arrive at the Kankai temple, a Hindu ceremonial center deep within the protected area, near its core zone, the national park. The place is festooned with lion images.

Kankai is dedicated to the goddess Durga, consort of Shiva, in her avatar as Kankeshwari. The *vahana* (steed, mount) on which Kankeshwari rode to battle against the forces of darkness and barbarity, according to Hindu belief, is a lion. That accounts for the prominence of lions in the temple decor. The whole compound is a tiered set of structures built on a hillside, like a squarish wedding cake, with cells for the temple priests, temporary lodging for pilgrims, various other practical buildings, and a set of stone steps leading down from the upper courtyard to the main shrine, an elaborate little edifice approached through a four-columned archway. Atop the archway sits an image of Durga as Kankeshwari, flanked by two lions.

The shrine itself is a small, enclosed room, a sanctorum, partly occluded behind a drawn curtain and guarded by a two-foot-high painted statue of a crouching lion. From the shrine doorway stretches a marble patio, where devotees and other visitors are allowed to assemble beneath a domed, mirror-inlaid ceiling. From the patio, a few more steps lead down to a secondary shrine. This one is devoted to Shiva, represented there with his own *vahana*, his personal beast of conveyance: a bull. Shiva's shrine, like Kankeshwari's, is topped by an arch-domed roof, within which stands another sculpted lion. Each of these iconic felines looks tame, dutiful, stalwart. Around the entire compound, offering a slightly ironic subtext for anyone pausing to notice it, is a high stone wall topped with barbed wire—that is, a lion-proof fence. Unlike other lion-bedecked Hindu shrines, Kankai lies in the midst of lion country.

Despite the barbed wire and the sprawl of concrete outbuildings, the temple compound is a pretty place in a beautiful locale, overlooking a branch of the Shingodah River. Gulmohar trees, sometimes called flame-of-the-forest, rise above the courtyard into parasol canopies of scarlet-orange blossoms that seem to glow in the sun. Bougainvillea are also in bloom, and fig trees shade the steps down to Kankeshwari's shrine. The shrine domes are trimmed in bright saffron orange and topped by flags of the same color, the sacred Hindu hue, handsomely echoed by the gulmohars. An orange windmill twirls slowly above the courtyard. Outside, a sign in English greets visitors: NATURE IS THE ONLY BOOK THAT TEEMS WITH MEANING ON EVERY PAGE. People come here to read it, or at least to look at the pictures.

That's the attraction for Chandrakant Panna, a late-thirtyish man in an airy white shirt and white trousers, with whom I fall into conversation. Mr. Panna grew up in a village not far beyond the PA boundary. For twenty years he has lived in Bombay, but he returns each year to visit family and, sometimes, brings them to offer *puja* (prayer observance) at the forest temples. Over his shoulder I see his elderly father, frail, grizzled, wearing a beatific smile, along with several younger women wrapped in bright saris. They plan also to visit the Banej temple, some miles southeast of here, Mr. Panna tells me. Banej is a similar compound, one of four major temples (Kankai, Banej, Tulshishyam, Patla Mahadev) that remain as inholdings to the protected area, encompassed within its boundaries but not fully subject to its rules. Having questioned me politely about what the devil I'm doing here at Kankai myself, Mr. Panna offers a tip: If lions are my interest, I should go to Tulshishyam. It lies in the eastern half of the sanctuary, where the landscape is more open and dry. That's the place, he advises me. For lion watching, try Tulshishyam at sunset. Lately there have been four lions—no, wait, he says, *five*, plus a leopard—in the vicinity. Mr. Panna doesn't elaborate, but I imagine those cats attracted to the temple by errant cows or goats, or by deer drawn to the garbage midden—or maybe they come to stalk the occasional human straggler. Dreamy, distracted pilgrims might be easier prey even than the little Maldhari girls mentioned by Brook Fox.

What about you? I ask. Why travel all the way to Kankai for offering puja? "Very calm and silent place," Mr. Panna replies. "Surrounded by jungles and trees."

It is that. It's a blessedly sylvan contrast to Bombay. But some observers are concerned, with all due respect to their Hindu co-religionists, about the booming popularity of Kankai and other forest temples, which have made Gir less calm and silent. Ravi himself, a dutiful Hindu, says, "I have nothing against temples per se. But they tend to become business operations. And to expand." Kankai has certainly expanded. The temple's history goes back four hundred years, long before the designation of Gir as a wildlife sanctuary, before even the founding of the Babi dynasty of nawabs. When the area received protected status under Indian law, the temple's preexistence was recognized by a sort of grandfather clause. In 1974 it got a limited charter of rights from the Gujarat Forest Department, including the right to exist where it was, on a small patch of land, and to welcome pilgrims, who would enjoy road access and use of the river. Since that time, the temple managers have quietly added nineteen unauthorized buildings and erected the stone perimeter wall, which almost triples the land area claimed. Visitation levels have sextupled, from eight thousand pilgrims a year to more than fifty thousand. Traffic along the sanctuary roads has increased accordingly, as have overnight stays by pilgrims, accommodations to support them, garbage, firewood cutting from the forest, and noise. Prayer chants are now broadcast through a public-address system, and on the marble patio outside Kankeshwari's shrine there's a kettle drum with an automated beating mechanism, standing ready to deliver inspirational cadences, like a percussion track on a Hammond organ. Only Kankeshwari herself knows what the real lions think of all this.

Ignoring Mr. Panna's suggestion about the lion-watching opportunities at Tulshishyam, we go from Kankai instead to Banej, the same place he's headed for. At the temple there, set within another oasis-like patch of gallery forest, we see more images of the lion in its various mythic roles. One of these, a small painting, strikes me as strange. It hangs on the wall of a shaded veranda overlooking a tranquil, dammed pool in the river. On the pool's gentle surface floats a pumice stone, airy and smooth, to which a young boy directs our attention, calling it a "swimming rock." Turtles scull back and forth in the clear, tepid water, like swimming rocks themselves. Respectfully, we take off our shoes before entering farther, but the concrete steps leading up toward the shrine have been roasted by afternoon sun and are too hot to walk on, at least for an infidel like me. My soul may be thickly callused but not my soles. We pause instead at the veranda, where a man serves us sweet tea in shallow steel saucers. Having long since emptied my canteen during the eight thirsty hours since we set out, I welcome any gulp of liquid. Am I giddy with dehydration, or is it simply the spell of a Hindu holy place that makes everything here at Banej seem disjointed, mystic, a little woozy? While we slurp our tea, I study the strange painting.

It shows a sadhu, a bearded holy man clad in a loincloth, guiding a two-wheeled bullock cart heaped with firewood. The bullock itself, liberated from its traces, lopes away in the background. Buckled into the cart, in its place, is a lion.

What's the story behind *this*? I ask Ravi.

Haven't a clue, he says.

He shrugs it off, but I can't stop wondering. *Panthera leo persica* as a draft animal? Maybe it's an ancient scene from the Vedas. Maybe it depicts an event in the life of some Hindu equivalent to Saint Francis of Assisi. Anyway, I find it spooky. A lion harnessed to a load of dead trees feels too much like an allegory of the future.

11

A quarter century ago, the ecologist Paul Colinvaux published a limpid little book titled *Why Big Fierce Animals Are Rare*. Within the noisy jostle of one publishing season it came and went quietly, a minor offering from Princeton University Press. But it took hold in certain circles, sold steadily over the years, and acquired an excellent word-of-mouth reputation among those people (including me) who operate in the border zone between professional ecology and the general public. What made Colinvaux's book durably valuable was his knack for posing simple and seemingly ingenuous questions, the answers to which turn out to be not only interesting but also fundamental to understanding the dynamics of life on Earth.

Why are there so many different species? he asked. Why do common kinds of animals tend to remain common and rare kinds to remain rare, year after year, notwithstanding that they're all breeding as fast as they can? Why are some animals big and others small? Why are there no trees in the Arctic? Ecology, Colinvaux wrote, is "the science that reasons why." Any serious attempt to explain basic facts about animal or plant species—their shapes, their abundance, their habits—necessarily leads to a deeper investigation of the environment they live in, the resources they depend on, and what they do to acquire those resources. Having worked for years as a researcher and a teacher, Colinvaux wanted to reach out beyond the corridors of his profession and share a taste of that deeper investigation with nonscientists. So he paused from other duties to write this brief book full of fruitful, transcending questions, among the best of which is the one underlying his title: Why are large predatory animals invariably rare?

The first step toward an answer is to focus not on the matter of rarity but on the matter of size. "Animals come in different sizes," Colinvaux wrote, unabashed about stating the obvious, "and the little ones are much more common than the big." Take a typical patch of northern temperate woodland. You'll find insects, small birds, foxes, hawks, and owls. "A fox is ten times the size of a song bird, which is ten times the size of an insect. If the insect is one of the predacious ground beetles of the forest floor, which hunt among the leaves like the wolf-spiders, then it, in turn, is ten times bigger than the mites and other tiny things that they both hunt." The factor of ten between adjacent size classes is only a crude approximation, but to a great degree it holds true, and not just in temperate woodland. In the sea, diatoms and other algae represent the tiniest class, planktonic plants taking their energy straight from the sun; planktonic animals such as copepods, feeding on the algae, are about ten times larger; larger still, by about another order of magnitude, are the shrimp and the small fish that eat copepods; then come the middle-sized fish, predators such as mackerel, gobbling quantities of small fish and shrimp; finally, feeding on mackerel and whatever else seems worth a snap of their jaws, come the sharks and the killer whales. The scale is x, $10x$, $100x$, $1000x$, $10,000x$, roughly. Look at a tropical forest or an Irish bog, Colinvaux advised, and you'll see the same pattern. "It is an extraordinary thing but true that life comes in size-fractions which, for all the blending and exceptions that can be found by careful scrutiny, are remarkably distinct," he wrote.

From there he turned to the corollary fact, equally important and almost as obvious: "Animals in the larger sizes are comparatively rare." Of diatoms there are many many many times many, while of mackerel there are merely many. Of sharks there are few. Why? What determines that larger animals, especially predaceous ones, should be so rare?

In earlier centuries and millennia, this was often attributed to the Creator's direct intervention, producing an ineffable tranquility known as "the balance of nature." Herodotus voiced that idea in the fifth century B.C.: "And indeed it is hard to avoid the belief that divine providence, in the wisdom that one would expect of it, has made prolific every kind of creature which is timid and preyed upon by others, in order to ensure its continuance, while savage and noxious species are comparatively unproductive." Colinvaux's answer was less theological. He looked back not to Herodotus but just a few decades, to a pioneering ecologist named Charles Elton.

Elton was an Englishman, born with the twentieth century, who finished a degree in zoology at Oxford in 1922. Zoology at that time

was largely a laboratory science, focused on the morphology and phys-
iology of individual animals. University-educated zoologists of the day
tended to look down on the natural-history tradition of descriptive
fieldwork and instead spent their time measuring heart rates and
skull dimensions. It was ironic, Elton later wrote, that these new fashions in zoological research had been triggered by the ideas of Charles
Darwin:

> Darwin, himself a magnificent field naturalist, had the remarkable
> effect of sending the whole zoological world flocking indoors,
> where they remained hard at work for fifty years or more, and
> whence they are now beginning to put forth cautious heads again
> into the open air. But the open air feels very cold, and it has become
> such a normal proceeding for a zoologist to take up either a mor-
> phological or physiological problem that he finds it rather a discon-
> certing and disturbing experience to go out of doors and study
> animals in their natural conditions.

This wasn't surprising, Elton added, since there existed no real
tradition of training zoologists for fieldwork. In botany, by contrast,
the inclination toward fieldwork was strong, and botanists had
made greater progress than zoologists in the identification and classification of species—a crucial, basic enterprise, prerequisite to
deeper ecological understanding. Then again, Elton noted, botanists
enjoyed a certain advantage, "because there are fewer species of
plants than of animals, and because plants do not rush away when
you try to collect them."

Elton himself liked that cold, open air. In 1921, while still an under-
graduate, he had joined an Oxford expedition to the island of
Spitsbergen, four hundred miles north of Norway, where he collabo-
rated with a botanist on a survey of arctic plant and animal communities. Those communities were relatively simple—high latitude, hard
conditions, few species—and therefore lent themselves well to ecolog-

ical analysis. Elton continued his fieldwork there during return expeditions in 1923 and 1924. Around the same time, based on his Arctic experience, he became a biological consultant to the Hudson's Bay Company, an arrangement that gave him access to trappers' records and other fur-trade data going back two centuries. Those data, along with his Spitsbergen work, allowed him to begin studying the population dynamics of northern animals.

He looked at a number of species—arctic fox, red fox, lynx, snowshoe rabbit, muskrat, marmot, and lemming. He was intrigued by the matter of their relative abundance and by the changes in absolute abundance from time to time. Population sizes of predators, Elton saw, tended to fluctuate upward or downward in delayed response to fluctuations among prey. That seemed straightforward enough; an increased abundance of prey would support more reproduction and survival among predators. But what caused the prey populations to fluctuate? This was more complicated. Food supply, weather, disease? In 1927, after just three months of writing, Elton published his first book, titled *Animal Ecology*. At age twenty-six he had produced a classic.

Animal Ecology became a landmark in the transformation of ecology into a systematic science. Like Colinvaux's *Big Fierce Animals*, it's written with plainspoken clarity and grace. In spite of that, it was taken seriously by the professionals, and it's still cited in texts and journal papers on predation. Among its innovative contributions was a discussion of food chains.

"Food is the burning question in animal society," Elton wrote, "and the whole structure and activities of the community are dependent upon questions of food-supply." A "food-chain" as Elton described it (and spelled it, with a hyphen that has since been dropped) is a linear sequence of consumers and consumed. Plants derive their food from the soil, the air, and the sun. Herbivores eat plants. Carnivores eat herbivores. A "food-cycle" he defined as the interconnected aggregate of all food chains within a community. This may sound like familiar stuff, but in 1927 it was new (as attested by the *Oxford English Dictionary*, which cites Elton's passage as the earliest known occurrence of the term "food chain"). His "food-cycle" has been replaced by "food web," but the concept remains indispensable. Elton noted two important things about food chains, both of which were echoed by Paul Colinvaux: With each shift from one link to another, up the chain, body sizes are larger and population sizes are smaller.

Size, Elton recognized, is not to be taken for granted. He wrote a revelatory five-page section under the simpleminded heading "Size of Food." Size differences make food chains possible—and, to reverse that, food chains make size differences necessary, especially among animals that eat other animals. Each stage in the chain transforms smaller food particles into larger units, thereby making the food conveniently available to still larger animals, who couldn't cope with the smaller particles. "There are very definite limits, both upper and lower, to the size of food which a carnivorous animal can eat," Elton noted. The normal carnivore can't take prey above a certain size, because the carnivore isn't strong enough, ferocious enough, or adept enough to catch and kill such formidable victims. Swallowing capacity is another limitation. Many predators, lacking teeth that cut or tear, have no alternative but to gulp their prey whole. "At the same time," Elton wrote, "a carnivore cannot subsist on animals *below* a certain size, because it becomes impossible at a certain point to catch enough in a given time to supply its needs. If you have ever got lost on the moors and tried to make a square meal off bilberries, you will see at once the force of this reasoning." The lower limit, he explained, is also influenced by the local abundance—the sheer density—of potential prey at a small size. When mice are pestilentially common, foxes can live entirely on them; when mice are at middling abundance, foxes eat rabbits. Somewhere between the upper size limit (too big to kill or to swallow) and the lower size limit (too small to bother with) is an optimal food size for each predator species, and that's the size generally eaten.

To this pattern, of course, there are exceptions. Elton mentioned one, an anomalous case that suggests how crucial the size consideration may be: "Man is the only animal which can deal with almost any size of food, and even he has only been able to do this during the later part of his history." Dexterous fingers, opposable thumbs, artificial weapons, and cutting tools have vastly widened the dietary options of *Homo sapiens*. Baleen whales can feed on plankton, killer whales can prey on baleen whales, "but it is only man who has the power of eating small, large, and medium-sized foods indiscriminately." Elton chose to overlook several other ingenious species, such as the chimpanzee and the brown bear, that take food in a great range of sizes. Still, his point stands: Food size is a crucial matter for *almost* every animal that doesn't use tools.

Upper and lower limits also govern the range of sizes that any animal can attain. Those limits dictate still another: a limit on the number of links that any food chain can include. From plant to herbivore to first-order carnivore and onward, Elton observed, a food chain usually comprises fewer than five links. At each link, there must be a certain reproductive surplus adequate to support the predatory toll taken by the link above. If there's no such surplus, the food chain is unstable. In a stable chain, the surplus is made possible by another difference between each link and the next—a difference in reproductive rate. Little creatures multiply faster than big ones. So a hundred million diatomic cells may reproduce quickly enough to sustain their number and, at the same time, to satiate a million copepods; those million copepods may in turn support ten thousand shrimp, which may support a hundred mackerel, which may provide a single shark its occasional meal. But what creature eats the shark? None does—except, sometimes, that ingenious tool-using species, *Homo sapiens*.

Elton dubbed this situation "the Pyramid of Numbers," for its mathematical resemblance to a step pyramid. Ecologists nowadays call it the Eltonian pyramid, in his honor. At the top of the pyramid—the terminal step—is a population of predators not generally subject to predation themselves. "Finally, a point is reached at which we find a carnivore (*e.g.* the lynx or the peregrine falcon) whose numbers are so small that it cannot support any further stage in the food-chain," Elton wrote.

Why so? "Finally, a point is reached," he says. Yes, okay—but why is the point reached *here* and not *there*? For all the merit of Elton's book, his explanation of this end point is unsatisfying. Why does the Eltonian pyramid narrow to a terminal step? Why doesn't it rise upward endlessly, each step ever narrower unto infinity? Why are there no super-monstrous predator species that eat tigers or sharks?

12

I've been pondering that little mystery, both with and without the assistance of experts. It evokes some interesting hypothetical images. *Panthera rex*, the two-ton felid. *Carcharodon magnificus*, the shark that's as big as a blue whale. *Ursus humongous*, the elephantine bear. They don't exist, but why couldn't they? There is no lion-eating predator, stomping thunderously across Africa and Asia, armed with claws as long as pitchfork tines and teeth as big as tent pegs. Why not? Why does a food chain top off where it does?

The next crucial insight came fifteen years after Elton's book, from a student-and-mentor pair of the highest scientific caliber. Raymond L. Lindeman and G. Evelyn Hutchinson belonged to different generations but were united by their interest in the ecology of lakes. They studied the relationships between eaters and eaten. From close scrutiny of small lacustrine systems, full of sunfish, tadpoles, backswimmer beetles, and other small creatures, they derived large truths. Lindeman, during his graduate work at the University of Minnesota, had investigated a little body of water called Cedar Bog Lake. Hutchinson meanwhile was at Yale University, building quietly toward a reputation as the world's leading limnologist. They developed similar ideas independently, then combined their thinking in the early 1940s during a fellowship period spent by Lindeman at Yale. Unfortunately for ecology, the time was cut short when Lindeman died of an illness in 1942. He was twenty-six. His most notable work was a dense eighteen-page paper, titled "The Trophic-Dynamic Aspect of Ecology," which appeared posthumously in the journal *Ecology*. Hutchinson called it "the major contribution of one of the most creative and generous minds yet to devote itself to ecological science."

On the subject of food chains, Raymond Lindeman had picked up where Elton left off. In the 1942 paper he translated Elton's pyramid into the concept of trophic levels, each level representing a stage in the flow of energy through an ecosystem. By "trophic" he simply meant nurturing, or energy-availing, as a way to characterize the relationship between one level and another. The term "trophic levels" would later become common parlance in ecology.

Energy, as Lindeman and Hutchinson both recognized, was the resource commodity by which food chains could be quantified and understood. The fundamental energy input to any ecosystem (except ocean-vent communities) is sunlight. Plants capture a small fraction of that energy, converting it to molecular fuel by means of photosynthesis; then they burn some of the fuel in the course of their own life and growth, leaving the rest available in their body tissues. Herbivores eat plants, and of the energy thus consumed, some goes to support living activity, some goes into growth. They breathe. They run, swim, or fly. They radiate heat. They build muscles, organs, bones. They sprout antlers, molt their feathers and grow new ones, secrete shells. In the course of those processes, another large fraction of the original energy supply is irretrievably lost. Entropy: poof, sizzle, gone. The remainder stands available to the next link of the chain, the primary carnivores. It's physically embodied in the herbivores' living tissues, mainly as protein and fat. And so on, with a still-smaller fraction made available to secondary carnivores, in their turn. All of this can be figured in calories. That's how Raymond Lindeman figured it.

Compare the total quantity of usable energy from one trophic level (say, the herbivores) to the total quantity yielded at the next level (the primary carnivores), and you get a ratio. Lindeman called this ratio the "progressive efficiency" of a trophic relationship. It won't surprise anyone to hear that energy transfer along a food chain, like most other activities in life, is not highly efficient. Plants generally capture around two percent of the available solar energy, though under optimal conditions they might get as much as eight percent. The rates for herbivores and carnivores are harder to measure but probably range up to about ten percent. At each level, far more energy is used up or lost than is captured as food by the next level. It's another version of the Eltonian pyramid, but instead of ever-smaller populations near the top, it's ever-smaller amounts of energy remaining available. Finally, there's just not enough left to feed a hypothetical population of *Panthera rex* or *Carcharodon magnificus*. So those creatures don't exist. The base of the Eltonian pyramid isn't wide enough to allow its ratio-determined steps to rise that high. It's barely wide enough to support the real creatures who live atop it—such as *Panthera leo* and *Carcharodon carcharias*—at such low population densities that their collective survival can be difficult.

Decades after Lindeman, Paul Colinvaux cited an ancient Chinese proverb reflecting this ecological and thermodynamic reality: Each hill shelters only a single tiger. The largest predators are spread thinly on Earth because energy, in forms they can harvest, is limited and broadly dispersed. They must travel distances. They can't afford congregation. They need to hunt and compete desperately. They must be bold, prudent, stealthy, opportunistic, and lucky. Their meals are few and far between.

Elton's pyramid, Lindeman's energy analysis, and Colinvaux's gloss on them help explain why, after thirteen hours driving with Ravi and Mohammad through the Gir forest and its bordering areas, I haven't seen a single Asiatic lion. Big fierce animals are inherently rare.

13

But that's not the only factor keeping the lions invisible today. We've been on roads. We've been noisy. We've moved far and fast, seeking a wide-angle glimpse of the ecosystem at the expense of deep focus. For serious lion watching, we'll need to ditch the vehicle, escape the judicious restraints imposed by the Gujarat Forest Department on casual visitors to the ecosystem, and get out into the forest on foot.

We exit the protected area at a checkpoint south of Banej and return toward Sasan along a public road, heading west into the dusty glare of sunset. This is a different sort of light from the subtle sepia of morning. It's thirstier, more bleary, and redder, like a bloodshot yolk. Another day's exudation of multitudinous human striving has been added to the atmosphere since sunrise. Here on the south perimeter of the Gir Wildlife Sanctuary and National Park, where conditions of soil and moisture seem to be more hospitable than in the north, we see sugar cane fields, peanut crops standing bushy and green, and a considerable expanse of mango orchards, shapely trees in neat rows laden with heavy fruits that dangle like soap-on-a-rope. The mango industry seems to be booming, Ravi observes. Mango is a cash crop for which there's growing demand, and it travels. Eat it fresh, or make a chutney, or pack it in boxes for shipping to Bombay, Delhi, overseas. The local variety, called Kesar, is pungently sweet and so creamy-smooth you can scoop out the pulp with a spoon. I've eaten my share, bought cheaply from street stands in the village. Notwithstanding its cost to the landscape, I can't help admitting it's the best mango I've ever tasted. I suppose that makes me complicit.

And each time another hectare of land is cleared, for mango orchards or crops, either legally or illegally, the forest gets smaller. The lion population becomes more constricted. The prospect for conflict rises. The prospect for long-term survival of *Panthera leo persica* declines. This brings to mind still another reason why large predators, big fierce animals, are a rarity on our planet. They have a high incidence of going extinct.

THE MUSKRAT
CONUNDRUM

14

It was almost by accident that Ravi Chellam came to the study of Gir's lions. Born in the southern state of Tamil Nadu, he grew up in Madras with hopes of becoming a physician. But because of his caste identity (as a Brahmin), and despite a good record in school, he found himself excluded from medical training by the reservation system—India's version of affirmative action, designed to compensate for two millennia of caste discrimination. So he went to work as a marketing executive and immersed himself outside the office in serious amateur cricket. Lanky and quick, he was a right-handed bowler with a nifty fastball—nifty enough that, during one of his years with the Grand Prix Cricket Club, he was the top wicket taker in the state. He got a decent offer to play semi-pro cricket for a corporate-sponsored team, and at one point he even toyed with the idea of turning pro.

But by his mid-twenties Ravi found that the comfortable job, the high life of a young executive, the city of Madras—and even the falling wickets—didn't satisfy him. Meanwhile he had gotten involved in conservation, initially as a volunteer for the World Wildlife Fund and at nature camps for Indian youngsters. He had no inkling, at first, that wildlife biology could be a profession. Then, in 1985, he joined the Wildlife Institute of India as a doctoral student. Presented with a

choice between two possible dissertation projects—he could partici-
pate in an elephant study, or he could start something on the Gir
lions—he chose the lions. They were more severely endangered and
therefore more urgently interesting. But while he was back down in
Delhi for a visit, awaiting the paperwork necessary to clear the bureaucratic path for his study, a third possibility arrived in the form of a
telegram. "Come and join the snow leopard project" is how Ravi
remembers it.

The snow leopard. Famed, mythic, and scientifically enticing.
Beyond the inherent attractions, too, there was some pressure from the
institute for him to accept. But . . . four years in the Himalaya? *No way*,
he thought. *I'm a southern boy, a Tamilian, I like it hot*. "Snow-leopard
habitat was like the Arctic Circle to me," he says, recollecting the decision years later. The Gir forest, with summer temperatures reaching 125
degrees Fahrenheit, was more to his taste. So was the lion, which as a
research topic had its own sort of hotness. He made his first trip to Gir,
a short reconnaissance visit lasting roughly a week, in December 1985.

That visit was almost discouraging enough to make him reconsider
the snow leopard. Like most new graduate students, he was long on
enthusiasm, short on practical experience, resources, and clout. The
responsible officials of the Gujarat Forest Department welcomed him
coolly. He had no vehicle. No research authorization, not yet. He had
hoped to establish some working contacts, but he didn't get far. To the
little bureaucratic nawabs who controlled the Gir Wildlife Sanctuary
and National Park, he was young, transient, academic, easily ignored.
He did receive permission (sufferance, that is) to walk around. Yes, he
could do that—and would he, please, try not to get eaten by a lion. He
made little forays into the forest, seeking a feel for the place, acquaint-
ing himself with the community of species. Chital were conspicuously
abundant, but there were also the sambar and nilgai, the chinkara, the
langurs and wild boar, the peafowl and flycatchers and lapwings and
mynahs and so many other birds. He saw that this was a teeming,

diverse ecosystem, not just a compound for a few hundred endangered cats. He had outfitted himself in full camouflage gear, with a knife and a canteen on his belt, but the camo couldn't conceal that he was a greenhorn. He recruited a young boy to guide him. The boy was ten or eleven years old, knowledgeable but taciturn; his sole piece of field gear was a stick, a sort of shepherd's staff, suitable for coaxing buffalo home after a day's grazing. With the boy's help, Ravi found plenty of lion sign—tracks, scat piles, kills. He heard lions roar at night. But he didn't lay eyes on one.

"It was getting to be my last evening," he recalls, "and I had still not seen a lion. So I was beginning to have second thoughts." His unease was based on the most practical of concerns: A graduate student needs not just a subject, not just a research question, but a decent prospect of gathering data. Without data, no dissertation, no degree. For a field biologist interested in behavior, data derive from observation. Ravi didn't want to spend years chasing an animal so rare or so wary it was invisible.

That evening at dusk, as he and the boy walked along a dirt track, Ravi heard grunting, then saw four lions emerge from the brush. They were subadult females, he recognizes in retrospect, though he couldn't have said that at the time. Oblivious to him, they roughhoused amiably among themselves. Maybe they were sisters. Their age, their gender, their possible family relationship, their easy mood—none of these registered on Ravi so vividly as did certain more elemental facts: "They were lions, they were making noises, I was on foot. And I definitely felt threatened." While he struggled to get hold of himself, the young lionesses noticed him and, wary but calm, turned to withdraw. The whole encounter lasted no more than a couple of minutes. Throughout that time, the little boy with the stick stood watching casually.

To the boy, a pride of lions was a familiar reality of the local landscape—nice to look at, amusing on a slow day, formidable, testy, but hardly more dreadful than a diesel locomotive rumbling by on its tracks. You had to stand clear and mind your manners, but there was no cause for undue alarm. Meanwhile the postulant wildlife biologist in the camouflage suit was, by his own later testimony, "rooted to the ground in utmost fear."

Then he remembered his camera. The four lionesses, highlighted by a setting sun, their faces glowing golden, would have made a lovely photo. Instead all he got was an ass portrait of the last of them as she retreated into the forest. "That's the way I began," Ravi says.

By the following March his fieldwork was under way. He'd found a place to live in the village of Sasan, just outside the protected area, but spent as much time as possible in the forest. His plan was to study the lions' predatory habits, to get a rough profile also of the prey populations, and to learn whatever he could about the lions' social organization and habitat use. For the predation work, his method entailed collecting scats and analyzing kills. By closely examining the hairs extracted from a washed and dried lion turd, he could tell what species of animal had been eaten. By locating kills before they were totally consumed and determining the species, sex, age, and physical condition (from teeth and bone marrow) of a victim, he could learn more about the lions' dietary proclivities. Those data were easily quantified. He supplemented them, whenever opportunity allowed, with observational notes on the lions in action. That's what he was doing one night, toward the end of the first year, when he had his closest encounter with *Panthera leo persica*. It supplied vivid evidence that Gir's lions are not like lions anywhere else.

He was documenting a predation event. The prey was a sambar stag. It had been killed on a steep slope and dragged downhill to be stashed in dense brush just above a streambed, where vultures and crows wouldn't pick it clean. While Ravi and his tracker collected basic data from the carcass, they noticed that the lions who had made the kill—a group of adult females with cubs—were lingering nearby. The stag was a big animal, sizable enough to offer the lions a day or two of intermittent feeding. Aha, Ravi thought, here's a chance for observations. So he returned to the site with water, food, and his sleeping bag, prepared for a stakeout. To minimize human presence, he sent his tracker back to sleep at the nearest Maldhari ness.

A cool, moonlit night. Ravi stayed awake, listening to the crunch of bone as the lions fed, watching their dusky shapes come and go. When they climbed down into a rocky gorge to drink, he followed cautiously. "It was fairly interesting," he says blandly. After daylight, with the sambar still not fully eaten, he decided to resupply himself and return for another night.

On that second night he lay awake until two A.M., cozily zipped into his sleeping bag, his back propped against a *Syzygium* tree as he made notes. Then he dozed off. Several hours later he woke with a shock of realization—that he was sleeping in the forest with lions all around. Oops, a reckless lapse; gotta be more careful. Then he felt weight on his legs. He looked down into the face of a lion cub that had curled itself onto him for a congenial snooze. Ravi's first thought was, *Where's your mother?*

He looked around. No sign of the female. "So I kind of wiggled my toes, to gently coax the cub away from the sleeping bag. And he obliged. I can still remember the startled expression on the cub's face." The little lion circled behind the *Syzygium* tree and peeked back at him curiously. Ravi snapped a flash photo at that moment, capturing mostly tree trunk. The cub, having realized that this flash-faced stranger wasn't its mama, ambled away to find her. And the mother herself? If she knew where her youngster was and whom he was cuddling with, she apparently hadn't cared.

This was more typical of Ravi's experiences than anomalous. "I did all my work on foot," he says. "When I was in Africa last year they refused to believe it." His counterparts among African lion biologists made plain that *their* lions didn't grant any such indulgence to foolhardy bipedal interlopers. "They're not used to people on foot," Ravi says of the African animals. "They'll just come and charge at you and kill you." But for one reason or another the attitude of *Panthera leo persica* toward humans seems to be notably different from the attitude of *Panthera leo leo* in Kenya or Tanzania. Ravi mentions another occasion, when a full-grown male seated just a few yards from him let loose a roar so forceful that he felt the lion's spittle on his face. Was this a threat? "No. He was just roaring." It's a distinct element of the vocal repertoire, Ravi explains, not to be confused with growling, which is quieter but more hostile. "Roaring is a signaling system to other males. To friendly males it's to say, *Here I am.* And to unfriendly males it's to say, *Here I am, bugger off.*" Because the Gir lions seem so disinclined to view humans either as potential threats to their own safety or as prey, the people who spend time afoot in the Gir forest—Maldharis, Forest Department functionaries, researchers such as Ravi, even villagers who sneak in to poach firewood or fodder or fruit—generally arm themselves, if at all, with nothing more than a stiff stick, as Ravi's boy did, or a kuwadi.

15

A kuwadi is what Mohammad Juma holds ready, on another afternoon in the forest, as Ravi and I squat amid the dry gravel of a streambed, not many strides from a lioness with three cubs.

We're accompanied this time by a young American photographer named Michael Llewellyn, whom I've dragged out here in connection with a magazine assignment. My purpose and Ravi's is to observe the lioness and her brood from a discreet distance, seeing as much as we can while committing minimal intrusion. Michael's purpose is slightly different: to capture strong images. Because he's an artful photogra-

pher with a very particular style—he does fashion and offbeat portraiture more often than wildlife photography—he eschews telephoto
lenses. Didn't bring a single big zoom-tube out here to India. He
works with Hasselblads, for Christ's sake. Have you ever tried to photograph a lion while gazing down into the viewfinder of a Hasselblad
C? The first problem is that a person has got to get close.

The cubs seem content to cooperate. They're as big as cocker
spaniels, and floppy with cheerful innocence. Their faces are fuzzy,
their paws are huge. Their legs and bellies are spotted, the usual pattern among infant lions, fading into tawny brown on their upper bodies. While Michael sidles closer and closer, coached and vigilantly
attended by Mohammad, Ravi and I chat quietly.

These cubs are about four months old, Ravi figures, born at the start
of the dry season. A litter of three isn't unusual, he tells me. In general
he has seen no evidence, either in reduced reproductive rate or birth
defects, that the population bottleneck at the turn of the century has
caused genetic problems due to inbreeding. Litter size has been good;
there are no kinked tails; no sign, so far, of extraordinary susceptibility
to disease. "The idea that this population is doomed is just wrong,"
Ravi says. Still, the bottleneck has presumably left the lions with low
genetic variability, meaning that their vulnerability to a microbial epidemic could be high. How serious is that danger? Very serious, as a
minatory lesson from Africa suggests.

In the Serengeti ecosystem, a variant of canine distemper virus
(which usually affects only canids) appeared in 1994, and this variant
proved lethal to cats. Within less than a year it killed about one thousand lions, a third of the Serengeti population. Those Serengeti lions
had presumably enjoyed much greater genetic diversity, and therefore
better collective resilience, than the few hundred surviving animals at
Gir. If the same sort of disease struck *Panthera leo persica*, the entire
population might be wiped out.

As though she can hear this disquieting thought, the lioness shifts, rising from her spot in the shade of a *Syzygium* tree. Michael freezes. She withdraws a few yards, moving into deeper shade along the streambed's far bank. "Something's bothering her," Ravi says. The cubs follow, and Ravi peers carefully. "She's got a kill."

It's a chital, stashed in cover. One cub gnaws at the deer's rib cage, sharing a snack with his mother, while the others rest. After a few moments the female pauses, lifts her head, and gazes straight back at us, now all of sixty feet away. In a glance you'd never distinguish her from an African lioness—she has the same golden face, creamy chest, black marks on the backs of her ears. A male of the *persica* subspecies usually has a shorter mane than his African counterpart, especially on top, leaving the ears more prominent. Both males and females of *persica* also have a telltale keel of skin down the center of the belly; a similar belly keel shows sometimes, but seldom, among African lions. These are subtle indicators, brooking exceptions, and noticeable only to an expert like Ravi, who has stared hard at many individual lions. Apart from such external clues, it would take genetic lab work to sort Gir lions and African lions into distinct groups. Even at close range, in daylight, I can't spot the belly keel on this female. Unperturbed by our presence, unflattered by our attention, she resumes eating.

"You can see how comfortable she is," Ravi says. Some Indian biologists, especially those who study tigers or leopards, are inclined to belittle the Asiatic lion as unnaturally docile. Ravi's own dissertation adviser, an eminent field biologist named A. J. T. Johnsingh, used to tease Ravi about working with those "tame" cats down at Gir. "But hell, she's a wild animal, she's got a chital," says Ravi. "She's not gonna commit suicide." He means that avoiding unnecessary conflict is judicious, not craven, for a predator. A broken tooth can be a dire misfortune. A gash sustained in a silly fracas can become infected. The extraordinary calm of the Gir lions, their stolidity, their sensibleness, is no proof that the subspecies has been enervated by persecution and inbreeding. "I'm a little touchy about that," Ravi admits.

Anyway, this female hasn't lived free of harm, as evidenced by a dark, naked scar on her left shoulder. She has a notch in one ear, as though clipped by a tooth. Ravi would use those marks to identify her, he says, if he were still doing fieldwork here. Apart from the nicks and the scars, she looks to be in good condition—no more than six years old, canine teeth intact, just reaching her prime. She commands a nice territory, including this stretch of streambed, with its gallery forest and its late-season pools. The shade and the water should be attractive to thirsty chital—especially in the hard weeks before the monsoon—and therefore tactically valuable to her as a hunter.

Those hard weeks are presently at their crescendo. The dry upland forest is so brown and leafless it seems cauterized by fire, and even the stream's shaded pools are nearly dry. But the Forest Department has made provision against that problem. Knowing that the lioness will be thirsty after her meal, Mohammad walks up the hillside to start a hidden pump. We hear the pump begin to chug, and then well water comes gurgling down through a fat hose. Slowly, her water hole is replenished.

But even at Gir, and with the luxury of tap water, a lioness is a lioness. There are limits to her tolerance; there are signals. One comes when Michael, shifting position for a different camera angle, turns his back to her. Suddenly she's crouched on her haunches, ready to move.

"Whoa," Ravi warns. "That is *not* a relaxed position."

Mohammad, trusting soul, has left his kuwadi with me. Now I'm supposed to be the enforcer? Not likely. I can't imagine using this blunt little ax on a lion, not even to save Michael from attack—in fact, not even to save his film. Someone could get hurt.

Fortunately for us all, I'm not put to the test. The lioness moves toward the water, not toward the paparazzo. She stops, hesitant, waiting for the pump to fill her pool. A crow makes off with a gobbet of chital. When Mohammad cuts the engine and silence returns, she strides onward across the gravel with loose-jointed assurance. Crouching on her haunches, paws down, elbows out, she leans forward to drink.

Michael and Ravi are now positioned at the opposite edge of the pool, just thirty feet from her across the water. Ravi reads her body language. Michael, gazing into the viewfinder, wants that leonine face filling his frame. With the next shutter click her head jerks up.

She doesn't like the noise. But she tolerates it as a minor distraction. After another drink she sits up and shakes her head, clearing droplets from her whiskers. She strolls back to the shade and lies down. When a cub approaches, she lets it sprawl across her forelegs. The cub, like a toddler served spaghetti, has slimed its face with chital gore; now mama begins cleaning it with her tongue. She squints contentedly. The cub tilts its head back in ecstasy.

Then she rolls over, exposing her tits, and invites that cub and another to nurse. They clamber aboard, gently jostling for position. At roughly this point she becomes affixed, in my memory, with a provisional label: the Happy Lioness. She makes eye contact with us, as though to say, *Haven't you men got any work of your own to do?*

16

It's a halcyon scene, yet it shouldn't be sentimentalized. Harder times lie ahead for the Happy Lioness and her cute little cubs.

If the youngsters manage to survive for another couple of years, they'll reach their own age of independence as their mother cycles back into estrus. Where will they go then? They'll need food, they'll need water, they'll need secure cover, and to meet those needs they'll need territory. The mother won't surrender her own—not happily, not to male offspring (though she might share territory with her grown daughters). What then? Gir is already full of lions. What's an unsettled animal to do? Finding space, finding safe harbor and sustenance, is difficult in an overcrowded ecosystem. Charles Darwin offered a famous simile on that theme in *The Origin of Species*: "The face of Nature may be compared to a yielding surface, with ten thousand sharp wedges packed close together and driven inwards by incessant blows, sometimes one wedge being struck, and then another with greater force." It's crucial to remember, he noted, that every organic being faces a struggle for existence, that each seeks to survive and to procreate, that many more offspring are born than the habitat can support, and that "heavy destruction inevitably falls either on the young or old." How does a juvenile lion manage to wedge himself in among the others?

The social system of Gir's lions is unlike that in East Africa, where open savanna habitats and big-bodied prey, such as wildebeests and zebras, allow for cooperative hunting and food sharing by large groups. Those groups include even adult males. Beyond their primary role as decorative studs, the African males offer some secondary social utility to the female-based prides—defending cubs against hyenas while the lionesses are hunting, for instance—but generally they live relaxed lives, in association with this or that pride, and mooch off the females. At Gir, circumstances are different. The forest is dense, the chital are petite, so cooperative hunting is less advantageous. Pride size is smaller. The association between sexes is more tenuous. As in Africa, males may form coalitions among themselves—consisting of three or four animals who hunt and travel together and who may jointly defend a territory—but a male seldom loiters with a pride, except at the time of mating. And if a young male doesn't find some buddies, he's utterly on his own. He might fight skirmishes to win a good place within the ecosystem, or he might wander out beyond the protected area into human-dominated terrain, there to face other forms of peril, or he might attempt to eke out a marginal existence amid the squash of competition. Various sorts of dangerous temptation and delinquency await him as alternatives to starvation.

One such temptation is domestic livestock. For that, a young lion needn't go far. Just moments downstream from where the Happy Lioness has been nursing her cubs, for instance, we come upon a dozen domestic buffalo, lapping water from a concrete tank under the watch of several Maldharis.

The Maldharis stand ready, if a lion should attack, to take action. Their first recourse would be to shout and throw stones; if that didn't stop the cat, they might bonk him on the nose with a stick. Pushed to extremity, they'd try to crack his forehead with a kuwadi. But for now they look serene and impassive. I suppose guarding livestock in lion country—this way, the old-fashioned way—is a Maldhari equivalent to serving in the army: long stretches of tedium punctuated by terror. The difference is that their hitches last a lifetime.

17

Domestic bovines, and the people who herd them, have never lived an easy existence within the Gir forest. Thirty years ago, as I've mentioned, the lions fed mainly on livestock. Killing cows and buffalo was routine, and not just for rootless young males. This was documented by a Canadian graduate student named Paul Joslin, who did fieldwork at Gir in the late 1960s for a doctorate at the University of Edinburgh.

From his microscopic analysis of almost five hundred lion scats, Joslin found that "some 75 percent contained hair of domestic stock." It was a high figure but not surprising, Joslin added, "considering the preponderance of domestic stock which had been shown to exist in the sanctuary." During the dry season, when outsiders from surrounding villages brought their cows and buffalo to graze in the sanctuary, that preponderance reached a nine-to-one ratio, domestic herbivores versus wild ones. The resident Maldharis owned about 20,000 head of livestock; the outsiders drove in another 25,000, or, during a severe dry season, as many as 70,000. Somewhere between 45,000 and 90,000 hungry cows and buffalo, then, competed for Gir's grass. The place was a big feedlot, complicated by the presence of trees and lions. According to Joslin's counts and calculations, domestic buffalo constituted fifty-three percent of the total population of large herbivores, and cows another thirty percent. Chital, the most common wild ungulate, represented just eight percent of potential prey animals on the hoof.

Joslin also gathered information, both from Maldharis and from outsiders, about lion attacks. He learned that cows suffered heavier predation than buffalo, possibly because buffalo were more belligerent when threatened, possibly also because cows usually found themselves at the front of a moving herd, the first to walk into any lion ambush. He learned that most of the lion-killed animals were Maldhari live-stock, presumably because Maldhari herds remained in the sanctuary twenty-four hours a day, whereas outsiders' animals came for daytime grazing and were herded out at dusk. The Gir lions tended to be noc-turnal hunters; when they killed livestock, it was generally by skulk-ing into a village near the protected area, under cover of darkness, or by jumping over the thorn fence into a ness. The government of Gujarat had established a compensation program for stock losses, hop-ing to keep people from poisoning lions in retaliation. But the program carried conditions that few Maldharis or other herders understood, and many potential claimants didn't even bother to apply. At best, stock owners had to wait months for payment, and when a claim was rejected, they seldom got an explanation why. The reported cases of lion poisoning averaged one per year, Joslin learned. How many poi-sonings went unreported was anybody's guess.

Joslin's other interesting discovery was that, in many cases, lions got no chance to enjoy the meat of a cow or a buffalo they had killed. Instead they were driven off by the animal's owner, who guarded the carcass until a professional skinner could arrive. Hides were valuable, and any herder who had just lost a cow or a buffalo wanted at least the consolation of selling its skin.

But since the task of flaying an animal was so unsavory to Hindu sensibilities, hide harvesting had become the special service of certain Dalit people from surrounding villages. Dalits ("Untouchables," under the old terminology of the caste system, or as Gandhi preferred to call them, Harijans, "Children of God") throughout India had been relegated by tradition to such unsavory jobs as butchering animals or cleaning latrines. Stripping hides from lion-killed livestock fell in the same category. Sometimes a hide collector also bought the carcass for its meat, if that hadn't gone rotten and the transport distance wasn't far. And if the hide collector spurned it, the meat would go down the gullets of whitebacked vultures, since the lions themselves had by then abandoned hope of getting it. The vultures, sedulous scavengers with nothing to prove, were more patient about waiting their turn.

This arrangement was a boon to hide collectors roundabout Gir, who derived a considerable fraction of their trade from salvaging lion-killed livestock. It was a boon also to *Gyps bengalensis*, the whitebacked vulture. *Gyps bengalensis* is a rangy, raucous creature, about which India's most authoritative bird book says: "Large gatherings collect at animal carcasses with astonishing promptness and demolish them with incredible speed. The obsequies are attended by a great deal of harsh screeching and hissing as the birds strive to elbow themselves into advantageous positions, or prance around with open wings, two birds tugging at a morsel from opposite ends." Despite that voracious moxie, whitebacked vultures can't tear cowhide with their beaks. If a lion-killed cow was left intact, as Paul Joslin noticed, they could only nibble at the soft areas (eyes, ears, mouth, nostrils, anus) or at any gaping wounds. But if the cow had been skinned—peeled like a banana by the helpful Dalit—a gaggle of vultures could pick the skeleton clean within half an hour. Even a nine-hundred-pound buffalo, left flayed, would be stripped of flesh in a jiffy by these birds.

In almost a quarter of the cases Joslin investigated, the lions that killed a cow or a buffalo didn't get a single bite of its meat. In another twenty-two percent of the cases, they got a tantalizing taste but no more. Yet they supported themselves in large part from the half share of their livestock kills that they managed to keep, and the loss of so many carcasses to Dalits and vultures merely required them to kill more than they needed. It was an altogether weird ecological economy, in which resident Maldharis, outsiders in search of free pasturage, lions, hide collectors, and vultures all profited from a gross overabundance of domestic livestock set to feed in the Gir forest. Facing such competition, the wild ungulates languished. The ecosystem as a whole began showing signs of deterioration.

As the 1960s ended, two other researchers dovetailed their work with Paul Joslin's. Toby Hodd was a British range ecologist who studied the effects of domestic grazing on soil condition and plant productivity within the forest. Stephen H. Berwick was an American, a grad student from Yale, who for his doctoral project investigated the food habits and population dynamics of Gir's wild ungulates—the chital, the sambar, the nilgai, and three other species. Together these efforts (with support from the Smithsonian Institution and the Bombay Natural History Society, and complemented by several other studies) became known as the Gir Ecological Research Project. They were intended to produce an integrated analysis of the Gir ecosystem that might answer two urgent questions: Are the lions failing? And if so, why? These questions had arisen from the latest lion census, in 1968, which suggested an alarming decline.

Methodology is a sensitive issue when it comes to censusing lions in the wild, and for this effort the Gujarat Forest Department had used a combination of methods, old and new. The old method entailed counting distinct sets of pug marks seen at water holes and other likely sites during a forty-eight-hour period. The problem with this method was in determining which sets were indeed distinct, rather than merely

multiple tracks from a single animal. The chosen solution was to measure length and width of individual paw prints, on the assumption that
every lion had paws of which the precise proportions were unique.
Forest guards and hired laborers did most of the actual measuring, by
way of small splints of bamboo broken to match this or that dimension.
Each pair of splints represented the length and width of a lion's right
front paw. Such pug measuring and counting had been used for the
first formal census in 1936 and, with minor variations, for subsequent
censuses in 1950, 1955, and 1963. According to the 1963 census, the total
lion population within Gir and in a few peripheral patches of habitat
was 285. That was almost exactly what the 1955 census had found. The
similarity of the two results, eight years apart, encouraged (but didn't
necessarily justify) confidence that the method was reliable. Based on
bamboo-splint calculus, the lion population appeared to be stable.

Still, accuracy could always be improved. For the 1968 census,
Forest Department officials decided to supplement the pug-mark tally
with a visual count based on baiting. Live prey animals were tethered
at strategic sites in the forest, and two observers were stationed at each
site, counting the cats that came to kill and to feed.

If this combination of methods was more accurate, it turned out also
to be less encouraging. The calculated total in 1968 was just 177. The
lion population seemed to have fallen by almost forty percent within
five years. That was disturbing news—especially so since, in 1965, the
Gir Wildlife Sanctuary had been established expressly to protect
Panthera leo persica.

Paul Joslin, arriving around this time, made a series of population
estimates himself, using several different methods, and concluded that
the Forest Department's figure of 177 was about right. He also noted
that most of the population decline had occurred outside the protected
area. The PA itself hadn't lost nearly as many lions as those peripheral
patches of habitat. Was that a consolation or not? It depended on
whether you took a narrow bureaucratic perspective or a wide ecolog-

ical one. Part of the reason for the decline, Joslin concluded, was that the peripheral patches had shrunk or disappeared. Lion habitat had become human habitat. With the Green Revolution in ascendance throughout India, bringing improved seed strains and artificial fertilizers (along with a naïve conviction that whizbang agricultural technology would cure the world of hunger), more and more acreage in southern Kathiawar was being cleared and plowed. The lion's realm was diminishing, as it had been for centuries. But now the diminishment approached a terminal threshold.

To that gloomy fact the research done by Joslin, Toby Hodd, and Steve Berwick soon added a gloomy corollary. The Gir ecosystem, supposedly safe within its statutory boundary, was in a sorry state of disrepair.

It had suffered, and was continuing to suffer, degradation of several sorts. For instance, small-scale commercial timber operators had been permitted to cut trees in some parts of the sanctuary; their main interest was teak, which the Gujarat Forest Department managed on a forty-year rotation. This yielded about 600,000 rupees ($60,000 at that time) in annual revenue but left stands of new-growth plantation teak and coppiced stubs in place of older, more diverse forest vegetation. The wild ungulates, which browsed on leaves and twigs, weren't very fond of teak, but the caterpillars of certain pestiferous lepidoptera liked it fine, and they occasionally exploded in huge outbreaks, eating teak leaves down to lacy skeletons. Fertile ground along the stream valleys, even inside the sanctuary, had been cleared and planted in crops. Some Maldharis had taken to gathering buffalo dung mixed with topsoil from around their nesses and selling it in bucketloads to outsiders, thereby exporting a fraction of the forest's recyclable protein, carbohydrates, fats, and minerals. Besides losing nutrients, the soil around each ness grew compacted from all that hoofed traffic, reducing its ability to hold and retain water. Overgrazing seemed to be changing the very nature of the vegetation community, as nutritious perennial grasses were replaced by less palatable annuals that could spring up from seeds each year. The hard grazing also invited erosion. Cut, eaten, plowed, trampled, scraped, flushed—the place was beset and bedraggled.

Were there just too many animals sharing Gir? Too many wild ungulates combined with too many head of livestock? Well, maybe it wasn't that simple. Steve Berwick's study of the food habits of chital, sambar, nilgai, and the other native herbivores persuaded him that those species lived almost exclusively from browsing, not grazing, at least during the critical period, the dry season. Competition between wild herbivores and domestic stock seemed to be minimal, and overgrazing was accountable to the latter. Cattle, he concluded, were destroying Gir. And as the forest went to hell, he wrote, so also would go its unique community of animals, "including, of course, the last of the Asiatic lions."

What could be done? Closing the sanctuary to outsiders' cows and buffalo would be an obvious first step. But what about the Maldharis, who'd lived within the sanctuary since before it *was* a sanctuary? This was a complicated dilemma, both in ecological and in human terms. "If the Forest Department could remove the Maldharis and their cattle," wrote Berwick, "the forest could be preserved, but the major source of food for the lions would disappear, and the Maldharis— an exceptional human resource—would have to be relocated." Relocation, he added ominously, would entail "unknown potential social consequences."

Evicting the Maldharis was a drastic idea, given their willful detachment from modern India. Where else would they go? What else would they do? Could they live without access to the wild landscape of Gir? Could the lions live without them? The government of Gujarat decided to find out.

18

No one knows just how long the Maldharis have inhabited the Gir forest. By one account they arrived during the late nineteenth century,
having begun their migration down from Rajasthan hundreds of years
before. They may have reached the Kathiawar Peninsula by the fif-
teenth century, but they don't seem to have landed in Gir until much
later, maybe 130 years ago, give or take a few decades. In other words,
although their culture is traditional, their presence within this ecosystem is probably not ancient—no more ancient, say, than the presence
of white ranchers and settlers in Montana and the Dakotas. One crucial difference, though, is that the Maldharis didn't displace a native
people already in residence. There was no series of abrogated treaties,
no Baker Massacre on the Marias River, no Little Big Horn, no Dawes
Act, no Wounded Knee. The Maldharis merely occupied space left
vacant by the last bands of Kathi outlaws. They found an ecological
role—living peaceably in the forest, raising cattle for dairy—that was
new to the area.

Who are these people, known conveniently but opaquely as "the
Maldharis"? That question has puzzled me ever since I first saw a
group of such men, in their kadias (the blousy shirts) and their chornis
(the peg-leg trousers) and their white knit hats, tending buffalo in the
forest. It hasn't been satisfactorily answered by the various ethno-
graphic studies I've found. Harald Tambs-Lyche, in his book on traditional Kathiawar cultures, places them among the *ter tansali*, the
"thirteen castes," a cluster of closely allied groups that have played varied but conspicuous roles in the region's history. Preeminent among
the thirteen castes were the Rajputs, that lineage of warriors famed for
their bravery and chivalric airs, who spilled out of Rajasthan into
Kathiawar and elsewhere, conquering, establishing princely courts,

causing magnificent temples to be built and heroic poetry to be written, setting a fine style. Of the remaining dozen castes that recognized the Rajputs as first among equals, three specialized as pastoralists: the Ahirs, the Bharwads, and the Rabaris. Another caste, the Charans, combined pastoralism and poetry, gradually raising themselves to privileged status as the court bards of the Rajputs. Those four—the Ahirs, the Bharwads, the Rabaris, and the Charans—accounted for much of the early Maldhari population of Kathiawar. Originally they were nomadic, moving their herds here and there across the peninsula, searching for open pasture or for harvested crop fields in which their animals might be welcome to munch stubble. The Maldharis didn't do any farming themselves, but in exchange for pasturage they could deliver manure. As the landscape became more owned and constricted, many of them took to other occupations. Some settled in uncrowded areas, especially forests, establishing semi-permanent camps that they could relocate every few months or years as grazing conditions or water supplies required. That became the pattern at Gir. Most of the Maldharis here are either Rabari or Charan.

According to one etymological hypothesis, the name Charan derives from *char*, meaning "to graze." An alternate view is that it comes from a word meaning "to spread," as in spreading news, spreading tales of glory. Charans have a myth of their origins that, given later history at Gir, seems almost prophetic. The story goes that the goddess Parvati, Shiva's consort, made a little idol or doll from what is delicately called "the dirt of her body," into which Shiva himself put the breath of life. Shiva commanded the fecal doll-man to serve as a herder, responsible for an incongruous quartet of animals— a cow, a serpent, a goat, and a lion. As recounted by a scholar named R. E. Enthoven, keeping peace among the four beasts was difficult: "The lion attacked the cow and the serpent attacked the lion, but the herdsman by the gift of some of the flesh of his arm quieted them and brought them to Shiva, who in reward gave him the name of Cháran or grazier." Trying to intervene between cows and lions, a Maldhari occasionally still sacrifices his own flesh.

According to Enthoven, the mythical rise of that humbly created herdsman was paralleled by an actual rise of the Charans into favor among Rajput princes, achieved on the strength of their poetic talents and their fine appearance. They were "a tall good-looking fair-skinned tribe. The men are like Rajputs, strong and well made," Enthoven noted in 1922. "They wear the moustache and long whiskers, and in Central Gujarát they wear the beard." Charan men dressed in blousy cotton, as they do today, and "a Rajput-like turban or a piece of cotton cloth four cubits long wound around the head." Affluent Charan women, sometimes the men too, adorned themselves with silver anklets, gold necklaces, gold earrings. Besides being poetic and natty the Charans had a knack, like the Irish, for eloquent cursing. They became valuable to the Rajputs as bards, keepers of genealogies, formidable (and presumably amusing) cuss-outs, spreaders of princely fame. Charan women were admired as smart, forceful, even sacred, the sacredness supported by old stories of goddesses incarnated as Charan females. Despite all this flash, the Charans retained their association with cattle herding, and some of them continued to live it. You could think of them—Charan men, anyway—as the earliest cowboy poets. You could think of Charan women as . . . well, there's no adequate analogy, so: Hindu goddesses in bangles churning buffalo yogurt into butter.

The Rabaris are proud and handsome too, though they have no equivalent to the Charan claim of sacredness. Enthoven called them "strong, tall and well-made, with high features, large eyes and oval faces." On the other hand: "The men are dull and stupid, but the women are shrewd and intelligent." Although it's wonderful to be reminded that "strong, tall and well-made" is in no way inconsistent with "dull and stupid," Enthoven was probably unfair to the men. Everyone agrees that Rabari women play a dominant role in family and business decisions. Sigrid Westphal-Hellbusch, a German scholar I mentioned earlier, wrote that the women bargained fiercely with merchants and kept the men out of debt. During her research on the Rabaris, Westphal-Hellbusch found that "it was often the women who decided, by signs or words, whether the men could give information or not, and often enough they let go of all politeness and refused the men their speech, only to take over themselves. Not seldom with the remark, that the men were too stupid to answer." Yet these ill-respected men came from a sporty lineage of camel jockeys, tactical wranglers who once provided support for military expeditions up in Rajasthan.

The Rabari myth of origin traces them back to one Sambal, who was Shiva's own camel driver. Shiva gave Sambal three celestial damsels as wives and then, once the poor man had fathered a sizable family, directed him to vacate the heavenly realm and go live outside. Hence the name Rabari, which derives from *rahabári*, meaning "he who lives outside." A variant explanation is that they were "goers out of the path," in the sense that they had originally been Rajputs themselves but didn't marry Rajput wives, and so were obliged to live outside the Rajput aegis. Both etymologies seem plausible, and today the more traditional Rabaris still live outside—outside of villages, outside of society, within only the thorn fences that circumscribe their forest nesses. Others, as we'll see, have recently been forced to live outside in a different sense—outside the forest itself—and therefore to put themselves back inside a conventional agricultural economy. It hasn't been easy.

Maldhari life at Gir was hard enough already. At the time Paul Joslin and Steve Berwick were studying the ecology of the lions and their wild prey, Berwick's wife, Marianne, did her own study—on the ecology of the Maldharis themselves—using roughly the same conceptual framework. How did they support themselves? What was their impact upon the landscape? What was their role in the food chain?

She focused on a sample area near the center of the forest that encom-passed about thirty nesses, gathering information by interview, ques-tionnaire, measurement, and direct observation. The questions for which she wanted answers ranged from basics to deeper essentials. How many people lived in each ness? What were the rates of birth and death? What was the age profile of the population? Sex ratio? Health habits? Infant mortality? How much energy in the form of raw calo-ries, and how much in protein, flowed through the community? How much energy did the Maldharis expend in the course of their activi-ties? How were they *doing*? wondered Marianne Berwick.

Not very well, she found. Their life expectancy at birth was only twenty-four years, compared with forty-one for the Indian population overall. Beyond infant mortality, there were sharp drops in survivor-ship at several other stages of life—notably, for females, in the child-bearing years and the menopause years. Maldhari adults tended to lose weight during the monsoon, when water and fodder were abun-dant but life was harsh in subtler ways. Lactating females didn't get enough protein. Others got enough protein but not enough calories. Nobody got enough green vegetables. "Data indicate that the *Maldharis* are nutritionally and medically at risk and are living very close to their resource base," she concluded, "with only a thin line sep-arating them from economic disaster." By the time she published her paper, twenty years later, one form of disaster had already struck. It was called resettlement.

Following the lion census of 1968 and Joslin's preliminary work, a conviction arose among the key players in Gujarat that something had to be done. After six decades of recovery, the lion population seemed to be shrinking again. The Gir ecosystem, overexploited and abused, was in decline. High officials began warming to the idea of a national park. Joslin presented a paper to the International Union for the Conservation of Nature and Natural Resources (the IUCN, the world's most august conservation body) at a meeting in New Delhi, in

which he outlined the problems at Gir and discussed possible reme-
dies. It's a complex situation, he warned. One shouldn't be too quick to
assume that banishing livestock will fix it. Although nobody likes to
see cows in a wildlife sanctuary, "there is no reason to suppose that
lions cannot live on cattle just as well as on wildlife, and possibly better. If cattle were removed suddenly, with the objective of allowing the
wild herbivores to recover, the lions might starve or emigrate." Joslin's
presentation helped focus a wider trend of concern. Committees met,
reports were written, letters were sent—including a letter from Guy
Mountfort, international trustee of the World Wildlife Fund, to Prime
Minister Indira Gandhi. Mountfort enclosed a detailed proposal for
upgrading the Gir sanctuary to a national park, closing the boundary,
eliminating timber extraction, ousting Maldharis and their cattle from
a core area immediately, and phasing them out of the whole sanctuary
over time. These various political, scientific, and bureaucratic nudges
produced action. On January 17, 1972, the Gujarat government issued
Resolution WLP-1971/P, specifying the construction of a stone wall
around the sanctuary, the establishment of gated checkpoints at the
sites of road entry, and the resettlement of 845 Maldhari families—that
is, all of them—outside Gir.

    This effort became known as the Gir Lion Sanctuary Project. An
official account of the project was published by the government of
Gujarat in 1975. That booklet touted a linkage between the new measures at Gir and an ancient Indian tradition of nature conservation that
goes back more than two millennia. Most famously, there were the
edicts promulgated on stone pillars by the emperor Asoka in 242 B.C.,
several of which decreed protection for animals and forests. Asoka's
guidance had often been invoked in connection with conservation at
Gir, since one of the surviving Asokan pillars is nearby, at Junagadh.
Dating back even earlier than Asoka's edicts was a document known as
the *Arthashastra*, written by an imperial adviser named Kautilya around
300 B.C. Kautilya had been a sort of Machiavelli to the first emperor of

India, Chandragupta, and the *Arthashastra* was his hardheaded treatise on governing effectively and maintaining power, roughly equivalent to *The Prince*. Among other practical advice, it proposed the establishment of *abhayaranyas*, or "forests free from fear," for the sake of preserving game animals. "In these protected forests," the Gir project booklet asserted, "there was strict supervision and certain mammals, birds, and fish were fully protected. If these animals were vicious, they were to be entrapped or killed outside the sanctuary, so as not to disturb the rest." Such *abhayaranyas* may have been, as the booklet claimed, the forerunners of national parks. But the *Arthashastra* itself offers no explanation of just how an interdiction against animals that are "vicious" ("dangerous" in the edition I own, translated by L. N. Rangarajan) in a "forest free from fear" might apply to a population of lions. And, of course, every animal is dangerous to the creatures that live below it on the food chain.

A perimeter wall, as specified in that government resolution, was built—just high enough to stop an ambling buffalo but not a leaping lion. A ban against outside livestock was declared. Even with the wall, even with the ban, incursion by outsiders' cattle didn't cease entirely. But those two measures were straightforward, useful, and uncontroversial compared with evicting the Maldharis.

The resettlement began in 1973. According to the official plan, each Maldhari family would get eight acres of cultivable land and a home site of 725 square yards, plus free transport for their possessions and the dismantled materials of their hut. They would also receive 2,500 rupees as a grant, for reconstructing the house and their lives, and another 2,500 rupees as a loan. The first ninety families were moved to an area called Chotila, about thirty miles west of the forest. The land was plowed for them at government expense. Nobody seems to have asked the crucial question: Were these people, with their ancient tradition as semi-nomadic herders, willing or able to become farmers? The answer would have been: Not very.

Within four years, about two hundred families had been relocated, but the cost was unexpectedly high—more than what had been budgeted for the entire resettlement operation. Impeded then by a shortage of money and by mixed reactions among the Maldharis, the program sputtered and slowed. Without ever being officially canceled, it effectively ended in the late 1980s, by which time almost six hundred families had been moved. Roughly three hundred families remained behind, on fifty-four nesses within the sanctuary, though none of them at its very center. That core area, which in the meantime had been declared a national park, was completely free of Maldharis, if not free—like the *abhayaranyas* as imagined by Kautilya—from fear.

Even the Forest Department, which had addressed a nearly impossible situation with good intentions and limited means, acknowledged in a later report that the Maldharis were "apprehensive" about any further relocations. The report noted, in a bland understatement, that "some relocated families could not do well in agriculture and dairying." From other evidence it appears that the land they had been offered was poor, with inadequate water, and that the various marginal subsidies proved unsatisfactory. Failing as farmers, they became field laborers for other small farmers or moved into villages to work for wages. Youngsters went off to distant cities. Tradition fell derelict and proud independence dissolved into despair. Ravi Chellam saw the aftermath during his own years at Gir. He described it, with his typical bluntness, in a co-authored 1993 paper: "The Maldhari translocation programme has failed and the translocated Maldharis have been reduced to a state of penury."

A graduate student named Shishir Raval later interviewed displaced Maldharis, as well as other residents of the Gir vicinity, for his dissertation in landscape architecture. He found a recurrent tone of gentle, plaintive nostalgia. "Our heart is disconcerted near the city," one person told him. Another said, "Gir means [our] heart." Several characterized the place as a sort of mother, nurturing and sustaining. The victims of

resettlement not only missed their forest but also loathed their new circumstances, as reflected in comments such as "If [they] throw us back
into the Gir, that will be better." When asked to imagine the future,
some sounded optimistic, or at least stoic: "Millenniums will change and
everything will turn over. Gir is unique; she has such qualities that she
will survive." A gloomier outlook was that "even with protection the
Gir will degrade; when there won't be trees, there won't be people who
used to gather under their shade—it will all be forlorn!" Another person said coldly, "If there is no jungle, we won't be there; the lions won't
be there." Among the various sad voices, Raval captured this:

> When the sky gets dark with heavy clouds and thunders, the big trees
> sway in the wind, the peacocks sing, I pine to run back to the Gir.

As appropriate for a doctoral project, Raval framed his work in dry
academic terms, setting himself to consider "perceptions of resource
management and landscape quality" at Gir. In other words, he under-
took to objectify the subjective. He had interviewed more than sixty
people, including about two dozen Maldharis. His dissertation, sub-
mitted to the University of Michigan in 1997, offers 327 pages of
orderly scholarship and three pages of starkly reported pathos.

19

One evening in a village called Haripur, some miles south of the protected area, I hear similar testimony myself. I've come here to meet a
Maldhari elder known as Khima Bhai, "Brother Khima," who carries
his own bitter memories of the forced resettlement program. In a
little dirt courtyard just off a stone-walled alley, we sit on *charpoys*

(string-woven cots of the old style, favored throughout India as portable furniture) beneath a bare bulb under the night sky, drinking tea thickened with buffalo milk and sugar.

The tea is served scorching hot, poured onto flat steel saucers that allow it to cool quickly. This is the traditional Maldhari style. We support our saucers delicately from beneath, in tripod position, with fingertips along the rim where they mostly escape the heat. It's like touching a low-voltage wire. This tea-balancing trick is familiar to me from earlier doses of Maldhari hospitality, but still it takes effort. Steady the saucer, cool the tea, sip the tea—sip it *loudly*, to show appreciation—don't rush, by all means enjoy it, and then . . . again I grab my notebook. Ravi Chellam keeps the talk flowing and functions as translator. Languages intertwine and besprinkle one another—Hindi, Gujarati, maybe a bit of Sindhi—but I can't even tell who's speaking which. I watch faces. I hear tones. Another man, an influential Maldhari known as Ismail Bapu, with whom Ravi is close, has brought us to this courtyard and arranged the meeting, and he occasionally inserts comments. Ismail Bapu is a large, charming fellow with his own stories to tell, his own intimate relationship with the lions and the leopards of Gir, and no shortage of opinions about the Forest Department. But tonight the featured witness is Khima Bhai. Several younger men sit on the stoop, respectfully silent.

Khima Bhai is a thin, handsome man in his seventies, with a long white mustache that seems to have been cantilevered with wax. He wears the traditional kadia and chorni but, unlike some of the crisp outfits I've seen on Maldharis in the forest, his are wrinkled and grimy. Despite the dirt on his clothes and the shame of his reduced circumstances, he bears himself with the dignity of a Vatican monsignor. His full name is Khima Dula Rabari.

He was forced out of the forest in the early 1970s, he tells me. The last ness he occupied was Sajia, which he shared with three brothers and their families. He and the brothers kept cows, buffalo, a few camels, about fifty head in all. Lions were less abundant in those days, he thinks. Anyway, no great problem. Sometimes a lion attacked but we could deal with it, says Khima Bhai. As for relocation, he says, that wasn't a new idea. The talk had been going on a long time, twenty years. There had been earlier schemes and offers from the government. We'll give you sixteen acres if you clear it for agriculture, they said, start raising crops, help feed the population of India. It was the Green Revolution business. But—growing crops? We weren't farmers. We didn't like the look of that. Besides, we knew farmers were very poor, Khima Bhai says. Us, we were independent, we were *graziers*. We wanted to stay with livestock. Then we were told, You *must* move. Not offered any choice.

Khima Bhai's voice rises. His anger gathers heat at the recollection. How many times in twenty-some years has he revisited this old injury? Not enough to extinguish the pain.

Move, move, move, they were saying. The government. All right, so Khima Bhai moved. Others who refused were beaten up.

Eight acres they gave us, he says. Eight acres at Chotila, which is forty-five kilometers away. Along with the land, we were supposed to get wells for irrigation, schools for the children, health care. None of that happened. The grass was poor. And not enough of it. The rains were unreliable. Eight acres, eight of *those* acres, didn't support a herd or a family. All my livestock died, he says. Half of us moved back here, but now we work as field laborers, he says, *laborers*, the word stressed with mortified disgust. Maybe we keep a few goats. Buffalo? Buffalo, no. Buffalo are expensive. I wish I could afford to do that, he says, but I can't.

Having confessed poverty, Khima Bhai directs one of the younger men, his son, to scoot down the alley to another house—the son's house, I gather, where Khima Bhai himself lodges humbly, the fallen patriarch—and fetch back a fresh batch of tea. The first round was supplied by our courtyard host, a member of Ismail Bapu's family, but dignity as well as generosity compels Khima Bhai to reciprocate. The son reappears with a small pitcher of steaming tea, again thickened with sweet buffalo milk and seasoned with nutmeg, which we receive on our tin saucers. The other bit of unspoken logic behind such tea service by the saucer-load, besides heat dissipation, is that it allows hospitality to be doled out in very small portions; tea is expensive to a Maldhari family. Muttering appreciation, I savor mine slowly, trying not to poach my tongue. The buffalo milk is rich and smoky. Ismail Bapu announces, in case anyone failed to notice, that this round is courtesy of Khima Bhai.

Meanwhile our talk turns back to the lions, and to the rigors of raising livestock in predator country. Yes, occasionally there were losses, says Khima Bhai, and not just among the old and weak animals. Sometimes you would lose a good buffalo. So be it. Wordlessly, Khima Bhai indicates his acquiescence to that inexorable reality—but I choose to press him a little. If you had the chance now to move back into the forest, I ask, and live there as a Maldhari, facing the various jeopardies as you know them, with lion attack and lost livestock and injury to yourself always possible, what would you say? I phrase this in the conditional, well aware that it's a stupid hypothetical question. He has no such chance and he's not going to get one.

Ravi translates Khima Bhai's answer: "All of us accept."

20

But they don't all accept—not the hypothetical temptation of a return to the forest, not the premise that resettlement has been unrelievedly bad. On the following night I hear a parallel story, with different particulars and inflections, from a middle-aged Maldhari known as Lal Bhai (his full name is Lala Boda Rabari), who was born on a ness in the forest and has adjusted rather contentedly to life outside it. Lal Bhai, I learn, is a bullish believer in both the inevitability and the value of change, his belief reconfirmed by the fact that five years ago, after thirty years of childless marriage to one wife, he married another, a sixteen-year-old girl, who has proved satisfactorily fertile. Now he has two little daughters, with a third child coming. Lal Bhai himself is fifty-eight, a heavyset man with a white handlebar mustache (no less stiffly pomaded than Khima Bhai's) and the chest of a retired long-shoreman. He lives with his young wife, his old wife (who remains part of the family as a sort of honored pensioner), and his two little daughters in a modernized ness beside the highway along the west margin of Gir.

The thorn fence that once circled this ness has been replaced by a stone wall. A power line brings in electricity. Outside the wall, cars and trucks flash past on their way to the big city, Junagadh. Lal Bhai summons tea and, when it arrives, passes round the steel saucers and does the pouring himself. Then, seated cross-legged on a charpoy, he answers my questions. Yes, they have running water here also. An electric pump pulls it up from the well. The perimeter wall was built six months ago with funds from an international "eco-development" grant. Lal Bhai is grateful.

He was born in the forest, yes, at a ness alongside the Hiran River. Since then he has lived at five or six different sites. Maldharis do that, they move. He lived sixteen years at Patariala, a ness within what is now the national park. When the park was created and all Maldharis were required to vacate, he went to Dudhala, still within the sanctuary. The Forest Department helped him move his possessions. Eight years ago, from Dudhala, he came here. There are twelve families, about three hundred head of livestock, not enough grazing in the dry season but otherwise it's good. The Forest Department allows them to harvest grass from another part of the sanctuary. They've been promised six hundred limestone blocks for rebuilding their houses. They may also get a flour mill. Lal Bhai has reduced his own herd to four buffalo and seven cows, from which each day he takes sixteen liters of milk. He buys cottonseed in bags as a feed supplement, and he sells fresh milk, not ghee. Sixteen liters, that brings two hundred rupees a day. For selling milk, says Lal Bhai, I must be near a road. It was an economic decision. And there are other advantages. For instance, the children can go to school.

He wants his children to be educated, to have a wider range of opportunities, to seek employment. The world is changing so fast. Look what people have accomplished. Lal Bhai has been quite happy with his Maldhari life, yes, but in a changing world the new generations must discover new roles and new forms of happiness. His little daughters, let them be educated and go outside the Madlhari tradition. If he should have a son, well, then the son himself would choose between stockherding and whatever else. Not everyone can be a Maldhari. Say there are ten children—only two can stay on the ness. The others must go, they must find different ways and places to live. Otherwise there would be too many people and too many cattle. Though he doesn't put it in such terms, Lal Bhai is acutely aware that Maldharis, like lions, are constrained by the carrying capacity of their particular niche within their particular ecosystem.

The lions themselves seem to be fewer than in bygone times, as gauged by his terms of measurement. (He disagrees, that is, with Khima Bhai and the official census.) Either they're fewer or, maybe, not so bold. Lal Bhai's terms of measurement are: How long does it take for an unwanted cow, staked out in the forest for the sake of her compensation value, to be killed? Recently someone put out an old, dry buffalo. She grazed alone there for fifteen days, *fifteen days*, before being taken. To Lal Bhai it's testimony that the lions of Gir aren't what they used to be.

He himself has never been attacked by a lion, no, though he has had his dicey encounters. One evening at dusk he was walking back to Dudhala from the village of Sasan, carrying a sack of onions and potatoes. He saw some lion cubs sitting in the road. He was alone, with no flashlight, no stick, no kuwadi. Of course he knew that where there were cubs, there would be a lioness. She revealed herself, growling, as he approached. Lal Bhai made a noise, trying to shoo her off. Instead the lioness charged. So he threw his bag of vegetables at her. Evidently this was a vegetable-eating lioness, because she diverted to the sack, seized it, and ran off. Lal Bhai made his escape. Returning later, he found a few onions but they were ruined. Nobody wants an onion upon which a lioness has slobbered.

Lal Bhai shares these thoughts with us in confidence. Sitting close, he leans closer, like a manicurist or a young priest in the confessional, and talks straight into his translator's face. As he makes a point, as he tells a story, his hands move through the air like big, slow moths. "There's one thing in the lion I've seen," he says. "If you don't run, if you stand your ground, they will never take you." Not even if they're frothing at the mouth, he adds. Of the Gir forest he says, "Slowly, slowly, it is being cut. If there is no forest, there will be no lions either." That's a familiar theme, almost a truism, as heard also among Shishir Raval's informants. The difference with Lal Bhai is that he declines to lament. All is fungible, in his view; all is flux. That's the world, period.

I'm not sure what to make of this opportunistic codger, with his young family, his electrified ness, his brisk fatalism that doesn't conform to my own biases or expectations. All I can say confidently is: He's a brave man, unafraid of lions and unafraid of the future. I wonder if he has thought about staking his old wife out in the forest.

21

At the Dudhala ness inside the sanctuary, where Lal Bhai once lived, now lives a man named Amara Bhai, a bantamweight fellow with warm brown eyes, narrow features, an unearthly smile, and a bulbous little disfigurement to the tip of his nose, as though it were once bitten off and reattached. I've come to solicit his testimony about lions because I've heard that he has a vivid personal story.

An ancient fig tree shades the yard at Dudhala. Stacks of firewood stand beside large steel drums holding water pumped from a nearby well. Although it's late afternoon when he welcomes us to the ness, with yellow light slanting low through the forest, Amara Bhai is freshly shaved. (Earlier in the day, when Ravi and I encountered him on a road and set this appointment, his gaunt cheeks were covered with white stubble.) Receiving far-traveled visitors from places such as Dehra Dun and America is, evidently, a spiff-up event. He wears a white turban and heavy cylindrical earrings that, if not brass, might well be gold. He has girded and primped to receive us and our nosy questioning. But first there must be tea. Amara Bhai passes the saucers.

We sit on charpoys beneath a corrugated tin roof, Amara Bhai talking, several other men listening, while children loll on nearby cots, gaping at us curiously, and women in bright-colored skirts go about their work, carrying water and firewood in the background. Four families, those of Amara Bhai and his three brothers, share the ness. He himself at age fifty-seven has only one child, a young boy, from among his three wives; how many others were born but failed to survive, he doesn't say. He and his brothers would like to see the children educated, yes, because if young people choose to leave the ness, they'll need something more than Maldhari skills. But there's a problem: no school. Small children can't walk into Sasan each day through a forest filled with lions and leopards. Nor is it desirable for them to live in the village, boarded out, too far separated from their parents. Maybe what's needed, he thinks, is a modest little school here at Dudhala or at another ness nearby. The nose injury? No, it wasn't a lion bite, Amara Bhai answers cheerily. His lion scars are less visible.

The first attack happened three years ago, during the monsoon. There were two male lions, possibly brothers. Earlier that morning he had heard them roar. As he was moving some buffalo out to graze, the lions struck, each seizing a different animal. Amara rushed in with his little ax. One lion ran off. The other released its buffalo, swirled, seized Amara Bhai by the leg, then dropped him and disappeared. No bones broken, fortunately. Amara Bhai was carried into Sasan on camelback, then by vehicle to the hospital at Talala. He spent fifteen days in the hospital, plus another month recovering. The government paid him compensation, though not enough to cover his medical costs.

The second attack happened just a year later. Again it was two males—Amara Bhai suspects they were the same two, a dangerous pair. This time in late afternoon he was bringing his animals home; just a kilometer from here, he paused to let them drink. The lions charged. One grabbed a two-year-old buffalo, and Amara Bhai, "being the Maldhari," as he says with a proud sense of station, went to its defense. He beams his ethereal smile, even laughs, while describing how the lion reared up, sank its teeth behind his right shoulder, released him, then turned back to the buffalo. Alone and bleeding, Amara Bhai found his way to the ness. Again hospital time, again costs. There is no punch line to his narrative, no ending at all, since the same two lions are still in the forest. His friends warn him, they say, Amara Bhai, don't give that pair a third chance to kill you. But what can he do differently? Let those lions *have* the buffalo if they attack, say his friends. But no, he can't afford to do that. Buffalo are precious and he's a Maldhari.

His two maulings within a year give Amara Bhai a certain authority on lion-human conflict, but he shows little inclination to leap from experience to opinionizing. I ask if he's scared. Of that malign pair, yes indeed, but not of lions generally, no. Lion attacks on people aren't a common feature of life in Gir, he says. They're accidents that sometimes happen. Aberrations. Is he angry about his injuries and losses? Is he discontent with how the Forest Department handles compensation? Angry, discontent—these complaints don't seem to fit his emotional categories. Despite the injuries, despite the ongoing risk, he's happy with his life in the forest. He would have no other place than Gir, and no other way than the Maldhari way of existence. Keeping livestock is what allows him to remain here, so he will continue to tend his buffalo, moving them from feed to water, trying each night to get them safely back inside the thorn fence, tolerating inescapable costs.

Is he concerned about changes that are coming? No. Having heard a question slightly different from the one I asked, he professes his deep, hearty indifference to the outside world and whatever changes it sees fit to inflict upon itself. The idea of changes transforming Gir— his own world, not just the world outside—doesn't seem to have occurred to him.

And about lions, how does he feel? "He says the animal is good," Ravi translates. "There's nothing wrong with the animal." Amara Bhai smiles.

22

Among this community of forbearing Maldharis, Ravi's old pal Ismail Bapu is an oversize figure. He's respected for his blunt, levelheaded opinions and sometimes called upon as a spokesman, despite an anomaly—he's not a Hindu from the Rabari or Charan lineages, not a Hindu at all, but a Muslim. His family has its own ancient, nomadic history tracing back somewhere into the Sind borderlands.

Ismail Bapu is fifty-six, a generous man of dignity and girth, with a round face and a grin that breaks wide beneath serious, watchful eyes. His unruly hair and his mustache are going gray, but his eyebrows are still black. The second of his names, Bapu, is an honorific that might best be translated as "Big Daddy." He wears the kadia and chorni as his daily uniform, along with a dusty red knit cap, though for festive events he replaces the cap with a white turban. Costumed otherwise, he could pass for a godfather in Little Italy.

Cagey and independent, Bapu has managed his relations with the Gujarat Forest Department, and with the larger world, deftly enough to avoid the extremes of exile, penury, and cultural capitulation. Besides keeping two dozen buffalo and a few cows, he does a bit of farming, and has a half-dozen men helping him, including his youngest son, a fifteen-year-old known as Bapu Mia ("Little Bapu"). His network of friendships spreads wide, and Ravi himself has been part of it—a close part, embraced like family—since the years Ravi spent at Gir doing research toward his doctorate. The bond with Ravi seems natural, given that they share a brusque self-assurance and a wry sense of humor, both in contrast to the stultifying, timorous careerism of the Forest Department bureaucracy. Bapu's current ness, known as Lakadvera, is a one-family compound that sits in a mixed landscape of forest edges, hills, rocky streambeds, and peanut fields just outside the Gir boundary. It's part of the ecosystem but not quite within the protected area.

Because we are with Ravi and Ravi is family, Bapu welcomes Michael Llewellyn and me for a night's stay. He has just finished killing a chicken for dinner when we arrive. Most labor around the ness is now delegated to the younger men or to his wife and daughters while Bapu supervises—a change from fifteen years ago, Ravi says, when Bapu was at his peak, robust and hardworking. But bringing a hatchet down on the neck of a bird is evidently still a task for the patriarch.

The house is small and simple, with mud-and-wattle walls, a little annex made of limestone blocks, and a roof of red tiles held in place by chunks of lava, which look like meteorites that landed there gently. The thorn fence encircles barely a quarter acre, including house and corral, and closes at a rusty metal gate hung between log gateposts. Just outside the fence, beyond a low stone wall lining a peanut field, we see a rotting rumen and some bone scraps, all that remain from one of

Bapu's cows, killed last week by a lioness with two subadult offspring.

Bapu saw it happen. The cows and buffalo were returning for the night, nearly home. This cow was young, but she'd been injured a year earlier, her hindquarters weren't right, and she moved a little slower than the rest. Before she could hop that little wall, the lioness had her. From where Bapu stood, peering over the fence, he had a clear view but could do nothing. When the lions had eaten their fill, he and his sons dragged the carcass away from the ness. No, Bapu says, he hasn't applied for compensation. Not on that cow. She was already lame, lion bait on the hoof, alas, and not giving milk; her value was too low to merit the hassle of filing. He didn't bother. Most people would have. But the compensation system is unfair, so never mind. Two months ago he lost several good buffalo to a pride of lions and in that case he did file, though he hasn't yet gotten any response. Compensation, if granted at all, generally falls well below the market value of the animal, he says. For a cow or a buffalo worth fifteen thousand rupees or more, you might receive four thousand.

For the killing of a person they pay a hundred thousand rupees, but of course that's another matter. Bapu hasn't lost any family.

The buffalo come parading home, nudged along by Bapu Mia, and Michael goes to photograph them in the good evening light. I follow for a glimpse of the milking routine as langurs call from their roost in an unseen tree. Then I rejoin Ravi and Bapu where they sit, on charpoys placed like lawn chairs outside the ness, enjoying the twilight smells and sounds. Bapu nods toward a wooded hillock in the near distance. Two male lions were up there last night, roaring for hours, he says. Because all his statements come translated by Ravi from Bapu's gumbo of Hindi, Gujarati, and Sindhi, I can't tell whether this particular one represents a felicity noted pridefully or a stockman's complaint. Probably it's both.

At dark we retreat inside the fence, closing the gate and repositioning the charpoys on a dry patch of ground between the buffalo crap and the veranda, where we'll sleep. But first there's dinner, a feast of curried chicken, millet flatbread, rice, raw onion, and bowls of rich, subtle buffalo yogurt, which is far better than any other yogurt I've ever tasted. We sit cross-legged on the dirt floor, cushioned by burlap mats, eating this tangy stuff with our fingers by the light of an oil lamp. The yogurt is served in little bowls with little spoons. After a smoldering curry it cools the palate even better than beer. Back outside, Bapu's middle son pours wash water for each of us from a pitcher. How was the food? asks Bapu. What's the word, I ask Ravi, for terrific, wonderful, great? *Saaru*, he says. *Boh saaru*, actually, would mean "very good" in Gujarati.

"*Boh saaru*," I tell Bapu.

We sprawl again on the cots, gazing up at a clear black sky to watch satellites winking across among the stars, while Bapu speaks of his various skirmishes with large carnivores and government agencies. "He has no negative feelings for the lions," Ravi translates, "in the sense that it's nature's way. The lion is not gonna eat grass. It has to have meat. And one packet of that meat is his livestock." Predation, Bapu knows, has always been part of life in the forest. It's ineluctable. The more contingent forms of aggravation, such as resettlement, grazing restrictions, compensation at below-market rates, are less easily borne. "His anger is with the Forest Department."

Bapu recounts a leopard incident reflecting a mortal reality of Maldhari life: that these five-foot-high thorn fences won't keep out a predator made desperate or deranged by starvation, injury, drought, or some other emergency. It happened during the monsoon, about twenty years ago. Although he has lived at other ness sites during the interval, back then he was right here at Lakadvera. On a drizzly morning, Bapu took the herd out and returned. His wife asked him to cut some vegetables. So he was sitting inside while his wife and sister dealt with the day's buttermilk. There was a banyan tree nearby—it's gone now, but it stood about where we saw the remains of his dead cow. Suddenly the banyan was filled with hysterical langurs. They'd seen the leopard coming. It streaked past the tree and leapt, sailing over the thorn fence. Bapu knew nothing until he heard his sister shout, "Panther, panther!" He thought she meant, Out *there*, outside the ness. When he stepped through the door it charged him, going for his throat. A big male, strong and quick. He caught one paw, pushed hard against the leopard's throat, and they danced.

His wife came to his aid, whacking the cat across its hind legs with a club. Bapu shoved it away, but the leopard hurled itself right back, clawing him, mangling his arm. He told his wife to get clear and then again he threw it off. The leopard's legs were now hurt, maybe broken. Bapu grabbed the club and walloped it on the head. He kept pummeling until it was dead. Then he walked into Sasan, several miles, and from there a Forest Department vehicle took him to the hospital at Talala. We'll go trap the leopard, said the Forest Department people, tell us where to look. You won't need to trap this one, Bapu said. They used fifty-two stitches to sew him up, Bapu recalls, but at least the Forest Department paid his medical bills.

Did the attack dampen your desire to live here? I ask.

No. "That was an accident. It's not likely to repeat itself," Ravi translates. Bapu carries no special abiding fear of leopards, or of lions either. "Death lurks in every one of life's turnings. If it has to be on you, it'll be on you."

This can't simply be the famously stoic Hindu sense of karma, since Bapu is a Muslim. As the talk dies, I lie comfy on my cot beside the corral, scanning idly for blinking motion among the constellations and wondering, more than idly, whether such fatalism is somehow general to India, specific to the Maldharis of Gir, or peculiar to Ismail Bapu, Amara Bhai, and a rare few of their compeers.

It doesn't seem to be shared widely by the non-Maldhari villagers surrounding Gir, who live separate from the forest but not always separate from its lions. According to that *Conservation Biology* paper on lion-human conflict, by Vasant Saberwal and his co-authors (including Ravi), more than sixty percent of the villagers interviewed were hostile to lions—and arguably they had reason, since more than eighty percent of the documented attacks occurred outside the protected area. A lion wandering through human turf is likely to be more unpredictable, more frantic, and therefore more dangerous than a lion in the forest. Most of the villagers complained that roving lions were especially threatening at night, yet these very people were obliged to do their irrigating at night, because electricity for their pumps was unavailable—diverted to factories—during the day. With the late-1980s drought and the sudden shortage of vulnerable livestock in its aftermath, the lions' aggressiveness had so increased that they were jumping fences and walls like Bapu's leopard, even trying to enter houses, in search of prey. Hard lives, for lions and humans both, had become harder.

Disappointingly for me, though not for Bapu, no lion visits his ness to shatter the tranquility of this particular night. No leopard comes flying over the thorn fence. The gentle wheezes and snorks of two dozen sleeping buffalo sound a restful rhythm just beyond our cots. The only pernicious animal in the vicinity is a rooster, on the rail near my feet, whose idea of dawn turns out to be half past four.

23

Of all the important patterns and trends described in the Saberwal paper, perhaps the most important is mentioned only in passing: "Many respondents (62%) reported that poorer villagers, who typically left their animals unprotected outside at night, lost substantially more livestock to starvation and lion depredation during the drought than did wealthier villagers, who could afford to build sheds and other structures to protect their animals." Boiled down: Poor people bore the brunt.

It's a general truth if not quite a universal one, relevant from Rudraprayag to Komodo to Tsavo, sometimes noted but seldom quantified or analyzed: Predation is costly and the costs are unevenly distributed. Large predators cause more material loss, inconvenience, terror, suffering, and death among poor people (specifically, poor people who live in rural circumstances within or adjacent to the habitat) and among native people adhering to traditional lifestyles on the landscape (who may be "poor" only in the sense that they aren't insulated by material wealth and have little political power) than to anyone else. Proximity plus vulnerability equals jeopardy.

This pattern isn't unique to humans and the alpha predators that sometimes menace them. It shows up in other predator-prey relationships at various points in the zoological spectrum. It emerges as a cardinal insight, for instance, from the published work of Paul Errington, an unassuming but percipient small-mammal ecologist who became an authority on predation during the middle decades of the twentieth century. Muskrats and minks in the wetlands of Iowa provided his window to larger understanding.

Errington was a consummate American Midwesterner, a Norman Rockwell exemplar of the wildlife biologist in a plaid flannel shirt. His research interests ran to prairies, marshes, little hardwood forests at the interstices of heartland agriculture, and to the birds and mammals that inhabit them. Raised in east-central South Dakota, he was a fur trapper and a hunter from a young age, learning to track animals and read clues to their life histories long before he grasped the formalized methods of science. He also learned the habit of keeping field notes. One of his earliest recorded observations, involving a mink, occurred on May 23, 1919, his last day of high school. To savor the new sense of freedom and the spring sunshine, young Paul went for a walk along a creek on the outskirts of town. Scanning the mud margins for animal tracks, he noticed a hole in the bank. It appeared to be a mink den, littered round-about with crayfish remains and mink scats. The scats themselves consisted largely of crayfish bits. Just inside the den entrance—as he reconstructed the moment, from notes and memory, forty years later—he spotted "a live crayfish, which I picked up. As I held it in my hand, a mink came out of the den hole, seized the crayfish from my fingers, and whisked with it back into the hole." Whatever that may have said about the food preferences or audacity of minks, it confirmed in Errington the thrill of close observation. The tracking skills also continued to serve him well, though his early ideas about predators and prey would change as he studied them at the population level.

During the late 1920s, Errington supported his off-and-on under-graduate work at South Dakota State College partly with wild meat and furs. "The game that I ate and the muskrats, minks, weasels, skunks and coyotes that I skinned," he wrote later, "made possible some things that might not have been possible otherwise." At the University of Wisconsin, in the early Depression years, he did a doctorate on the bobwhite quail (*Colinus virginianus*) and its predatory enemies. Then he moved down to Iowa State College, shifted his attention to the muskrat (*Ondatra zibethicus*), and began what would stretch into three decades of research on muskrat ecology among the wetlands of northern Iowa.

The muskrat is a semi-aquatic rodent. It lives in swamps and along the edges of ponds and streams, building its dens as vegetation mounds in shallow water or as burrows in muddy banks. Its native range includes most of the United States and Canada, and under good conditions it reproduces quickly, raising two litters, three litters, sometimes even four litters in a year. It eats aquatic vegetation such as cattail shoots, supplemented with frogs, clams, and other meat items as available. The mink (*Mustela vison*) is also semi-aquatic, with roughly the same native range and habitat preferences as the muskrat, but it's fully a carnivore, both by classification (it belongs to the mustelid family, within the order Carnivora) and by diet, feeding on small mammals, eggs, crayfish, frogs, and fish. Although a mink and a muskrat are similar in size—the average mink being slightly longer but less bulky than the average muskrat—minks prey voraciously on muskrats when they can do it without too much risk or trouble. The annual toll of mink predation on muskrats in mid-twentieth-century America, by Errington's loose guess, ran to millions of individuals. No one had quantified that toll precisely, but the perceived losses caused annoyance and anti-mink animus within the fur industry, because muskrat fur was a staple of the trade. The mink, though valuable as a fur-bearer itself, was a relatively rare creature, not easily harvested from the wild, and it seemed to be eating up inordinate muskrat resources viewed jealously by trappers and furriers. Errington's work corrected that illusion.

"Undoubtedly, muskrat flesh is one of the mink's favorite foods," he wrote. Under ideal circumstances (ideal for minks, anyway), a mink might live for months on a diet of nothing but muskrats. "Undoubtedly also, the mink is one of the most accomplished killers of muskrats among the wild predators frequenting north-central marshes and streams. Once a large mink succeeds in grappling a muskrat, that muskrat has little chance of escaping." What he meant by "grappling" was that the mink grabs the muskrat with its front legs and rakes it viciously with its hind claws while biting it around the head and neck. Even with those tactics, a mink generally doesn't dare attack a healthy adult muskrat in the safety of water or its den. Errington's evidence from the field showed that minks preyed mainly upon muskrats that were already somehow weakened or disadvantaged, either by muskrat-on-muskrat conflict or by the exigencies of weather or disease. His decades of sedulous observation, data gathering, and analysis convinced him that mink predation was unimportant as a limiting factor on muskrat populations.

What *did* limit population size, he concluded, was the muskrat's own social behavior within its environment. Muskrats hated crowding and, beyond a point, wouldn't stand for it.

Nowadays scientists would refer to muskrat population dynamics as a density-dependent phenomenon. *Ondatra zibethicus* is a territorial species. The muskrat's territorial instinct determines how many individuals can find food and adequate denning sites within a given area of habitat; also, the muskrat's social tolerance or intolerance determines how many litters of young in a given year will receive decent nurturing. Once the habitat is full and the threshold of crowding is reached, all other muskrats that may be born, or wander in from elsewhere, or find themselves handicapped by sickness or displaced from good territories because of age or some other problem, face a high likelihood of early mortality. That's what Errington saw, anyway. He called those excluded, dispossessed, doomed muskrats the "wastage parts" of the population.

A crucial but little-recognized distinction exists between how many muskrats are present in an area and how many are available as potential prey. Well-situated muskrats are effectively unavailable, Errington explained; they have good den sites, adequate food nearby, and routes of travel between den and food that don't expose them to attack. The "wastage parts" of the population, lacking those advantages, suffer most of the predation by minks. Errington figured that seventy percent of the muskrat flesh eaten by minks in central Iowa came from scavenging animals left dead or near-dead by disease or climatic emergencies, such as a drought or an especially severe freeze. Another sizable increment came from individuals, especially male muskrats, dispersing from winter dens into unfamiliar terrain in springtime, seeking new summer territories in which to breed. Still another increment was what Errington called "overproduced young." The weak, the homeless, the unsupportable offspring—these are the victim classes. By contrast, healthy adult muskrats holding good territories have little to fear.

Territoriality is what separates the haves from the have-nots. Holding a territory within good habitat strongly affects the life expectancy of an individual, as the total supply of such territories limits the size of a population in the presence of predators. "Proper consideration of this factor," Errington wrote, "calls for some modification of conventional views as to the struggle for existence, the ruthlessness of natural testings, and the nature of predation." Consideration of that factor also helps illuminate why the poorest villagers around the perimeter of the Gir forest have little appreciation of lions. No one wants to be among the "wastage parts" of the human population.

Is it inevitable that the costs exacted by alpha predators be borne disproportionately by poor people—in particular, by tradition-bound rural groups such as the Maldharis of Gir, the Udege of southeastern Russia, the shepherds of highland Romania, all nearly powerless and voiceless within larger national contexts—while the spiritual and aesthetic benefits of those magnificent beasts are enjoyed from afar? If not, then how should society redistribute the costs? How might we also redistribute the (material, as well as spiritual and aesthetic) benefits? That's what I call the Muskrat Conundrum. Paul Errington limned it without pressing the broader application. Decades after his death, it's a matter we cozier muskrats need to address.

LEVIATHAN
WITH A HOOK

24

Dangerous predators, of whatever species, are more easily admired from afar. Taking the Muskrat Conundrum as a starting point, a hard-headed humanist could argue that exterminating all alpha predators is basic to the enterprise of civilization, and that it's only distant, safe, hypocritical sentimentalists (like you and me) who might assert otherwise. This is a serious view that needs to be heard and addressed.

It was well articulated, back in 1973, by a biologist named Alistair Graham in a splendidly garish book titled *Eyelids of Morning: The Mingled Destinies of Crocodiles and Men*. The book was based on Graham's own field study of Nile crocodiles in northwestern Kenya, along the margin of what was then called Lake Rudolf. The lake (later renamed Lake Turkana) supported a large population of crocodiles and a small settlement of Turkana people. The people lived by subsistence fishing, spent their lives near and in the water, and knew crocodile attack as a dire possibility of daily existence. The crocodile species Graham studied, *Crocodylus niloticus*, is one of the two largest and most menacing crocodilians in the world. His project, begun in June 1965, involved killing five hundred crocodiles in order to measure them, determine their ages, assay their stomach contents, skin them, and sell their hides to pay the expedition expenses. It took a year, during which Graham was assisted by a small crew of Turkana and by a friend and (in Graham's description) "fellow crocophile," the photographer Peter Beard. The official product of the effort was a report to the Kenya Game Department. The unofficial product was *Eyelids of Morning*, a pastiche of photographs and artwork and text, which might be the most cantankerous coffee-table volume ever published.

Eyelids of Morning is presently out of print and has been available only intermittently for decades. But it remains an underground classic, well known and vividly remembered by a small audience as one of those books that stamp their images on the brain like an auto accident. Besides recounting Graham's crocodile work, it offers Beard's lurid and beautiful photographs of dead crocodiles, living Turkana, and the bleak landscape surrounding the lake. It's a rowdy mix of verbal and visual elements, festooned with snippets of history, old drawings and cartoons, crocodile myths and legends, crocodile kitsch, and the caustic cogitations of Alistair Graham on the larger subjects of wildness, man-eating predators, science, progress, and the seemingly quixotic enterprise of nature conservation in an ever-changing world. For instance:

> So long as one is constantly threatened by savage brutes one is to some extent bound in barbarism; they hold you down. For this reason there is in man a cultural instinct to separate himself from and destroy wild beasts such as crocodiles. It is only after a period of civilization free of wild animals that man again turns his attention on them, seeking in them qualities to cherish.

Just above that statement, on page 201, appears a black-and-white snapshot of two severed human legs dropped into a blood-spattered cardboard box, following their recovery from the belly of a crocodile.

The legs had belonged to an American named William Olson, a young Peace Corps volunteer, who was killed on the Baro River in southwestern Ethiopia during the time Graham and Beard were working at Lake Rudolf. Olson's fatal mistake was that while visiting a village on the Baro for a larkish vacation with some Peace Corps friends, he ignored warnings by local people and went for a swim. In the river somewhere lurked a large, brazen crocodile that had recently eaten a child, then also a woman, within full view of the village. For the locals, dependent on their river for practical uses, it was a dull, odious, quotidian danger. But that danger seemed too abstract to deter Bill Olson and his pals on a hot day. When the others finished swimming, Olson chose to linger behind, standing on a submerged rock and staring into the water. The attack happened quietly. One minute he was there, then he wasn't.

Apparently the crocodile had swum up from behind, bit hold of Olson's legs, and dived. Big crocodiles sometimes kill their larger air-breathing prey by drowning, a fact to which Graham may have alluded coyly when he wrote "they hold you down." The whole event, so gruesome and shocking as to seem newsy, drew passing attention in the "Milestones" section of *Time* magazine: "Died: William H. Olson, 25, Cornell graduate ('65) and Peace Corpsman since last June who taught science in the Ethiopian village of Adi Ugri; after being attacked by a crocodile while standing waist-deep in the muddy Baro River near Gambela, Ethiopia. Five fellow corpsmen heard Olson shout and saw the beast pull him under; next day police found and shot the crocodile." The shocking aspect that brought the press coverage, truth be told, was that Olson was an American, not just another among the dozens or hundreds of Africans who died from crocodile attacks the same year. Needless to say, the Ethiopian woman and the child who'd been killed earlier by the same crocodile hadn't achieved notice in *Time*. Not even Alistair Graham put them on record by name.

Olson's own memorialization, both in *Time* and in Graham's book, was blurred by a couple of small mistakes. His name was William H. (for Henry, according to alumni records at Cornell University) Olson, not William K. Olsen, as Graham had it. Apart from that point, Graham's version corrects *Time*'s misprisions. No one saw the beast pull Olson under, and no one heard him shout. He disappeared while his friends were occupied elsewhere; his fate was revealed only when his body reappeared in the jaws of a large crocodile surfacing downstream shortly afterward. And the crocodile wasn't shot by police; it was stalked and killed by an American big-game hunter who happened to be on the scene with his professional guide. These details were clarified by the guide himself, Karl Luthy, whose meticulous account Graham quotes. Luthy had been standing near the river when Olson disappeared, later helped drag the crocodile carcass onto a sandbank, and described what came next—opening the croc's stomach to see whether this was indeed the animal that got Olson. "We found his legs, intact from the knees down, still joined together at the pelvis," Luthy reported. "We found his head, crushed into small chunks, a barely recognizable mass of hair and flesh; and we found other chunks of unidentifiable tissue. The croc had evidently torn him to pieces to feed and abandoned what he could not swallow." The crocodile was just over thirteen feet long, not huge for its species but more than big enough to kill and dismantle a human. Karl Luthy admitted that he and his client, one Colonel Dow, had felt impelled to kill the croc not just because it was dangerous to the villagers but from some primal hunger for human-against-crocodile vengeance.

The Turkana whom Graham knew at Lake Rudolf had their own complicated attitudes toward crocodiles—a mix of loathing, practicality, and insouciance tempered by their loss of a few community members to crocodiles each year. These lakeshore Turkana were unrepresentative of the tribe overall, with its tradition as proud warriors and nomadic pastoralists. They had settled near the lake during a severe famine in the 1920s, when the Kenyan government had established a relief camp at a spot called Kalokol on the lake's western coast. The relief supplies continued after the famine conditions ebbed, so

Kalokol became a permanent village. It served as Graham's jumping-off point to more remote edges of the lake, and from Kalokol also he recruited his field assistants. Like other Turkana, the men tended to be slender and handsome and confident, the women lovely and stat-uesque and proud of their bodies, as Peter Beard appreciatively docu-mented in his photos. Unlike other Turkana, they were intimately acquainted with *Crocodylus niloticus*. One white man who lived among them, a sort of extension agent sent by the government to advise on fishing techniques, assured Graham that the Turkana "did not care if crocs laid hard-boiled eggs or swam upside down. In their world there were quite literally only two important facts about crocodiles: they were evil and they were edible." Graham himself found the Turkana to be happy, arrogant, mindful of their tribe's warrior reputation, ashamed of their own poverty and fish-nibbling existence at Kalokol, tough, introspective, petty, aggressive, insolent, wild, and yet "peaceful in a way that only contented people can be." Despite (or because of?) the contradictions, he seems to have admired them.

And he was fascinated by their ambivalence toward the crocodile, which blended nonchalance and contempt. In the dour animist the-ology of the Turkana, *C. niloticus* played a role as the punishing agent of a capricious God who was by turns benevolent and vindictive— like the Lord in the Book of Job, only worse. Such a God, as Graham saw it, "was indistinguishable from the devil." All of life's uncertain-ties were "basically the handiwork of the same spirit, who was devil-ish or compassionate according to his whim." Still, whimsy notwithstanding, this Turkana God seemed to reserve his ugliest afflictions for bad people, and crocodile attack was one mode of delivery. Graham's field assistants were sanguine about their own exposure, telling him that they felt quite safe wading in crocodile-haunted waters so long as their consciences were clear. "Theirs is not the contempt of familiarity, nor the apathy of resignation," Graham wrote. "It is the supreme confidence of an intact ego. To a man who has no doubts about himself an attack from a scaly, evil crocodile *must* be bad magic."

What made the bad magic seem worse were two factors. First, crocodiles are stealthy—amazingly so, for such large beasts. When stalking terrestrial victims at the water's edge, they remain invisible underwater until the last second, as that Ethiopian croc did to William Olson. Strong as they are, quick as they are, they take people by surprise, not just by quickness and strength. The second factor is that they don't just kill humans—they eat them. "The fear of being eaten by an animal," Graham argued, "is much greater than the fear of merely being killed by it."

Why does man-eating make a murderous animal—any murderous animal, but especially a sneaky, reptilian one—seem more horrific? Graham offered a baroque theory. "One of civilization's imperative taboos is against cannibalism; little else arouses such fear or loathing. And we do not easily distinguish, emotionally, between a human eating a human and an animal eating a human." The fear of man-eaters, he claimed, is entangled with the fear of becoming a man-eater oneself. "Such fears generate volcanoes of rage and terror. Nor should we be so culturally arrogant as to suppose ourselves too far removed from cannibalism. Given the conditions of physical and mental anarchy that accompany war, for example, or other cataclysms, cannibalism soon reappears." Given their proximity to people along lakes and rivers, and their genuine menace, crocodiles almost inevitably became burdened with imputations of malice and depravity, according to Graham. Without quite explaining, he added: "It is around the matter of cannibalism that the symbolism of crocodiles twists and writhes. To be eaten by a croc is to be consumed forever *by evil*. One forfeits all hope of immortality. One's soul is irrevocably Satan's, one's body is dung."

Much of *Eyelids* is in that vein, boldly discomfiting but overheated. Although I agree with his take on the redoubled horror of man-eating,

I don't find his cannibalism argument persuasive. My own view is simpler and less arch: that the extra dimension of dread derives largely from ancient concerns about funerary observances and the deceased's prospects in an afterlife. Respectful, decorous disposal of the mortal remains has been important across virtually all times and cultures. When Hector met his ugly end outside the walls of Troy, the ugliness was intensified not by some flesh-eating predator but because Achilles dragged his dead body behind a chariot. Bury the corpse, cremate it, put it up on a platform to be picked clean by birds, pile rocks over it like the cairns of Komodo, even cook it and eat it yourselves (as the Foré people of highland New Guinea did, until medical researchers alerted them about kuru disease, transmitted in undercooked brain), but by all means don't leave it to be gnawed at by leopards or hyenas. People hire morticians to cosmeticize the cold remains of their loved ones. Why? For the sake of dignity and closure. Soldiers make a high obligation of rescuing not just their wounded but also their dead from the battlefield. Why? To give peace to themselves, to the next of kin, and in some sense to the fallen comrades. Achilles was consciously flouting that piety, not just mindlessly overplaying his fight against a dead opponent. Seeing the corpse of a friend or a relative being chewed on and swallowed by a crocodile—or a lion or grizzly bear or tiger— must carry much the same flavor of desecration. Graham's cannibalism hypothesis is plausible enough, then, but it's not necessary as an explanation of why man-eating predators seem more abominable than other agents of death.

The Turkana of Lake Rudolf dealt with such abomination, according to Graham, by equating crocodile mayhem with divine justice. That helped open some emotional distance. "They appear to be quite callous about victims of attacks. But in fact they are not being cruel or cynical. They are simply taking it for granted that the victim was an evil person—why else would a croc have attacked him?" The attack, the killing, then the tearing and swallowing, represent the claiming of a mortgaged soul, in the way Mephistopheles claimed Faust's. To be swallowed by evil, Graham wrote, "in the nursery logic of our unconscious, is to *be* evil, for all practical purposes." It's a reversal of the apothegm *You are what you eat*. What eats you, you are.

How much of this fevered theorizing should we trust? Well, Alistair Graham was a white wildlife biologist, not a Turkana elder or even an anthropologist. He spent just a year at Lake Rudolf, during which he was very busy killing crocodiles, skinning them, pickling their organs, and extricating himself from episodes of shipwreck, marooning, and grounded airplane. His characterizations of Turkana spiritual beliefs are sensationalistic, dogmatic, and a bit vague. If he made any concerted effort to poll a sizable sample of informants, rather than simply learning what he could from his field crew and other casual Turkana contacts, he didn't mention it. The views expressed and the impressions reported in *Eyelids of Morning* are always interesting, but they shouldn't be mistaken for judiciously collated fact.

His crocodile data, on the other hand, seem solid. Graham and Beard killed their quota of five hundred crocs, of which the last three were lost to science during a final boat-sinking adventure on the wind-stirred lake. Among the other 497 animals, 278 were females ranging up to eleven feet long and 202 were males up to sixteen feet long (the remaining 17 carcasses evidently went unmeasured and unsexed). Both the large crocodiles and the smaller ones seemed to live mainly on fish, supplemented by any other meaty items (from a stork to a human) they might catch. They hunted at night, mostly in shallow waters near shore. The hunting must have been difficult, as reflected in a high incidence of empty stomachs among the specimens Graham took. The young grew roughly fifteen inches per year until they reached sexual maturity around age six; then the rate slowed. A big male might weigh half a ton and live to be seventy. A female could produce up to a thousand young in her lifetime. By Graham's estimate, the entire population comprised about 14,000 individuals (not counting those less than a

year old), possibly the largest assemblage of Nile crocodiles then surviving in Africa. But only 5,500 of the total were adults, and that
seemed puzzling.

Graham suspected that the population was overcrowded. He figured that the annual crop of hatchlings averaged another 14,000 but
that very few were reaching adulthood. Among the adults, he found
evidence of stunting. Body size at the onset of sexual maturity was
relatively small, compared with that of other known crocodile populations, and so were the maximum sizes attained. Mature females in
particular were small, and therefore they produced fewer eggs. Then
again, Graham noted, stunting itself was a relative concept. The population seemed to be an old one, long undisturbed. There had been
little hunting by humans, and apart from humans an adult crocodile
has few enemies. It might simply be that this crocodile population, as
Graham had gauged it, was right for the limited resources (just so
much food, just so many nesting sites) of the lake, and that the
reduced individual size and fecundity represented forms of attunement to those limits.

But attunement to limits, however it might affect *Crocodylus niloticus*, is not such a binding imperative upon *Homo sapiens*. The ever-
increasing squeeze of humanity could already be felt, or at least
foreseen, in that remote part of Kenya. At the end, it seems, Alistair
Graham grew disconsolate over the meaningfulness and likely application of his own research. "Our knowledge of crocodiles ultimately
was of potential value only to those far from Lake Rudolf who, feeling
overcrowded, needed more resources, more ideas, more space—simply
more." He'd begun hearing talk, presumably from Nairobi, about setting aside the lake's northeastern shore as a national park, a prospect
that pleased him not at all. He scorned the notion of preserving wilderness by locking it up within regulations, and he guessed that any such
scheme would mean eviction of the Turkana herders up in the hinterland. Change was coming inexorably to the lakeshore Turkana too.

And what would our increased knowledge of crocodiles do for *them*? The incompatibility of men and predatory carnivores remains; our findings could not alter that. Knowledge dispels the *evil* of crocs— for those who bother to acquire it—but our facts would not change the Turkana's outlook. For them, crocodiles would remain evil, hostile denizens of the lake. Nevertheless, no Turkana would ever attempt to exterminate crocs. They do not *hate* them. It takes a civilized, cultured, overcrowded man to hate crocs, or love them, or exploit them, or exterminate them.

And it takes the patience of Job to let them be.

One of the dilemmas outlined by Graham remains only more pertinent thirty years later. He warned, remember, that "there is in man a cultural instinct to separate himself from and destroy wild beasts such as crocodiles." From this axiom flows a list of hard questions. Doesn't separation inevitably mean eradication? Isn't containment (as in a national park that's too small to support a viable population) just another word for captivity? Is it possible to separate *Homo sapiens* from the dangerous inconvenience of alpha predators—around a lake, along a river, in a forest anywhere on the planet—without separating those predators from the habitat they need for a continuing existence in the wild? Can we have them at all if we're unwilling to suffer among them? Finally, in answering that one, whom do we mean by "we"?

25

Late in the day on April 25, 1977, two crocodiles were quietly released into a creeklike channel of the Brahmani-Baitarani river delta on the east coast of India. They were small animals, each barely one meter long, though the species to which they belonged, *Crocodylus porosus*, attains larger maximal sizes than any other crocodilian, larger even than *Crocodylus niloticus*. Both of the little crocodiles were female. Maybe they would survive and grow big. Maybe they would find males, eventually, and mate. That was the hope. *Crocodylus porosus*, commonly known as the saltwater crocodile (or sometimes the estuarine crocodile), though it lives in brackish and fresh water also, had nearly disappeared from the Brahmani-Baitarani delta, mostly because of heavy hunting and mangrove cutting during previous decades. Now it was being put back.

Another three crocodiles were released the next day. A week and a half later, ten more were turned loose in the same channel. The releases had been scheduled to precede the start of the monsoon by a margin of months, so that these young crocs might adjust to their habitat before the waters rose and everything got crazy.

The Brahmani-Baitarani delta is a maze of river branches, narrow channels, and murky sloughs that converge, split, arc around, swing back, and interlace through a large area of flood plain and mangroves on the coast of the state of Orissa, roughly a hundred miles northeast of the city of Bhubaneswar. It's one of the largest mangrove formations in India, comparable to the Sundarbans area at the delta of the Ganges, farther north along the country's eastern shore. Unlike the Sundarbans, which is famed for its man-eating tigers, the Brahmani-Baitarani delta is tiger-free and little known outside India, but its biological richness is spectacular. The mangrove forests, mud banks, grassy clearings, and swampy swales harbor Indian rock pythons, king cobras, jungle cats, fishing cats, and one of India's biggest concentrations of barheaded geese. A vast heronry, spread across treetops, attracts giant wading birds of several species, including tens of thousands of openbill storks. Olive ridley turtles come ashore to lay eggs by the thousands among the sands of the southeastern beachfront. Lunker specimens of the water monitor, the heftiest lizard this side of Komodo, bask on mangrove limbs and scuttle through the understory.

Warm, wet, tangled, the Brahmani-Baitarani delta is a good place for aquatic reptiles on a bigger scale too. At one time, not many decades ago, it supported a thriving population of saltwater crocodiles.

Although the water drains generally southeastward from its sources among the modest highlands of north-central India, emptying eventually into the Bay of Bengal, within the delta it flows every which way, meandering north, south, even bending back westward before the final turn and spill-out to the sea. The small channels are nameless, but the big branches have their labels on maps, one name giving way to another—confusingly, like urban boulevards that change names at each major intersection—wherever the water forks or makes a sharp bend. Besides the Brahmani and the Baitarani, there's the Dhamra River, the Hansina River, the Maipura River, and several others—including, near the heart of it all, the Bhitarkanika River.

Around the delta's edge are human settlements, protected from river surge and tidal bore by an ambitious, desperate network of dikes. The crucial line of diking, a five-foot-high wall of hand-piled clay, is known as the saline embankment. *Here*, people have stalwartly declared, *is where the slosh of the sea ends and dry land begins*. But the sea, in combination with the river system whose high-water pulses sometimes collide with tidal sloshing, doesn't always respect that line. The terrain behind the embankment is richly besmeared with fertile soil, thanks to flooding and silt dumping by the rivers, and offers excellent ground for rice farming in paddies, so long as the salt water can be kept out. It's a zone of tenuous opportunity, hazard, and flux.

During the late years of the British empire, much of this land was held by the Raja of Kanika under the *zamindar* system, a neo-feudal arrangement whereby absentee landlords paid franchise fees to the British for the privilege of milking rents from the landless peasantry. At the time of India's independence and partition, in 1947, a wave of Bengali immigrants came down into northeastern Orissa from the border region, where the Indian state of West Bengal now stood in

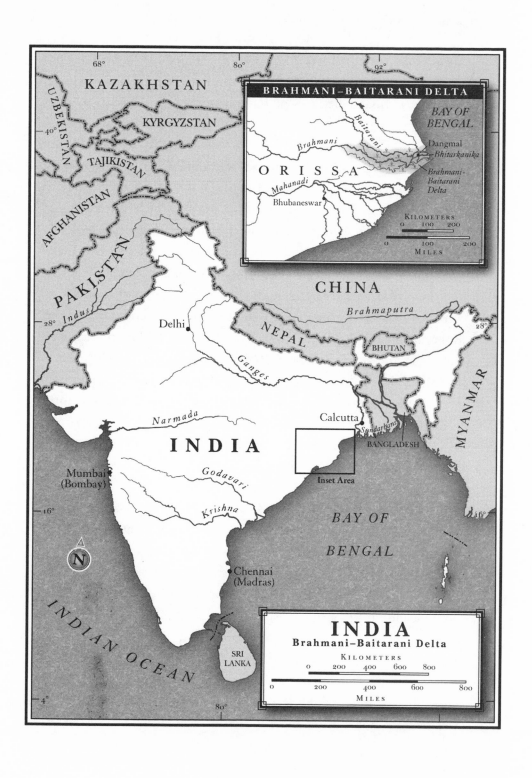

BRAHMANI–BAITARANI DELTA

Baitarani
Brahmani
BAY OF BENGAL
O R I S S A
Dangmal
Bhitarkanika
Brahmani–Baitarani Delta
Mahanadi
Bhubaneswar

KILOMETERS
0   100   200
0   100   200
MILES

68°   80°   92°

KAZAKHSTAN
UZBEKISTAN
KYRGYZSTAN
40°
TAJIKISTAN
AFGHANISTAN
PAKISTAN
Indus
CHINA
28°   28°
Delhi
Brahmaputra
NEPAL
Ganges
BHUTAN
MYANMAR
Narmada
I N D I A
Calcutta
Sundarbans
BANGLADESH
Mumbai
(Bombay)
Godavari
Inset Area
Krishna
16°   16°
BAY OF
BENGAL
N
Chennai
(Madras)
INDIAN OCEAN
SRI
LANKA

INDIA
Brahmani–Baitarani Delta
KILOMETERS
0   200   400   600   800
0   200   400   600   800
MILES

4°
80°

uneasy contiguity with East Pakistan. The zamindar system was abolished soon afterward, but not before the Raja of Kanika had leased some ten thousand hectares of forest land to local Oriyans (Orissa natives) and Bengali settlers. The industrious lessees cleared mangroves, improved dikes, and built more paddy fields, fishponds, and villages. What remained of the forests went into revenue management by the government of Orissa. Another wave of Bengali refugees arrived during and after the Bangladesh war (in which East Pakistan, with India's help, won its independence from Islamabad) of 1971. Some acquired land rights and some merely squatted. Eventually there would be a hundred villages with a total human population of almost four hundred thousand in the immediate region. But at its center was a muddy mangrove wilderness, fecund and diverse, embraced by the Bhitarkanika River. This parcel, not yet tamed for people and rice, was known as the Bhitarkanika Forest Block. By a government decree issued in 1975, an area encompassing that untamed core plus a portion of half-settled landscape roundabout became the Bhitarkanika Wildlife Sanctuary. The sanctuary's founding rationale was to conserve—or, more accurately, to reestablish and *then* to conserve—the saltwater crocodile in one of its last refuges of good habitat on the Indian subcontinent.

Besides being the largest crocodilian and among the most ferocious, *Crocodylus porosus* is the most broadly distributed. That distribution, stretching from the western Pacific to the Indian Ocean, once included much of coastal India. Thousands of miles eastward, the species thrived in estuaries and rivers draining northern Australia, New Guinea, and the Philippines. Well adapted for open-ocean dispersal, it even turned up on islands as remote as Palau and as far east as the Solomons and Vanuatu. It lived throughout Indonesia and southeastern Asia and had colonized the Andaman and Nicobar islands, almost four hundred miles offshore from Malaysia in the Bay of Bengal. It was found also in Burma, the Sundarbans, down the east coast of the

Indian subcontinent, on the island of Sri Lanka, and barely around India's southern tip onto the Malabar coast. In short, the saltwater crocodile was an abundant, widespread, and very successful tropical reptile, whose closely similar ancestors dated back eighty million years and whose primordial presence was an important factor to early humans throughout the Indo-Pacific region as they defined their relationship to the natural world. Then, after millions of years of success, *C. porosus* began to fail.

It coped poorly with the pressures of predation and competition exerted by modern humanity. By instinct and necessity it nested along riverbanks, but riverbanks were exactly where people chose to settle. Also, having the misfortune of smooth, handsome belly skin that could be tanned into premium leather, it suffered badly from hunting, especially from the new style of night hunting, practiced with rifles and motorboats and battery-powered spotlights. After World War II, the saltwater crocodile became an international market item as well as a staple of local diets and local mythologies. Its distribution shrank and its reproduction was impeded as human commerce turned estuaries into harbors and as human settlements sprawled along rivers everywhere. Mangroves were cut and cleared, the muddy silt beneath them "reclaimed" by diking and planted in rice. As late as 1939 a Ceylonese herpetologist had written that *C. porosus* "flourishes along the shores of the Bay of Bengal." That changed fast. By the early 1970s, it was extinct on the south Indian coast. A few individuals could be found in the Sundarbans, and a small population survived in the Bhitarkanika area of the Brahmani-Baitarani delta.

Then, under India's Wild Life Protection Act of 1972, crocodile hunting was banned. The export of crocodile skins was prohibited. *Crocodylus porosus*, as well as the two smaller species of native Indian crocodilian—the mugger, *Crocodylus palustris*, and the gharial, *Gavialis gangeticus*—received full statutory protection on Schedule I, the act's list of most-threatened species. But the Wild Life Protection Act required ratification by each state, plus enforcement measures, all of which took several years. Given the current scarcity of *C. porosus* and the ongoing loss of habitat, even well-enforced protection probably wouldn't have been enough. Wildlife officials recognized that, at least in Bhitarkanika if nowhere else, the species had a chance to recover. But it would need help.

26

H. Robert Bustard was a crocodile expert who had worked in Australia and Papua New Guinea (the eastern half of the New Guinea island, which until 1975 was held as a territory of Australia), surveying wild populations and developing management recommendations. Along the Fly and Sepik rivers of Papua New Guinea, two of the wildest stretches of *C. porosus* habitat remaining anywhere, Bustard found many areas where crocodiles had been nearly extirpated. Twenty years of commercial hunting, first by expatriates and later by local Papuans outfitted with flashlights and guns by expatriate skin buyers, had claimed most of the large crocodiles and many of the medium-sized and smaller ones. Throughout earlier times the native people had killed crocodiles, but not so efficiently or abundantly. Harpooning was the traditional method. They also caught crocodiles with nets and hooks, as they caught fish. Along the Sepik in particular, according to Bustard, hooks were commonly used. For old and wary crocs, hooking was sometimes the only method that worked. Bustard himself, in eight nights of scanning the banks of the Sepik, saw only six crocodiles. Between the new-style shooting and the old-style hooking, he concluded, the big breeder crocs were being eradicated.

Crocodiles grow slowly, even in optimal habitat. A female saltwater crocodile reaches sexual maturity when she's about seven feet long, at roughly age twelve, and a male matures at about eleven feet, after some sixteen years. The danger in New Guinea was that the populations of *C. porosus* in major rivers would lose all their mature adults, and hence their capacity to reproduce. Along some sections, that might already have happened. Bustard therefore recommended that besides setting an upper size limit on legally killable crocodiles (to protect the breeders, wherever they survived), the territorial government should initiate restocking. Collect eggs from the wild, hatch them, and rear the youngsters for a year in captivity, he advised; then, after they've passed through the vulnerable life stages at which their natural mortality rate tends to be very high, release them back into the rivers. Plant juvenile hatchery crocodiles now to get, eventually, big wild crocodiles.

The goal of his proposals, as Bustard explained in a 1969 journal paper titled "A Future for Crocodiles," was not just to help New Guinea's crocodile populations recover and survive. That was a short-term concern. Over the longer term, it was equally crucial to arrange circumstances so that local people would have reasons for *wanting* them to survive. "Crocodiles do not readily arouse public sympathy," he noted dryly, "and the best argument for conserving them is their economic value."

Their value at that time, as calculated from the export of untanned skins, totaled 470,000 pounds sterling per year. With a sensible program of management, Bustard argued, the annual yield might exceed £2,350,000. How to achieve that? He listed four ways: by increasing the total quantity of crocodile skin harvested; by increasing the average size of individual skins, since a large skin is more valuable per unit area than a small one; by increasing the quality of harvested skins, through careful attention to skinning and salting techniques; and by establishing a tannery in New Guinea, so that value could be added before export. Teach people to skin the animals deftly and salt the skins well. Encourage close contact among hunters, tanners, and buyers, so that the first two groups know the realities of the market. Enforce a lower size limit as well as an upper one, so that young crocodiles have time to reach optimal size. And grant local people proprietary tenure—by way of land ownership, exclusive harvest rights, whatever—over the crocodiles that inhabit their waterways. Bustard recognized that, so long as crocodile hunting went unregulated along publicly accessible rivers, it would represent another instance of the familiar dilemma that the ecologist Garrett Hardin had recently labeled the Tragedy of the Commons: Resources owned by no one get overused and abused by everyone. "Clearly," Bustard wrote, "there is no incentive to build up crocodile populations on a local river-swamp complex if anyone can come and shoot them." Better to let the local people exercise ownership, in some sense, by way of an exclusive though regulated harvest right to the local crocodiles.

Bustard also suggested that the territorial government establish two prototype crocodile hatcheries, one on the Sepik River, one on the Fly, and that these hatcheries be used to train Papuans for running similar operations in their own villages. The hatcheries would be supplied with eggs from the wild, for which local collectors would be paid one shilling per egg. For each crocodile hatchling released back into the wild after a year's growth, the local people would get another nine shillings.

Bustard's message, in sum, was: Manage the crocodile populations for a sustainable harvest, and make sure that a sizable cut of the action goes to the folks who live in the habitat. There were two excellent rationales for such an arrangement. First, it would discourage poaching, over-harvesting, and the inclination to exterminate all crocodiles. Second, it was fair. Enduring the risk of crocodile attack in their home waters was a service to society, to the territorial economy, for which people should be paid.

These proposals, Bustard added, "could be applied in many areas of the world where once large crocodile populations are now seriously depleted." Someone in India must have been paying attention.

In 1974, now under the auspices of the Food and Agriculture Organization of the United Nations, Bustard was brought to New Delhi. The original terms of his visit stipulated a six-week consultancy to assist the government in establishing a crocodile farm at the New Delhi Zoo, aimed at raising animals for the eventual harvest of their skins. If that farm succeeded, similar projects might be developed elsewhere in India. Upon arrival, though, Bustard learned that "the problem was of greater magnitude, involving conservation of India's three crocodilian species as well as development of an industry." The gharial, a long-snouted, fish-eating crocodilian that lives in cool rivers draining from the Himalaya, was sorely endangered because of habitat loss, hunting for its skin, and fishing operations using nylon gill nets, which tend to drown crocodiles as unintended victims. The saltwater crocodile and the mugger had been hurt by the same factors, though not quite so badly. Bustard did another broad-stroke field survey like the one in New Guinea, visiting ten different states to assess the population status of all three species in their various locales and to formulate recommendations for conservation and commerce. His report was optimistic on the economic prospects, and dismissive of concern about how local people might react. "Indian villagers do not fear crocodiles," he claimed, "nor do they kill for the sake of killing. Both crocodiles and local people co-exist with minimal conflict. The villagers do not attack the crocodiles and the crocodiles very rarely disturb the villagers." That may have been true where and when he looked, but it suggests an underestimation of the importance of rare, grievous, and provocative exceptions. Bustard compounded the affront with a statement that could only have come from a fly-in, fly-away consultant: "It is well known that man is not within the preferred prey range of crocodiles and that the occasional attacks, except for the aberrant crocodile which becomes a 'man-eater,' are the result of mis-predation." *Mis-predation?* It's a ridiculous term, especially as applied to an animal of such varied and capricious dietary proclivities. Bustard failed to appreciate that, for a victim, or for the next of kin, getting gobbled in a case of mistaken identity is no more acceptable than getting gobbled on purpose.

What obscured the genuine (if marginal) dangers and dreads was his golden vision of the potential contributions of crocodile farming to the Indian economy. Because crocodiles are native to the tropics and the subtropics, he noted, tropical countries had a natural advantage in commercial crocodile farming. Crocodile hides were much in demand for fashionable leather products, and prices seemed likely to continue rising. The current price for saltwater crocodile skins, he reported, was thirty-nine rupees (about five dollars then) per inch of width across the belly. A smallish skin was therefore worth five hundred rupees, serious money in 1974. Bustard imagined a future, twenty years on, in which India might be producing a half million skins annually, worth 260 million rupees. If there was concomitant growth in the capacity to manufacture fancy leather goods from domestic skins and export them in value-added form, the annual total might go to 600 million rupees.

Where should such efforts begin? Bustard saw particular promise in Bhitarkanika, where he had visited for a week in early June. "The area is prime saltwater crocodile habitat in which the species was formerly abundant," he wrote. The crocodiles of Bhitarkanika had been subjected to trophy shooting during the Raja's time and had later suffered from harpooning by Muslim hunters who came into the delta from the west. The harpooning, done at night using lights, had severely reduced the number of large crocs but not killed them off entirely. Bustard figured that they would recover if protected and sensibly managed.

Bolstering the wild population would be necessary but not sufficient. Conserving the habitat against further encroachment would also be important. And there was a third factor: human attitudes. "In order to retain the co-operation of the local people, so essential for a project of this kind, the crocodile management programme should result in real material benefit to the people. This would be possible by tying conservation in the sanctuary to commercial crocodile farming at the village level," Bustard wrote. As in New Guinea, so in India—crocodiles might be tolerated, despite the danger and inconvenience, if they paid their way. In the Bhitarkanika area alone, he suggested, a combination of crocodile farming and sustainable harvest from the wild might eventually yield twenty thousand skins annually.

A year later, in response to Bustard's recommendations, the Orissa government established the Bhitarkanika Wildlife Sanctuary, with its special mission of rescuing *C. porosus*. Fishing was banned within the sanctuary, to protect crocodiles against drowning in nets and against poachers functioning under fishermen's cover. Mangrove cutting was prohibited. The government set up a Saltwater Crocodile Research and Conservation Centre within the sanctuary. Robert Bustard himself became chief technical adviser for India's crocodile-conservation efforts, there and elsewhere.

A research officer from the Forest Department, Sudhakar Kar, worked closely with Bustard during the early years of the project (and would remain with it long after Bustard left). Among their first tasks was a more thorough survey of the sanctuary, during which they found only twenty-nine adult crocodiles, six subadults, and sixty-one juveniles. The adult tally was low enough, but the extreme scarcity of subadults seemed even more worrisome. Presumably it reflected high mortality of two- and three-year-old crocodiles in fishnets before they reached the subadult stage. Any hope for a reproducing population, therefore, would require protecting those precious young crocs until they grew big enough to breed, and releasing more captive-reared youngsters into the wild. The two little crocodiles released on April 25,

1977, and the thirteen soon afterward, were the first graduates of Kar and Bustard's nursery.

They were virtually shooed out the back door. Near the channel into which they were liberated sat a village called Dangmal.

27

Apart from the addition of crocodiles, Dangmal doesn't seem to have changed much since 1977. Tidy mud-and-wattle houses, with hand-rounded corners and thatched roofs, front on a dirt footpath that constitutes the central avenue. Big-leafed vines twine across the roofs like strings of Christmas lights, radiantly green against the brown thatch. There are hand pumps for water, picket fences of bamboo and palm, freshly washed saris flapping dry on the fences, and, beside some of the houses, little ponds. Squarish and mud-banked, the ponds seem to be hand-dug, presumably for raising edible fish. Beyond are the rice paddies, stretching off toward a low horizon capped by the saline embankment. It's late November when I arrive, end of the monsoon, and the rice harvest is in progress.

I've driven up from Bhubaneswar with a grad student named Bivash Pandav, whose work on sea turtles hereabouts will eventually bring him a doctorate through the Wildlife Institute of India. Bivash is a brisk, incisive, mid-twentyish fellow from a middle-class Oriyan family. Dressed jauntily for this field expedition in jeans, a Waikiki T-shirt, and an Australian bush hat, he resembles a young Jesse Jackson trimmed out to go bass fishing in Wisconsin. His own research season doesn't begin for a few days, but meanwhile he has generously agreed to show me the ecosystem. We check ourselves through the formalities at a Forest Department office on the western edge of what's now Bhitarkanika National Park (recently upgraded from a sanctuary) and hitch aboard a Forest Department patrol boat heading into the mangroves. It's a large, motorized river launch christened *Baula*, meaning "Saltwater Crocodile" in Oriyan.

We descend the brown Baitarani River on an outgoing tide, then turn back upstream along the Dhamra. The world of roads, trucks, cars, bicycles, bullock carts—the world of teeming human India and its terrestrial modes of conveyance—disappears behind. The world we're entering, by contrast, is all water and mud and mangroves. We wind through smaller channels, Bivash and I seated in kingly vantages on the upper deck, watching herons and egrets along the banks, kingfishers dipping from limb to limb like trapeze artists, rose-ringed parakeets flying overhead in raucous little flocks. We discuss crocodile-management problems at Bhitarkanika and the fact—which Bivash finds maddeningly wrongheaded—that there's still no legal harvest of crocodiles in India. No hunting permitted. No crocodile farming, despite Robert Bustard's urgent recommendation more than two decades ago. No commercial take of skins or meat whatsoever, not for export, not for domestic sale, not for local use. Not in Bhitarkanika, not anywhere. There should be, Bivash argues. This crocodile population has recovered. The reintroduction program has worked nicely, and now the habitat is full. Crocodiles are becoming victims of their own success; local people have started to hate them. Still the government prohibits harvest. Thousands of chickens and thousands of goats are killed every day in this Hindu country, he notes caustically, so why should there be reluctance to countenance the death of a few large reptiles?

Bivash himself is a fervent conservationist, but he sees crocodile protection with the cold clarity of a population biologist. Also, he knows the realities of the habitat-overlap zone, where dangerous though wonderful animals come in contact with poor, rural humans. "I'm telling you," he says, "no amount of nature education will help if your father gets killed, if your mother gets killed, if your brother gets killed by crocodile."

In late afternoon we're put ashore onto a wooden dock serving the Forest Rest House at Dangmal. This is where Bustard, Sudhakar Kar, and their colleagues hatched and reared crocodiles. A palm-lined lane leads up to the Rest House. Beyond that is a path to the rearing compound and, onward, into the village.

On this first evening there's time for only a short walk before dark, so I stroll out among the long, shallow fenced ponds of the rearing compound and find two puzzling remnants of what occurred here. One is a small crocodile looking lonely and abandoned—though steadied by a dim reptilian patience—in the first of the ponds. It almost seems to be waiting for Bustard's return, as though he were Godot. Except for this animal, the place is deserted. The other bit of opaque evidence is a painted sign, a sort of scoreboard, that bears the heading RELEASE OF CROCODILE IN NATURE. Below is a list of totals, year by year.

The first releases in 1977, according to this record, amounted to fifteen crocs. That matches the numbers reported elsewhere (a paper by Kar and Bustard in the Bombay Natural History Society's *Journal*) and includes the two pioneers I've already mentioned, freed into Dangmal's little creek on April 25 of that year. The program expanded robustly, despite some minor gaps and reversals, to a peak of three hundred crocodiles released in 1986. From there it declined to just four crocs in 1989 (what happened?), rose back in the early 1990s, and seems to have ended in a sizable, spasmodic gush with the release of 190 crocodiles in 1995. Then no more. Was the program finished, or just the scoreboarding? Fascinated by these ambiguous data and knowing that they somehow reflect a complicated history of conservation politics and reptile husbandry, I copy the full list into my notebook:

1977 – 15	1988 – 77
1978 – 80	1989 – 4
1980 – 30	1990 – 10
1982 – 75	1991 – 100
1983 – 50	1992 – 100
1984 – 100	1993 – 120
1985 – 200	1994 – 80
1986 – 300	1995 – 190
1987 – 123	

It wasn't so very many crocodiles, but it was enough, given the rate of survivorship for such meter-long juveniles, to bolster the population back to a level that seemed commensurate with available habitat. As of 1995, according to a later report by Sudhakar Kar, there were 660 crocodiles in the wild, of which goodly portions were subadults and adults.

Of course the numbers listed on this sign—most of them, anyway—are suspiciously round, so they must represent approximations. Why would the tally keepers have been obliged to estimate? If it was possible to record that first year's crop so precisely (two crocs on April 25, three on April 26, et cetera), why round off to the nearest ten or hundred in later years? And what was different in 1987 and 1988 that allowed such exceptional precision again? Was there a change in personnel? Was there a custodian of crocodiles during those years who cared fervently about each animal under his protection? If so, what became of him, and of the crocodiles, during the blighted years of 1989 and 1990? The scoreboard doesn't explain. And aside from the little crocodile in the pen, there's no one around for me to ask.

The rest of my information, during three days at Dangmal, comes from listening and watching. Out on the channels after dark in a small boat, Bivash and I spotlight roughly a hundred crocodiles in an evening. Most of them are juveniles, less than three feet long, but almost a dozen are sizable subadults or adults. Only the eyes and the nostrils dimple visibly above slick water as these animals float inconspicuously, alert to danger or food. Their eyes shine as orange dots in the darkness, reflecting our beam. The bigger crocs seem to spend their night-hunting sessions hovering not far off the bank, facing in. A careless chital or pig, at the water's edge to drink, would make a good meal. An unwary heron could get snatched—mostly feathers, legs, and beak, but crunchy and digestible nonetheless. A reckless human, poaching fish with a cast nest, might look like a prey item too.

Exploring farther by daylight, we find many large crocodiles basking on east-facing mud banks to catch the sun's morning warmth. One of them is a big fellow, sixteen or seventeen feet long, who glides down the mud and slips underwater as our boat draws near. It's uncanny that an animal so armored and heavy and huge can be, also, so liquid and discreet. I'm impressed—now *that's* a real bull crocodile—until half an hour later, when we reach a stretch of river to which Bivash has brought me, purposefully, for the chance of glimpsing a greater monster, a certain notorious behemoth. We approach, scanning the banks and the river expectantly. "Massive chap!" Bivash says, pointing, and now I see it too: a twenty-foot crocodilian barge floating gently in midstream. The creature's head is as big as a beaver. Its tail scutes protrude like a row of shark fins. With invisible, relaxed leg strokes, it turns in the current, catches an eddy line, circles. "It" is undoubtedly a he, since only males of *C. porosus* grow so large. And he has a name, Bivash tells me. Local people call him Mahisasur, meaning "Big Demon."

We hold the boat off at a respectful distance, admiring this beast. Let's not go any closer, Bivash suggests; we'll only disturb him, and he gets disturbed too much already. Good idea, I say.

Back at the compound, Bivash helps me persuade several Forest Department workers to recount their crocodile mishaps. Generally these tales reflect the discomfiting fact that when monsoon rains raise the river, fish and other animals come spilling through the dikes, and for weeks afterward a crocodile is liable to turn up *anywhere*. A tiny man called Nata, in a strap T-shirt and a dhoti, tells me (via Bivash's

translating) of the day back in 1995 when he walked down to a pond—
this one here, just behind the Rest House—to dip out a bucket of
water. A seven-foot croc erupted from nowhere and grabbed his arm.
Nata pulled, the croc released, and Nata fell backward, mangled but
alive. He shows me the scars on his forearm. Twenty-two stitches, a
month of healing, and still he has trouble with that hand. The same
crocodile attacked a village girl several days later. It was captured then,
Nata says, and released into the river.

"They don't have this concept of killing 'problem' animals here,"
Bivash explains, "they" meaning the Forest Department, responsible
for wildlife protection and control. An arm grabber in the pond is
viewed as a normal crocodile that merely needs relocating.

Does Nata fear for his children? Not particularly. Now if someone
enters the forest to poach fish, that person might be attacked, he says.
But if people keep to their proper byways, their houses, their village,
they will be safe. Without pressing the question of how that squares
with his own backyard skirmish, I slip Nata fifty rupees and offer
warm thanks.

Khaga is a middle-aged fellow, barefoot, in shorts, with a wide,
supplicant smile, who presently works as a boatman. He started with
the Forest Department as a laborer on the crocodile project, back in
Robert Bustard's time. Khaga's job was to look for nests in the wild.
During the first year they found only two. One of those held forty-
eight eggs, by Khaga's recollection (ancient memory with the precision
of legend), which yielded twenty-six live hatchlings in captivity. Those
hatchlings, plus another dozen scooped from the wild, were the
founder generation of the program. Dr. Bustard's idea, Khaga tells me,
was to develop the hatchery here and then help people begin farming
crocodiles in the village. But no, there has been no farming. Why not?
"Government is not taking any decision, so it's not happening," he
says, an answer that constitutes a primly articulated nonanswer. Khaga
himself has been chased by female crocodiles while collecting their

eggs, fair enough, but never attacked for no reason, and never bitten. The people who get hurt, he asserts, are those who go recklessly into the crocodile's habitat. Crocodiles live in the river and protect the forest; because of them, villagers are reluctant to sneak across the statutory boundary to poach fish, meat, or wood. Khaga's own attitude toward *C. porosus*, to hear him tell it, is tender. He has cared for crocodiles as his own children, reared them, handled them with respect, released them to the wild—so how can he hate them? He's only sad that the program has given nothing back to the people who assisted it. Him, for instance. Nowadays, a poor working man, hurt by inflation, he can't even buy a kilo of onions. Bivash earlier alerted me to the current onion crisis—low supply, high prices, heavily in demand as a staple for poor Oriyans. I express my sympathy to Khaga with a hundred rupees, enough for two kilos of onions.

Then Bivash and I walk the foot trail that leads through Dangmal and onward, toward a village called Khamar Sahi. It's a hot, dry afternoon, ideal for bringing in the rice. We see half-naked men, and women in bright saris, swinging hand scythes and laying armloads of rice stalks aside for bundling into sheaves. Other men carry sheaves up from the paddies—two big bundles per man, one on each end of a cross-shoulder pole. Threshed rice is spread to dry on grass mats, where frail old men and women do their part by stirring it with shambling bare feet. Cow dung, a valuable resource, is patted out like tortillas to bake in the sun. Most of these people work as day laborers, Bivash tells me. The zamindar system is gone, but the paddies are still controlled by absentee landlords. Fifty years after independence, half a day's drive from the state capital, and Dangmal doesn't even have electrification.

Farther on, we stroll into Khamar Sahi, where the sight of a white man is so unusual that children gape at me slack-jawed, without even hollering the usual cheery taunts and greetings that a visitor can expect along little-traveled Third World routes. I haven't seen many villages

so remote, so blessedly unconnected, that the quick-study kids couldn't
accost a stranger with at least two words of English—"Hallo, meester!"
But this is such a place. Bivash hazards that I might be the first white-
faced alien to intrude here since Bustard.

 We meet with twenty men in a small house that serves as a community center and basketry workshop. Light baskets nicely woven
from a sturdy grass known as *nalia* hang from the ceiling, and among
them, representing a whim or a totem, one basketry effigy of a crocodile. Friends and relatives have assembled, as requested, to give me an
earful about crocodile depredations. I hear of Ananta Jena, twenty-six
years old, father of two, who went to cut nalia in the forest and was
attacked while crossing a stream. The crocodile seized him around the
middle, took him down; it prevented his brothers from recovering the
body. I hear of Jeigun Nesa, a girl of sixteen, killed by a nine-foot croc
that had come in, with the monsoon waters, to lurk in her family's
pond. She went out to wash herself one evening; the crocodile caught
her hand, drowned her. Jeigun Nesa was unmarried, but plans had
been made. That crocodile remains in the pond even now, guilty of a
recent attack on a bullock that lowered its head to drink. The bullock
struggled free but needed a hundred stitches in its face. I hear of
Kartik Mandal, thirty-two, married, one son, who went with his
brothers and several friends to fish in a forest channel. His younger
brother, having cast a net, couldn't pull it in. He asked Kartik to
unsnag it. Kartik reached down, gave a little scream, vanished. They
thought he was diving to untangle the net, but he didn't come up.
Finally a very large crocodile surfaced, holding Kartik's body in its
jaws. The croc's mouth was so big, I'm told, that besides Kartik there
would have been room for someone else. This comment strikes me as
a shadowy hint of the unspoken question in everyone's mind: *Who
else—who will be next?*

 Again with Kartik Mandal, there was no recovering the body.
There was no retributive action by the Forest Department. There was

no visiting safari hunter with a high-powered rifle to oblige the village by making this crocodile his prey, and no doughty big-game guide to help retrieve the victim's remains from the croc's belly. Kartik Mandal, Jeigun Nesa, Ananta Jena—these people correspond to the nameless woman and child, on the Baro River in Ethiopia, who became blind footnotes to William Olson's story as told in Western magazines and books.

The account of Kartik Mandal's death comes from a tall, slender, angry man in a blue shirt and a blue sarong, one of the fishing companions who saw it happen. Having declined to sit on the floor with the rest of us, he stands near the door, interrupting other witnesses, speaking fiercely, smiling darkly, grandstanding, waving his arms to depict the length and the girth of one crocodile or another. His graying hair is combed back slick. From the flat expressions of the other men and their silence as he holds forth, I can't tell whether he's a community leader or merely an infamous loudmouth, but he's forceful, opinionated, and involved. Jeigun Nesa was his niece. Kartik was his friend. Ananta Jena's widow still lives here, he tells us, with her two fatherless children. The government should do something about all this. Should put up a fence, says the man in the blue sarong. Should somehow prevent crocodiles from invading their paddies and ponds. When I ask about farming of crocodiles, the other men dismiss that notion with nervous grins. Put crocodiles into their ponds *on purpose*? Forget about it. Sarong Man turns the subject back onto his earlier complaints. Crocodile farming? Farming might be okay, farming is fine, yes of course they would consider crocodile farming, but first the government must control these animals in the wild—in the river, in the fish ponds. The government must do something. Must relieve people of this fearful burden. I listen to the testy back-and-forth in Oriyan, study faces and tones, wait for bursts of translation from Bivash. There's a complex emotional chemistry in the basketry hall today, people unloading their aggrieved memories and their frustrations on an

American visitor who is keenly interested but, as they must realize, can do nothing to bring them a remedy. The very subject of crocodiles, for this community, is a sort of collective psychic abscess. I've poked at it, verified its tenderness, but I haven't drained it.

Afterward we're steered to the house of Ananta Jena, to whose brother I've given a hundred rupees for the wife and children. A hundred rupees is nothing from my wallet, zero, just scraps of oily paper, but it's a sum suggested by Bivash as appropriately scaled—not too large, not embarrassingly so—to express respectful sympathy from a passing acquaintance. The brother goes inside, delivering my piddling alms, and a shy teenage girl emerges to bow a thank-you, *namaste*, at the ghoulish white stranger who pays people for their sad stories. The widow herself remains inside, but Ananta Jena's elderly mother emerges to address me, whining and howling in grief that's either undiminished by time (the dozen years since Ananta's death) or well reenacted. What she's howling, Bivash reports, is *Who'll give me back my son?*

At twilight, Bivash and I go out again on the channels by boat. We count eighty-seven crocodiles, further evidence of the reintroduction program's success. Its failures are harder to measure.

28

In Australia, on the other hand, *Crocodylus porosus* is dealt with pragmatically. The city of Darwin, capital of the Northern Territory, is also the capital of a rambunctious crocodile-enhanced tourism industry and the nexus of a quiet but considerable trade in crocodile skins, crocodile meat, crocodile-parts souvenirs, crocodile tabloid journalism, crocodile natural-history documentaries, and—all right, if you insist on considering the *Crocodile Dundee* series—schlock adventure movies based on crocodile-cowboy mythology draped on an actor from down in Sydney, where there aren't any saltwater crocodiles. In the Northern Territory there are plenty, and quite a few too in Western Australia and Queensland.

Northern Australia also harbors an endemic species, *Crocodylus johnstoni*, sometimes called Johnston's crocodile or, more commonly, the freshwater crocodile. In breezy Australian style, which grows breezier with distance from the big southern cities, these two native crocodilians inevitably become "the saltie" and "the freshie." The freshie is a smallish, inoffensive creature with a broomstick snout, almost like a gharial. It lives in the upper, inland stretches of rivers that drain to the north coast, feeding on fish, frogs, crustaceans, even insects and spiders. The elongated snout is well adapted to such a picky, small-item diet, functioning more like a forceps than like the great slamming maw of the saltwater croc. Freshwater crocodiles don't bite hold of bullocks by the face or represent any grave threat to humans, and they mainly avoid the lower stretches of rivers where they might find themselves in competition with salties. Although a six- or seven-foot-long freshie is a powerful animal, well armed with teeth and not to be taken for granted, it's a meek country cousin compared with *Crocodylus porosus*.

Both species have been hunted for their skins, though *Crocodylus johnstoni* was always less highly prized because of the little bony plates (known as osteoderms to the scientists and, within the leather trade, as buttons) reinforcing the skin of the belly. When a crocodile skin is tanned, the osteoderms take on a different shade of stain and leave a rough irregularity to the touch. Among the factors that make the saltwater crocodile so desirable commercially is that, notwithstanding its great size, its ferocity, and the knobby toughness of its dorsal surface, its belly skin is supple, smooth, uniformly creamy in color, dappled nicely into many soft scales, and free of osteoderms.

The distinctness of the two species was a subtlety lost on early white explorers of northern Australia, who often referred to these big toothy reptiles as "alligators," a confusion still reflected in the names of the East Alligator, the South Alligator, and the West Alligator rivers, three of the best stretches of crocodile habitat in the Northern Territory. During a survey voyage along the north coast by the British ship *Beagle*, in 1839, a first lieutenant named J. L. Stokes found "alligators" abundant, and recorded that a certain sailor "had a very narrow escape from them. It appears that one of these monsters who had come out of the water at night, in search of food, found him sleeping in his hammock, which he had very injudiciously hung up near the water. The alligator made a snap at his prize; but startled at this frightful interruption of his slumbers, the man dexterously extricated himself out of his blanket, which the unwieldy brute, doubtless enraged at his disappointment, carried off in triumph." When that crocodile was later killed, a portion of the blanket turned up in its stomach, along with "the paw of a favourite spaniel, taken when swimming off the pier head." J. L. Stokes is memorable not just for his account of Australian explorations but for the fact that, during an earlier voyage of the *Beagle*, devoted mostly to similar charting along the coast of South America, he had shared a cabin with Charles Darwin. Back then, Stokes was just an assistant surveyor and Darwin was a young gentleman-naturalist, invited along to give the captain some dinner-table company commensurate with his class. Darwin proved himself egalitarian enough to mix with anyone, and evidently he and his cabinmate got along well, despite cramped quarters, because it was Stokes who, just a few years later, stepped ashore in a nice natural harbor at the northwestern corner of northern Australia and named it Port Darwin, after the friend back in England who wasn't yet famous.

J. L. Stokes had his own brush with a crocodile when he swam a small creek in order to hunt curlews along a far bank. He climbed out, nude, carrying a gun loaded only with birdshot, at which point "an alligator rose close by, bringing his unpleasant countenance much nearer than was agreeable." Stokes retreated, hoping to outswim the animal, and "reached the opposite shore just in time to see the huge jaws of the alligator extended close above the spot where I had quitted the water." We might suspect Stokes of overdramatizing—the image of jaws clapping closed as the man jumps clear of the water seems cartoonish—but the ill-advised swim, even today, is a contributing circumstance in many crocodile-attack cases. Learning caution, Lieutenant Stokes lived long enough to captain the *Beagle* himself.

Crocodiles were little more than a nuisance to the early white settlers in northern Australia, who came to mine gold, to establish cattle-grazing outposts (they called them "stations"), to impose their missionary Christianity on the Aboriginal people, or to help build a telegraph line up to Port Darwin, from where it would link Australia to Java (and from there to the rest of the world, including England) by undersea cable. Freshwater crocodiles could easily be ignored; salties, more assertive, were shot as dangerous vermin, or for sport, but not pursued for their skins. The impact on the wild populations was negligible. Aboriginal people had been living in this region for tens of millennia, of course, and the various language groups and clans—such as the Larrakia, whose homelands encompassed Port Darwin—had their own ancient relationships with crocodiles, which often entailed sacred mythic bonds that didn't necessarily exclude hunting crocs for the meat or collecting the eggs to be eaten. Because the Aboriginal population was always sparse, and because these people were closely attuned to the imperatives of long-term survival on a hard landscape, their forms of harvest tended to be sustainable. During the late nineteenth century and into the early twentieth, white hunters occasionally tried to sell crocodile skins, but there was no steady trade. Domestic buffalo, which had been introduced from Asia and gone feral in the northern bush, offered easier targets and a more familiar sort of hide for skinning, tanning, and marketing.

The boom years of crocodile exploitation began after World War II. Small boats with outboard motors brought hunters with rifles (mostly old .303s from the war) into the habitat; night shooting with spotlights increased the effectiveness of those hunters; and suddenly saltwater crocodile skins joined American alligator and Brazilian caiman as staples of the international trade in luxury leather. In fact, *Crocodylus porosus* emerged as the premium skin-yielding species among all crocodilians. Besides the absence of osteoderms, what made saltie skins especially desirable was their large number of rows of scales along the belly, each scale a pearly rectangle, like the polished ivory of a mahjongg tile. The skins were traded mainly through Singapore, and to the world market (which was indifferent to distinctions among animals from New Guinea, Indonesia, and Australia) they became known simply, misleadingly, as "Singapore small-scale crocodiles." Between 1945 and 1958, by one estimate, 87,000 saltie skins were exported from the Northern Territory.

That was enough to make crocodiles scarce. They disappeared almost completely from the more accessible tidal rivers, except for a few large adults that, wary of the vibrational thrum stirred into their waters by outboard motors, eluded shooting. They survived better in the dense, spring-fed swamps that merge with river drainages only during the wet season, and in the more remote rivers and swamps of Arnhem Land, a vast Aboriginal reserve encompassing the northeastern lobe of the Territory. But even in the swamps, even in Arnhem Land, there was crocodile hunting. Some Aboriginal people worked as guides, trackers, and skinners for white hunters, or hunted commercially themselves; Christian missionaries encouraged Aboriginal hunters and helped ship the skins. When a motorboat and a rifle were ineffective against a cagey old croc, hunters lay for it with traps or baited hooks. By 1959, the saltwater crocodile population was depleted across northern Australia. Then the trade turned to freshies.

Freshwater crocodile skin was worth only a third as much as saltwater skin per inch of belly width, but the market and the hunters who supplied it had grown hungry enough to make the lesser product commercially attractive. Besides, freshies could be taken in large numbers, especially at the billabongs where they congregate during the dry season, as the greater wetlands of the north constrict. Although the freshwater crocodile population suffered a few years of heavy harvest and disrupted reproduction, in this case unbridled crocodile killing bumped against a threshold of public resistance. Cattle-station owners grew annoyed at croc hunters who left skinned carcasses littering the landscape and befouling the waters. Others objected on more abstract grounds. *Wait a minute*, people said, *these animals are not only harmless—they're bloody unique to Australia!* The freshwater croc was the Aussie croc, no one else's. Both considerations made *C. johnstoni* just that much more popular than *C. porosus*.

In 1962, the state of Western Australia set a conservation precedent by making the freshwater crocodile a legally protected species. The Northern Territory followed that example in 1963. Queensland, the third of the northern states and the most curmudgeonly, held out. As a result, poachers killing freshwater crocs in remote parts of the other two states were still able, for a while, to sell their take in Queensland.

Hunting for saltwater crocodiles continued. The big individuals were mostly gone, but a few survived to breed, build a nest and defend it in some out-of-the-way swamp, and produce hatchlings. Meanwhile the focus shifted to smaller animals; even a two- or three-year-old saltie yielded a valuable little skin. Hatchlings and yearlings were collected, to be stuffed for the souvenir trade. There were no regulations, no limits, because the species was considered a pest. Some of the hunters were just dabblers, fishermen or other bush people who would take the occasional croc, even a marginal specimen, to make a few bucks. Serious market hunters were more selective. But together, unbridled, variously zealous or casual, they were putting themselves out of business. The keenest of them realized that and welcomed some form of regulatory intervention. In 1969 Western Australia protected the saltwater species, and the Northern Territory again followed, in 1971. Queensland was a different world, inhospitable both to crocodiles and to regulations. But in 1972 the new national government put through a total ban on the export of crocodile products, which effectively choked off the profit from crocodile hunting, even in Queensland.

Robert Bustard did a population survey of Western Australia's crocodiles at about this time. He also served as Australian representative on the newly formed Crocodile Specialist Group, a committee of experts that conferred under the aegis of the international conservation union, IUCN. Among biologists and conservationists around the world, concern over crocodilians was increasing. There were twenty-two living species, some of them widely distributed (such as *C. porosus*), some confined to small distributional ranges (such as the Philippine crocodile, *Crocodylus mindorensis*, and the Orinoco crocodile, *Crocodylus intermedius*) or meager remnants of habitat (the Chinese alligator, *Alligator sinensis*). By the early 1970s, many of those had been badly affected by decades of unrestricted harvest. Three million American alligators had been killed in Louisiana between 1880 and 1933. Seven million caiman skins had been exported from Brazil's state of Amazonas between 1950 and 1965. In the early 1960s there were twenty-five reptile tanners operating in the United States; one company alone tanned 1.5 million crocodilian skins per year. Export had always been a central fact of the trade, since most of the tanneries and most of the customers were in countries other than where crocodilians lived. Somewhere between five and ten million skins were shipped internationally each year. That changed abruptly in 1975, with the adoption of the agreement known as CITES (the Convention on International Trade in Endangered Species of Wild Fauna and Flora). Most of the world's crocodilian populations were listed on CITES Appendix I, prohibiting all international trade. Australia's population of *Crocodylus porosus* was initially listed on Appendix II, allowing some regulated trade, and then shifted to Appendix I in 1979. The shift to more absolute protection, according to one expert source, reflected concern over *C. porosus* in India and southeast Asia rather than a direct assessment of the Australian situation. By that time, ironically, Australia's saltie population had already come a distance toward recovery, thanks to the state prohibitions against hunting and the national ban on export.

The recovery was documented in a long-term, river-by-river survey initiated by an eccentric professor from the University of Sydney named Harry Messel. A confident man with a jut-jawed grin and a Hutterite beard, Messel was a Canadian physicist who found his way into research on Australian crocodiles by way of—as he tells it—a failed experiment in radio-tracking polar bears on the Arctic ice. When the radio transmitters in his bear collars didn't function, he saw a need to develop better technology for similar field studies around the world. In 1971, based now in Sydney and fascinated by crocodiles, he began (in collaboration with the Northern Territory government) a study of *C. porosus*, intending to fit crocodiles with new prototype transmitters. He established a field base at an Aboriginal settlement called Maningrida, on the north coast of Arnhem Land, and somehow found funding for a custom-built 125-ton research vessel (modestly christened the *Harry Messel*), a Cessna 206, several four-wheel-drive vehicles, and some smaller boats for crocodile catching. The project, as it developed, focused more on surveying the sub-populations in various northern rivers than on radio-tracking individuals for behavioral data, which at that moment in crocodile history was probably just as well. The numbers were down, and the question "How *far* down?" urgently needed answering. Messel took on junior collaborators, who would eventually help him turn out a long series of monographs and river-by-river reports. The first full-time biologist he hired was a young man, fresh from finishing his Ph.D. at another university, named Grahame Webb.

It was Webb who assumed chief responsibility for an intensive capture-mark-recapture program on the Liverpool and Blyth river systems, intended to give a relatively precise view of the age structure and size of those remaining sub-populations. During that study, Webb and his colleagues caught and measured 1,354 crocodiles.

Twenty-five years later, Harry Messel has retired from the scene but Grahame Webb remains, the most authoritative and influential voice on crocodile biology and management in the Northern Territory, and therefore also in Australia.

29

"When people said, 'Let's protect the crocodile,'" Webb tells me as we sit at his headquarters on the outskirts of Darwin, "it was like saying, 'Let's protect the *unicorn*. Let's protect the *dinosaurs*.'" He sniffs. "There wasn't any *around*." Legislating protection for crocodiles wasn't so controversial as it might have been, because most people seldom if ever saw one.

From that nadir, in the early 1970s, both saltwater and freshwater crocodile populations have rebounded robustly throughout northern Australia, thanks to four—or maybe five—crucial factors. First, the small number of wary, mature holdouts in the rivers and wetlands continued breeding. Second, crocodiles customarily produce large annual clutches of eggs (more than fifty per nest, for the saltwater species) and therefore can multiply quickly when conditions are good. Third, conditions *were* good during the 1970s and into the 1980s, because sizable areas of habitat remained unspoiled but relatively empty of crocodiles. This was quite different from the situation in India, where human population pressure had obliterated habitat while hunters were obliterating crocodiles. The Northern Territory harbored only 150,000 humans in an area ten times the size of India's state of Orissa, which harbored thirty million. For crocodiles along the Alligator rivers, or in the Liverpool drainage near Maningrida, or on the Adelaide, or the Mary, or the Blyth River and its big tributary the Cadell, food was plentiful. Nesting sites were available. Competition was low. Bingo: Reproduction surged, and a larger than usual fraction of hatchlings survived. The fourth factor, of course, was that well-enforced laws against hunting allowed hatchlings to become juveniles and juveniles to become adults.

The fifth factor is the most complicated—and the most arguable and distasteful, at least to some advocates of crocodile conservation. This one operates on a longer scale of time than the prohibitions against hunting, which were crucial in the short term but might not be adequate for the middle and distant future. In 1985, based on a proposal prepared for Australia by Grahame Webb and some co-workers, the CITES body reversed its earlier move and transferred the Australian saltwater crocodile from Appendix I to Appendix II, thus allowing the resumption of international trade. Australia was back in the business of selling crocodile skins. The downlisting from Appendix I assumed a commitment by the Northern Territory government to continue monitoring the wild population and to guard against any future decline. Webb himself, who by that time had parted ways with Harry Messel, became the principal scientist involved in the monitoring.

Grahame Webb is a sturdy man, now barely past fifty, with a squarish face, dark hair going gray, pale hazel eyes, and a set of adamant opinions about how saltwater crocodiles—or any species of large predator—should and should not be managed. "The worst thing that can happen to an animal is that it goes on Appendix I of CITES," he says. If you prohibit all trade, you dry up the flow of money that otherwise feeds back toward conservation, research, and conflict-mitigation programs, not to mention habitat protection that benefits not just the species in question but also many others. "As soon as you put it on Appendix I of CITES, and there's no commercial value at all—well then, you can kiss your ass good-bye."

Webb has studied crocodiles for most of his adult life. He has written books about crocodiles, including one novel, *Numunwari*, celebrating the majesty and menace of a huge old bull croc, sacred to the Aboriginal people who knew it, profaned by the white hunters who coveted it. He has built a career around crocodile research and management. He has been bitten by a crocodile—a female, defending her nest, who tore a gash down his calf that took months to heal—and dismissed that as a normal act of territorial behavior. Yet his passion was always for the research process itself, he says, not for *Crocodylus porosus* or for crocodilians as a group. In this he differs from some field biologists, who over time develop a deep emotional affinity for the species they study. Webb wants crocodiles to survive but not to be romanticized. He wants clarity. "If you talk bullshit to people about a big predator, you *will* be caught out," he tells me. "Some people would say, 'Ooh, crocodiles, they're not that bad.' But they *are* that bad."

This is my first face-to-face encounter with Webb, a casual chat on the veranda of the sprawling research-and-tourism compound he has named Crocodylus Park. Behind us, parents with children and out-of-town visitors pay an admission fee and stroll toward the fenced crocodile ponds outside, while others linger in the gift shop inspecting take-away goods: books, videos, postcards, stuffed baby crocodiles, hand-painted crocodile eggs, crocodile-foot back scratchers, crocodile-skin belts, crocodile-skin handbags and attaché cases, crocodile-skin keychain fobs, sharkskin wallets, barramundi-skin wallets, cane-toad coin purses, and dried crocodile feet in three sizes, sold loose and unembellished for whatever knickknack purposes the customer's imagination can invent. Webb wears a flowered shirt. Leaning back in ·his seat, he discourses with genial dogmatism on the necessity of devising economic arrangements that give local people some material incentive for accepting the presence of predators.

"You can't do it by education," he says, echoing Bivash Pandav. "Animals *eat* people." On later occasions when I talk with him, he'll sound a little more judicious, more sensitive to uncertainties and exceptions, but today he's giving me his core convictions at full strength. He's a blunt fellow by disposition, in the best Aussie style, but also a busy businessman as well as a scientist; possibly he hopes that one dose of his obdurate opinionizing will crook my snoot and he'll be done with me, soon and for all. "In the end, people are not gonna conserve something that has no use or value—that, in fact, has a negative use or value. It's not gonna happen. It never has, in the history of the world."

Webb's approach is contrarian and audaciously unsentimental. He rejects the view that endangered-species status, permanent bans on hunting, total trade closures, and righteous proselytizing—the latter aimed at producing, what, some sort of woozy spiritual awakening to the innocuous beauty of all creatures?—can achieve security, in the long run, for populations of flesh-eating beasts. Instead he puts his faith in the notion of sustainable use. Wildlife as renewable resource. *Use it or lose it* is the basic premise, encompassing many complex particulars involving habitat preservation and human attitudes. He consults on that subject around the world, under the aegis of a company he founded, Wildlife Management International. He plays a role in the IUCN's Sustainable Use Initiative. He runs Crocodylus Park not just as a tourism-and-education facility but as a research institution serving public and private clients, including some in the crocodile-skin trade. He believes that exploitation of wildlife populations at sustainable levels, yielding significant payoffs to local communities, combined with profit-based habitat protection that engages private landowners, is the necessary strategic approach to preserving big, dangerous, inconvenient animals. The case of *Crocodylus porosus* in Australia stands as a successful example.

The commercial trade here revolves around two types of institutions, loosely known as crocodile "farms" and crocodile "ranches." The earliest crocodile farms (that is, captive-breeding operations) in the Northern Territory, established around 1979, were founded with so-called "problem" crocs, trapped from the wild for public safety and put to breeding as an alternative to killing them. Ranching of crocs (that is, collecting eggs from the wild, hatching them in captivity, and rearing the hatchlings for skin and meat) began first with freshwater crocodiles and then, on a provisional basis, with salties in 1983. Given the high mortality of wild hatchlings and the high rate at which wild nests are lost to egg predators or flooding, this marginal off-take of eggs was expected to have no appreciable effect on the wild population; hatchling survival-rate would increase even as some nests were collected and total hatchling numbers went down. Later experimental work confirmed that expectation. The first skins were exported in 1987, two years after the CITES clearance. By 1989 there were three crocodile farms (operations generating some eggs in-house, in addition to collecting them from the wild) in the Northern Territory and one in Queensland, with revenues of almost a million dollars, and the trade was just beginning to grow. Landowners received permission to charge a fee for croc eggs collected from wild nests on their property, and those eggs fed the ranching cycle. The total wild-egg harvest increased from an annual level of 2,320 in 1984 to 29,000 in 1996. Other eggs were left to hatch from undisturbed nests, and those hatchlings renewed the wild populations. Within the Northern Territory alone, there are now about 70,000 saltwater crocodiles (not counting hatchlings of the year), according to estimates assembled by Webb and his colleagues. That's probably close to the population level before intensive hunting began. "It's been a win-win situation," Webb tells me.

Is the good news from Australia transferrable to other places and species? "I think it is," he says. "But you gotta be bold." If you've got

250 tigers in a population, he suggests, you might auction off the right to shoot two per year, and from the proceeds give everyone living nearby a handsome dividend. You might even sanction tiger farming, so that some of the big cats could be legally raised in captivity for the same vital products—bones, teeth, internal organs, skin, penises, whatever—that sell for pharmaceutical and charm uses on black markets in Asia. Of course, tiger farming wouldn't help wild populations except, maybe, indirectly—by meeting the demand for tiger parts and therefore undermining the illicit trade that incites poaching. There's still the problem of habitat. And there's the political problem: popular resistance, based on principle or emotion, to the idea of killing so much as one tiger, one lion, one bear for someone's profit. Even in Australia, many people opposed the notion of turning an endangered species into an exploitable one. Regarding the initiatives that came out of Darwin, Webb says, "We were bold enough, because we were a little territory and a long way from Sydney, to say, 'Stick it up your ass.'"

Boldness carried them one step further. In 1997 the Northern Territory introduced a limited, trial program allowing the commercial hunting of adult crocodiles in the wild. The underlying goal of that gambit was to increase benefits to landowners who preserved crocodile habitat. Seventeen crocs were taken, for their skins, from the Mary and Tomkinson rivers. The first two were harpooned by men of the Bawinanga Aboriginal Corporation, from the settlement of Maningrida on the north coast, the same place where Harry Messel had based his early fieldwork. The harpooning was assisted by a crocodile expert from Webb's operation, Wildlife Management International.

Several days after this conversation, I return to Crocodylus Park and pay the admission fee, in order to view Webb's compound through the eyes of a tourist. I gaze down at sequestered pairs of large male and female crocodiles from a walkway above their breeding ponds. I watch five dozen juvenile crocs in a fenced lagoon respond to the sound of a young man's voice as he alerts them to the 10:00 A.M.

feeding. The young man is a guide named Ben, wearing shorts and a neat Crocodylus Park shirt. Standing on a sort of dais, he throws meat into the lagoon, large bony chunks of what looks like short ribs of old horse. I see crocodiles splash, scuffle, gulp. When swallowing its prey, Ben explains, a croc must lift its head clear of the surface or else risk drowning, since a valve at the back of the throat admits both food and water. I hear him describe the high mortality of hatchlings in the wild, the land speed of a sprinting crocodile, the optimal age for harvesting skin. "We are not a farm," Ben says. "We are a research institution for crocodile farms." They are also a zoo. I take the loop walk through landscaped grounds and see the ostrich, the cassowaries, the cocka-toos, the green iguanas, the emu, and several species of monkeys. Back among the crocodiles, I read a note: "The male in this pen has been known to eat up to 17 whole chickens in one week." I stroll into the gift shop, browsing past the attaché cases and the back scratchers and the key fobs. I pause, tempted by a bad crocodile necktie, but decide it's not quite bad enough to be fun. Instead I buy a copy of Webb's novel.

Numunwari is an adventure yarn informed by experience and her-petology. It portrays the mingled destinies of a sympathetic white biol-ogist, an Aboriginal guide, and a gigantic, dignified, majestically amoral crocodile that eats innocent children and evil redneck jerks. The Aboriginal guide is a wise, placative man with a strong sense of his people's sacral history and a lifetime's knowledge of the Liverpool River. The eponymous crocodile is a twenty-five-foot behemoth lurk-ing in the waters upstream from Maningrida.

30

Maningrida sits at the mouth of the Liverpool River, on the coast of the Arafura Sea, two hundred miles east of Darwin. It's accessible by commercial flight, but since it's an Aboriginal settlement within the Arnhem Land Reserve, you need an entry permit from the Northern Land Council, which looks after native land rights. You also need contacts on the ground. This is no tourist destination, and there's nothing resembling a hotel. A lone stranger would walk the dirt lanes of Maningrida feeling as out of place as a crasher at a Mormon family picnic.

I've made my arrangements through the Djelk Rangers, a group of young men involved in community-development and traditional-knowledge projects under the Bawinanga Aboriginal Corporation. Bawinanga is not a clan or a language group itself but a portmanteau organization representing three distinct language groups—the *Ba*rrada, the Kun*win*ku, the Rembar*anga*, together making *Bawinanga*. The people of those groups live scattered among thirty-five tiny communities (known as "outstations") in the region around Maningrida. BAC supports various programs, including several that offer training, gainful enterprise, and a strong ethic of community responsibility to the Rangers—whose own name, Djelk, is translatable as "caring for country."

One of the Rangers' projects entails collecting crocodile eggs for incubation and later shipping the hatchlings down to Darwin for sale. They operate their own incubator, a climate-sealed little blockhouse within a metal work shed. During the wet season just past they collected nine hundred eggs and managed to sell 512 top-quality hatchlings at thirty dollars each. Five bucks from each hatchling goes back to the traditional landowner whose stretch of riverbank held the nest, and the rest covers expenses, other fees, and wages. The point is not to generate big profits but to create jobs and to nurture skills grounded in cultural traditions and sustainable uses of the landscape. This is year three for the Rangers in the hatchling trade. Next season their permit will allow them to collect two thousand eggs.

More recently they've also begun harvesting adult crocodiles from the wild. The word "harvesting," an agronomic euphemism that intrudes constantly into discussions of wildlife exploitation, means, in this instance, harpooning. Jab the croc, reel it in like a fish. The practice is ancient, though it lapsed for a generation, and the tools are a mixture of ancient and new. Harpoon points are now brass, threaded with 1/8-inch nylon parachute cord strong enough to stop a ten-foot crocodile like a smallmouth bass. The harpoon shaft is an old-fashioned pole of peeled eucalyptus. A pistol is necessary at certain moments or, less conveniently, a rifle. Also you want a high-powered beam, a stable boat, assorted ropes, and tape. Any chance, I've asked, of my tagging along on such a hunt? Well, that's complicated. We're not harvesting right now. There are seasonal constraints, plus certain issues of spiritual observance that must be addressed delicately within the community. Harpooning can be ticklish, even up here. But yes, well, it might be possible—sometime.

Meanwhile I want to meet not just young men like the Djelk Rangers but also people of deeper memory. Who are the sages, the elders, the key sources? Who bridges the chasm between old ways, mostly lost, mostly irrelevant during the years when there *were* no crocodiles, and current attempts to revivify those old ways within the context of laws made in Darwin and Sydney? Ask for "the Professor," Grahame Webb has advised me. Ask for old Jackie Adjarral.

I ask. I stop at his house, a cinder-block rectangle with a corrugated roof. I'm told that Jackie Adjarral has "gone bush." He'll be back tomorrow, maybe. Or maybe not. Settlements such as Maningrida, which was founded in 1957, are a late and tenuous phenomenon within Aboriginal life, exerting their gravitational pull through mate-

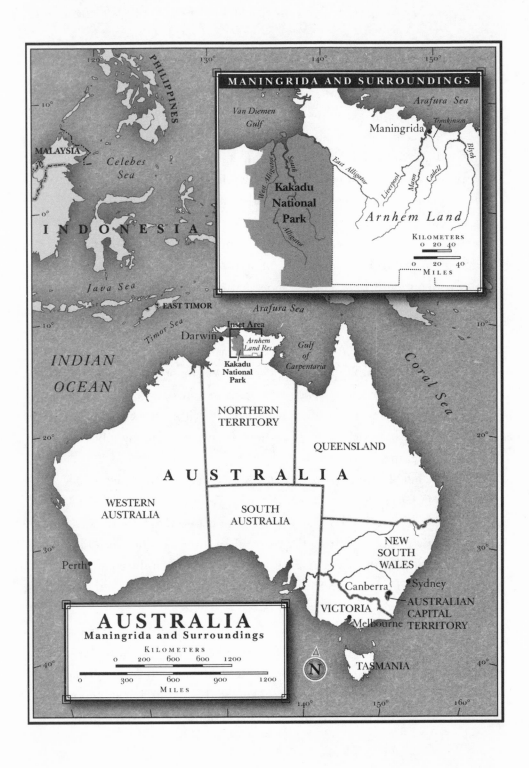

MANINGRIDA AND SURROUNDINGS

Arafura Sea

Van Diemen
Gulf

Maningrida

Tomkinson

Kakadu
National
Park

West Alligator

South Alligator

East Alligator

Liverpool

Mann

Cadell

Blyth

Arnhem Land

Alligator

KILOMETERS
0   20   40

0   20   40
MILES

PHILIPPINES

MALAYSIA

Celebes
Sea

I N D O N E S I A

Java Sea

EAST TIMOR

Timor Sea

Arafura Sea

Darwin

Inset Area

Arnhem
Land Res.

Kakadu
National
Park

Gulf
of
Carpentaria

INDIAN
OCEAN

Coral Sea

NORTHERN
TERRITORY

QUEENSLAND

A U S T R A L I A

WESTERN
AUSTRALIA

SOUTH
AUSTRALIA

NEW
SOUTH
WALES

Perth

Canberra

Sydney

AUSTRALIAN
CAPITAL
TERRITORY

VICTORIA

Melbourne

TASMANIA

N

AUSTRALIA
Maningrida and Surroundings

KILOMETERS
0   200   600   600   1200

0   300   600   900   1200
MILES

rial enticements but always opposed by the pull of the landscape (or "the country," as Aboriginal people call it when speaking English) and the millennia of experience binding them to it. Living amid concrete and steel in a government-subsidized village seems to be, for many of them, a form of camping; melting away upriver or into the forest is a form of returning home. For all I know, Jackie Adjarral may have gone bush for the season.

I spend two days among the Rangers, seeing their incubator, hearing of their various projects, learning a scant bit about their sense of connection to the country, and touring crocodile habitat on the Liverpool and its tributaries in their little aluminum dinghy. Half a dozen of them hobnob with me, but my particular hosts are two friends, Alister (Djalbalba is his Aboriginal name) James and Stuart (Yirawara) Ankin, each of whom wears a brown uniform shirt with an embroidered shoulder patch reading DJELK RANGER / CARING FOR COUNTRY. Alister is the quieter one, shy and hunkering, who embraces his boat-driver role with a grave sense of professionalism. Stuart, a boyish man with a soft ovoid face, a stubble beard, and a maimed right hand, is the jocular one, whose constant readiness to find humor or create some, to giggle, to tease in the gentlest ways, makes a nice complement to Alister's dourness. On one day's cruise upriver, we're joined by a younger Ranger named Cain Redford, bright and open-faced, more voluble. Our destination is the upper Tomkinson, a major tributary of the Liverpool. Branching left at the confluence, Alister steers us along its sinuous path through an area of floodplain meadows, tidal gullies, paperbark trees overhanging the brown water, mangrove thickets, soft muddy banks left exposed at low tide. "We took four crocodiles from here last year," Cain volunteers. Once the animal has been pulled to the boat, getting a bullet into its brain is crucial—otherwise there's a mad struggle. But the brain is a small bulb, encased beneath a thick, bony plate that's as sturdy as the roll bar on a Jeep. Miss it and the animal may perk up once you've got him underfoot, at which point your dinghy becomes a very crowded place. "Pretty dangerous," Cain says cheerily, "harpooning crocodiles at night."

We see crocodile after crocodile as we putter upstream, a seven-footer here, a five-footer there, another seven-footer, a few small ones, an eleven-footer, and more, each basking solitarily on its bank, then sliding away into the water or, in some cases, ignoring us as the boat passes. One of the seven-footers, dark-skinned and glum, declines to be rousted by our motor's low burble. Not afraid of us, eh? "Maybe he's tired," Stuart says. "Tired and *full*." A gourmand himself, Stuart often thinks his way around to the subject of eating. "Maybe he's asleep," Cain posits, "or in wonder world." Besides the crocodiles soaking up sunlight on their warm, comfortable mud, we see a few hefty specimens of what's locally called the mangrove goanna (a species of monitor lizard, *Varanus indicus*) recumbent on overhanging limbs. "We eat those," Stuart tells me, motioning at one goanna, and he greets another with the single word "Yum." He'll come back next week, he resolves, and shoot a mess of them for food. How do you cook a goanna? "Fire," Stuart says. Oh, you roast it?—what, on a stick? "Throw it in the fire," he says simply. Soon after, our riverbank lunch is done the same way: raw sausages tossed into campfire embers, scraped out, dusted off, and wrapped in white bread from the Maningrida store. An unfancy technique, but no ballpark bratwurst ever tasted better.

"So what do you think, David?" asks Cain. "About crocs here." I think there's quite a passel of them, I say, and that's gratifying. Are they this abundant throughout the Tomkinson and the Liverpool?

"Everywhere," says Cain.

"Everywhere," Stuart agrees.

On my third day in Maningrida I find "the Professor," Jackie Adjarral, a tiny black man of indeterminate middle age, in shorts and mustard yellow socks, changing a tire on a Mitsubishi truck at the community motor pool. With his fast, mischievous eyes and his rooster confidence, he reminds me of Chuck Berry. Speaking a quick, clipped pidgin that I can barely understand, Jackie agrees to accompany us on another jaunt upriver. His one condition, brusquely stated, is that there be "top-up"—that is, some payment for him afterward. Agreed. Wrenching down the last of the Mitsubishi's lugs, Jackie walks away into the shop as though bored or offended. He returns some minutes later, without explanation, and we depart. Small talk is not his style.

From the store we provision ourselves again with lunch groceries and, at Jackie's stipulation, a pack of Winfield Red cigarettes. Then we're off, crossing the estuary on a light chop and ascending, today, the mainstem Liverpool. It's a silty brown river bending its lazy way between mud banks and mangroves, like the Tomkinson but several times larger and with a greater reach. The Liverpool's headwaters lie many miles to the south, among the red-rock canyons of the Arnhem Land plateau—what people down here speak of as "the stone country." But the stone-country creeks and pools are mainly the province of freshwater crocodiles, and our interest is the lower Liverpool, within tidal range, where salties rule. We pass one other boat and then, winding on, share the river only with fish and reptiles. This is not a place, and these are not a people, for hammering lumber together unnecessarily, tossing up fishing cabins and getaway shacks every half mile. The high banks, thick with grasses and brush, shaded occasionally by a droopy paperbark, are left to the female crocodiles who build nests there, piling compost atop their clutches and then lingering to guard them. If it were otherwise, the croc population couldn't renew itself. Wild crocodiles need more from their environment than water to swim in and prey to eat; they need riparian frontage above the flood line for nest sites, because crocodile eggs either breathe or drown.

Jackie sits cross-legged on the dinghy's front deck. He's still as a hood ornament except when he acts as skipper, directing Alister's course with a wave of his hand. From one bank, a ten-foot crocodile slips into the river. Jackie waves the boat onward: Never mind that one, no big deal. Another croc, plunging past us toward the depths, accidentally bumps the boat, causing Jackie, calm but cautious, to pull in his dangling foot. When I pester him with questions, he responds tersely, though sometimes one recollection leads to another. Who taught you about crocodiles, Jackie? To my surprise, he answers, "Webb." Almost thirty years ago, he served as a field assistant to Grahame Webb's population work. As an afterthought, he adds that his own father showed him a few things too.

Jackie's father hunted crocodiles by harpoon, from a canoe. He would dry the skins in the sun and take them to Oenpelli, a settlement just inside the Arnhem Land border on the road to Darwin, to trade for supplies. Not cash, no, solid goods. His father would get tobacco, sugar, cornmeal, flour, clothes, and mosquito netting. His father killed many crocodiles. What's your totem animal, Jackie? "Crocodile," he says. Then adds, "We call it Mururrba," a name particular to his language group, the Gurrgoni. (Other language groups personify—and to some degree, deify—the saltwater crocodile as Bäru, Dinbal, and Ginga.) If Mururrba is Jackie's totem animal, does that mean he's forbidden to kill crocodiles? No. And his father's totem animal? Crocodile also. Yes, his dead father's spirit might now dwell within one of these very crocs on the river. Not a problem. Killing crocodiles is still acceptable so long as the protocols and rituals are observed. What are the rituals? Jackie isn't saying. No, he has never been attacked, never been bit by a crocodile. He *won't* be bit; he has protection, he says. Unless he makes "a mistake." A mistake such as . . . ?—but now Jackie goes silent. Protocols and rituals, mistakes to be avoided—that's private stuff, not fare for discussion with every prying white visitor who steps off the plane.

We stop to inspect a slick smear down one muddy bank, left by a crocodile as it tobogganed into the water. The mud, freshened daily by tide, is soft and grabby to a fully weighted foot. I clamber out, sink, and within two or three shoe-sucking steps begin to see logic in the low-slung crocodilian posture and the belly-scoot move—yes, now *that's* the way to locomote on such muck. Flanking the slide mark, four feet apart, are light footprints, palmate and clawed like a raccoon's.

"How big, you reckon?" asks Jackie. I gaze at the prints, imagining a crocodile large enough to fit their stride. About ten feet, I guess, trying to err on the short side so as not to seem breathless. Jackie reckons longer. Thirteen feet, maybe more. Alister, steadying the boat, says, "Let's get the hell out of here."

We ascend a small channel. As the afternoon passes, we amuse ourselves with more crocodile counting and then, when Jackie sees promising water, a bit of fishing. He tosses a big hooky lure by hand line, swirling it, looping it out like a bolo, dropping it niftily into slack water behind a drowned snag. Alister fishes too, but it's Jackie who boats two barramundi, one of them a meaty five-pounder. Engrossed in his fishing, content to be on the river and away from the motor pool, Jackie breaks the stillness occasionally with another memory. In the old times, he says, people caught crocodiles by hand. Just grabbed them right *here*—he gestures—around the hips. He knew of two brothers, relatives on his father's side, who would swim up alongside a submerged croc and snout-noose the animal where it lay. "Dem pella, pinish now." Those fellas, they're finished—that is, dead. He recalls another fella, a man named Peter, who made the error of grabbing a large crocodile by its tail. It spun back on him, counterattacking. Peter's brother had to paddle up here, retrieve his body, kill the croc for revenge. "Come and roast 'im." Peter's next of kin ate the crocodile but discarded the skin, possibly (though Jackie doesn't say) as required by spiritual observance. Grabbing, noosing—nobody catches crocodiles that way anymore, Jackie says. It's dangerous.

This stretch of the Liverpool also stirs some nostalgic memories of his work with Grahame Webb. They would build a trap, Jackie recalls, just a gated little stockade of stakes driven deep into the mud. Bait it with a freshly shot wallaby. Crocodile enters, chomps the wallaby, flips a latch, and the gate falls closed. The falling gate triggers a transmitter that tells Webb, back in Maningrida, there's been a catch. They return upriver by boat, snout-noose the crocodile using a long pole, tie its legs, blindfold its eyes, then get measurements and do their other biologizing. Release it, jump the hell out of the way. These were big crocs, Jackie says, sometimes twelve, fourteen, seventeen feet long. Jackie himself is no bigger than a jockey. His beard is gray, his hair is a globe of grayish-black curls. He sits on the bow, smoking his Winfield Reds, remembering.

Low tide, he says, is the best time to hunt crocodiles. Low tide during dry season. For collecting eggs from wild nests, on the other hand, wet season is the time. "You come back," Jackie tells me, indifferently, not imagining that I will.

31

But first other travels and other crocodiles intervene. For instance, I've been welcomed to bring my curiosity out to the eastern extremity of Arnhem Land, along the Gulf of Carpentaria, where a people known as the Yolngu have struggled stubbornly for a century, against several forms of intrusion, to maintain their connectedness to the sacred sites, creatures, and meanings of their landscape, including crocodiles.

Isolated as they are, and infused with exceptional strength of character, the Yolngu have managed to preserve an ancient, complex culture that spiritualizes every aspect of their lives, from birth to burial, from marriage to fishing, and from the freshwater creeks and billabongs of the hinterland to the bays, offshore islands, and coastal waters of the gulf. These are the people who venerate saltwater crocodiles as living embodiments of a hero or deity—Bäru, also called the

Ancestral Crocodile—from the creation era. This era, known to them as Wangarr, was a primordial time during which not just Bäru but other animal-human personages such as Gawangalkmirri (the Stingray), Mäna (the Ancestral Shark), Marrpan (the Sacred Turtle), Warrukay (the Ancestral Barracuda), and Mundukul (the Lightning Snake) acted out their rivalries, truces, and other epic adventures, animating the landscape and seascape of eastern Arnhem Land with sacred narrative and eventually delivering this world into the stewardship of the Yolngu.

Decades of study by some devoted anthropologists and linguists, and the abstruse publications they have produced, testify to the fact that Yolngu mysteries are rich with intricacy but don't open themselves readily to outsiders. Yolngu art, especially in the form of bark paintings, is much admired throughout Australia and sold to collectors internationally. Yolngu music and dance are powerful, fierce, and fabulous. But even beyond those distinctions, the Yolngu have a special place in modern Australian history. The fervor of their attachment to the landscape, in all its geophysical, biological, and mythic dimensions, drove one of the pivotal episodes in the Aboriginal land-rights movement. The legal documents speak of *Milirrpum* v. *Nabalco,* but it's a sequence of events that is more commonly known as the Gove Land Rights case. Neither label is quite accurate. "Gove" is an Anglo name, applied to the Gove Peninsula of northeastern Arnhem Land in memory of one William Gove, an Australian airman killed nearby during World War II. The dispute in question might better be labeled *Yirrkala* v. *Colonial Hegemony*.

My plane ticket says I'm flying to someplace called Nhulunbuy. That's a small air-routing detail reflecting big grievances and unresolved tensions. Yirrkala, a mission settlement founded in 1934, is an administrative hub for the Yolngu people; Nhulunbuy is a mining town populated mainly by whites. Both communities lie on the Gove Peninsula, just east of Melville Bay, separated from each other by fifteen miles of two-lane blacktop and a chasm of cultural disaffinity. The airstrip is halfway between, but the airline books for "Nhulunbuy," presumably because more of its passengers from Darwin are headed there, on industrial rather than Aboriginal business.

To many Yolngu people of the coast area, even Yirrkala is just an artificial if useful nexus, a relatively bustling place that they visit for supplies, services, or social festivities, while they ground their lives more sedately on "homelands" (their term for outstations, tiny compounds embracing a few families each) in the bush. In Yirrkala is a school, a clinic, and the Buku-Larrngay arts center, serving as a gallery for many bush-based painters. Yirrkala isn't the sole capital of Yolngu social life, but it's one of several. Nhulunbuy, by contrast, exists in support of a large bauxite mining and exporting operation that was imposed on the Yolngu people thirty years ago.

Bauxite is the principal ore of aluminum, a material of no value whatsoever within the traditional Yolngu economy. Exploration during the 1950s revealed a huge bauxite deposit underlying the red soil of the Gove Peninsula. The Yolngu might as well have found themselves lamentably situated atop uranium, petroleum, or gold. Permission to extract it was granted, without Yolngu input, in 1963. To understand the context within which crocodiles collided with aluminum on Yolngu lands, though, you need to glance back a few centuries, to a time when international resource-extraction pressures involved nothing more than fish, trepang (a form of edible sea cucumber), and pearls.

Northeastern Arnhem Land lies only twelve hundred miles from Sulawesi, the big spider-shaped island east of Borneo, along a hopscotch route across what was once known as the Malay Archipelago (nowadays Indonesia) and the Arafura Sea. Sailing ships could swing from island to island without crossing more than a few hundred miles of open water. That geographic circumstance put Yolngu territory within range for commercial voyagers from the port of Macassar, on Sulawesi's southwestern tip. Long before Europeans colonized the

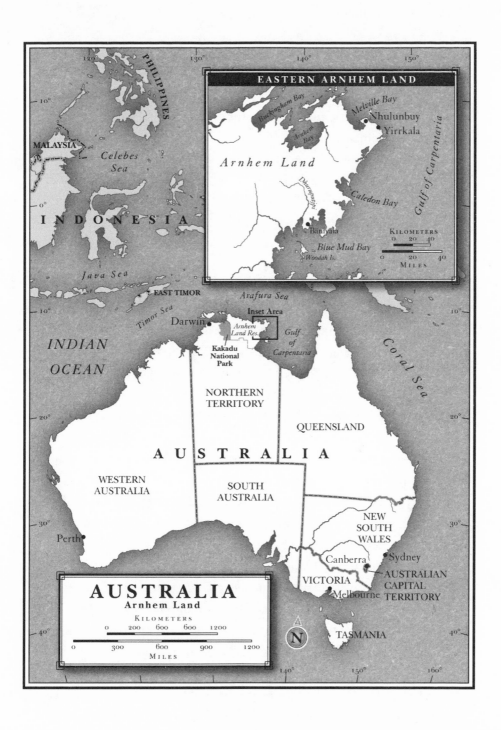

PHILIPPINES

MALAYSIA

Celebes
Sea

I N D O N E S I A

Java Sea

EAST TIMOR

Arafura Sea

Timor Sea

Darwin

INDIAN
OCEAN

Inset Area

Arnhem
Land Res.

Kakadu
National
Park

Gulf
of
Carpentaria

Coral Sea

NORTHERN
TERRITORY

QUEENSLAND

A U S T R A L I A

WESTERN
AUSTRALIA

SOUTH
AUSTRALIA

NEW
SOUTH
WALES

Perth

Canberra

Sydney

AUSTRALIAN
CAPITAL
TERRITORY

VICTORIA

Melbourne

N

TASMANIA

AUSTRALIA
Arnhem Land

KILOMETERS
0   200   600   600   1200

0   300   600   900   1200
MILES

Malay region, Macassan traders had begun making adventuresome journeys on the seasonal monsoon winds, sailing eastward as far as Aru, a little cluster of islands off the New Guinea coast, in search of bird-of-paradise skins and other natural treasures. They also came down to Arnhem Land and even rounded the Gove Peninsula, discovering rich sea beds there, along the western edge of the Gulf of Carpentaria, where they could dive for (or barter for, if others would do the diving) trepang, pearl shell, and pearls. The Macassans became welcome as annual visitors among the Yolngu. They were congenial trading partners who recognized Yolngu proprietorship of the coastal seascape as well as the land, respected Yolngu mores, and gave satisfactory payment in tobacco, rice, knives, cloth, or other luxury goods. Communication seems to have been adequate for practical purposes, though to what degree the Macassans learned one of the Yolngu languages, or the Yolngu picked up Malay, or both, is hard to say. Anyway, there was enough trust between these peoples that, when the monsoon ended and the Macassans rode the westward winds home, a daring Yolngu occasionally went with them. If he was lucky, the Yolngu traveler returned to his own home on a later voyage, telling tales of stepping-stone islands (Tanimbar, Leti, Timor), wide seas (the Arafura, the Flores), the port of Macassar itself, and a peculiar, light-skinned tribe of strangers (the Dutch) who by that time had founded their own trading presence in the Malay Archipelago. Who be *they?* the Yolngu abroad may have asked. They be *Hollanders*, was the answer. From the word "Hollander" derived (according to one etymological guess) a muted variant, *balanda*, that became the generalized Aboriginal term, used throughout Arnhem Land, for white folks.

Matthew Flinders was a balanda of the English-speaking variety, the first British sea captain to survey the north coast of Australia, almost forty years before J. L. Stokes arrived in the *Beagle*. When Flinders's ship rounded the cape near what is now Yirrkala, in 1803, he found a half-dozen Macassan praus already there. Within a century, the white governors of colonial Australia had grown jealous of the Macassan traders, some of whom dodged customs duties or supplied alcohol to their Aboriginal contacts. Toward the end, as they were being squeezed out, the Macassans themselves committed abuses all along the coast—breaking commercial trusts, offering more booze and less value, less respect. In 1904 the colonial Minister for Customs took steps to choke off access for Macassans, and by 1908 they were gone. What replaced them, quite disagreeably for the Yolngu, were white Australian officials, white traders, and a rough bunch of Japanese who had come for trepang. The balanda were bad enough, but the Japanese trepangers, at least sometimes, were worse.

Under this new regime, there was more strife—fights, accusations, killings, and further killings in revenge. From the perspective of officials in Darwin, the natives were getting murderously uppity. From the Aboriginal perspective, it was guerrilla resistance to an occupying tyranny. Within the first decade of the century, two police patrols went out on "punitive" expeditions, one of which approached from the southwest, along the Roper River, and raised some lethal havoc between there and Blue Mud Bay. The other patrol came out of the northwest down to Caledon Bay, just south of the Gove Peninsula, where they harried the populace and shot several people. The most notorious of all Arnhem Land murder cases began in the same area, Caledon Bay, on September 17, 1932, when a group of Yolngu men killed five Japanese trepangers.

The triggering incident occurred when a local patriarch, Wonggu Mununggurr, approached the Japanese men to complain about their behavior toward Yolngu women. The Japanese grabbed Wonggu and dunked his head in a bucket of trepang guts. You don't do that to a Yolngu elder. The Japanese also let off some gunfire, though evidently no one was hit. Three sons of Wonggu showed up the next morning and (at least according to an eventual verdict in court) speared the Japanese men as they ran away.

Nine months later, in June 1933, another police patrol came east from Darwin, this one to "investigate" those killings. It included a young constable named Albert S. McColl, who had recently mentioned in a letter to his brother, "I am going out with the party to Arnhem Land next month to look for the niggers who murdered the Japs some time ago." Having followed a trail of rumors and campfire smoke, the patrol eventually found itself in the thickets of Woodah Island, fifty miles south of Caledon Bay, stalking what seems to have been an almost randomly targeted group of Yolngu. The patrol had caught several Yolngu women, holding them in handcuffs and chains to bait their men out of the bush. The men approached, then fled, and the police gave chase. Shots were fired, spears were thrown. For reasons unclear, Constable McColl lingered behind. When the others returned a day later, they found McColl dead—speared through the heart. A man named Dhakiyarr Wirrpanda, one of whose wives had been among the hostages, was charged with the killing. The wife was a young woman named Djapirri.

During the various inquiries that followed, Dhakiyarr Wirrpanda disputed not the central event but the circumstances, claiming that he'd found McColl in the act of molesting Djapirri, and that McColl had fired a shot. At which point, said Dhakiyarr, yes, you bet, I speared him. The case was complicated by the fact that Dhakiyarr also admitted having helped kill two disreputable white men who had turned up at Woodah Island and taken their pleasure on several Yolngu women, again including Djapirri. Dhakiyarr may well have been a hot-tempered man, but he seems to have had grounds for feeling testy. He went to Darwin voluntarily, on an understanding that the whole conflict would be discussed as among sovereign parties settling grievances. Instead he was tried for murder, under a mangled variant of his name (*The King* v. *Tuckiar*), convicted, and sentenced to hang. His conviction was reversed on appeal to the High Court, and on November 9, 1934, he was released from jail. The next night he disappeared forever. According to a historian named Ted Egan, who reexamined all the old evidence and talked with surviving witnesses, it was "widely believed in Darwin that Tuckiar was shot by the police and dropped in Darwin harbour." But nobody knows.

I mention the case of Dhakiyarr Wirrpanda because it has larger meaning as an emblem of Yolngu resistance—proud, prickly, not always nonviolent—to violations of their people and their land. That resistance showed again in 1963, when the Australian government granted leases to an entity called the Gove Mining and Industrial Corporation for bauxite exploitation in the vicinity of Yirrkala.

32

The area at issue was within the Arnhem Land Reserve, created back in 1931 "for the use and benefit of the aboriginal native inhabitants of the Northern Territory." Aboriginal people in the Territory were not then recognized as citizens of Australia, and what might or might not be of "use and benefit" to them would be decided by white men in distant offices. The terms of assignment for reserve lands were modified slightly in 1953, by which time the value of strategic mineral resources (notably aluminum and uranium) had become harder to resist or ignore. Still, even the new terms specified that if "any part of a native reserve *has ceased to be necessary* for the use and benefit of the natives it may be severed from the reserve and, if mining takes place on the severed portion, royalties will be paid into a special fund to be applied to the welfare of the natives." The italics are mine, highlighting words that the Commonwealth of Australia soon chose to forget. On March 11, 1963, the leases were announced, under a cozy agreement among the government, the company, and the Methodist Overseas Mission

Board, which held a preexisting lease (for the mining of souls, if not of bauxite) at Yirrkala. The Yolngu themselves hadn't been consulted. Two days later, a parcel of 140 square miles was excised from the Arnhem Land Reserve.

The Yolngu response was a petition to Parliament, stipulating eight points of grievance and request, each expressed both in English and in one of the Yolngu languages, Gupapuyngu. For instance:

2. That the procedures of the excision of this land and the fate of the people on it were never explained to them beforehand, and were kept secret from them.

And:

4. That the land in question has been hunting and food gathering land for the Yirrkala tribes from time immemorial; we were all born here.

And:

6. *Dhuwala yulnundja mala yurru nhämana balandawunu nha mulkurru nhämä yurru moma ga darangan yalalanumirrinha nhaltjanna dhu napurru bitjarra nhakuna Larrakeahyu momara wlalanguwuy wänga.*

It was no ordinary paper complaint. The typescript was mounted on a panel of stringybark, and served as a centerpiece to a composition of iconic figures and motifs painted in Yirrkala style. It was signed by twelve Yolngu elders, including Milirrpum Marika, whose name would come to stand, in court records and historical accounts, for the collective set of all the aggrieved parties. This unusual document, which went to Canberra in August 1963, became known as the Bark Petition and remains in Parliament House today, publicly displayed, and much honored in the breach. Meanwhile, the mining began. The landscape near Yirrkala yielded to explosives and heavy equipment.

By 1968 the bauxite operation was in the hands of an international syndicate known as Nabalco, with a new forty-two-year lease. At that point the Yolngu filed suit in the Supreme Court of the Northern Territory, against both Nabalco and the national government, claiming that their land had been unlawfully invaded. The case eventually took its full, formal label as *Milirrpum and Others* v. *Nabalco Pty. Ltd. and the Commonwealth of Australia.* The plaintiffs of record were a handful of elders, representing both themselves and their respective clans—an important point, since clans are profoundly significant units within Yolngu culture, through which individuals attach themselves to particular areas of landscape and to the mythic histories that animate those areas. Milirrpum Marika sued for himself and the Rirratjingu clan. An elder named Mungurrawuy Yunupingu sued for himself and the Gumatj, a clan with its own special attachment to Bäru, the Ancestral Crocodile. When the case was heard in Darwin, a bright young man named Galarrwuy Yunupingu, Mungurrawuy's son, served as one of the interpreters, earning praise from the presiding judge for his explanations of Yolngu law. Galarrwuy, then fresh out of Brisbane Bible College, would later become one of the most formidable Aboriginal leaders in Australia.

In the meantime, bad news. The judge, Justice Blackburn, ruled against the plaintiffs. In Blackburn's view, it had come down to the crucial question of "native title." Did the Yolngu people hold a legally enforceable right to the land—a *title*, if not on paper then at least through common law and ancient possession? Or were they merely transients who happened to be there when white people arrived to initiate a preemptive regime of true ownership? Blackburn had listened carefully, though not very imaginatively, to the testimony. In his decision, he declared his impression that "the aboriginals have a more cogent feeling of obligation to the land than of ownership of it." He concluded: "It is dangerous to attempt to express a matter so subtle and difficult by a mere aphorism, but it seems easier, on the evidence, to say that the clan belongs to the land than that the land belongs to the clan." The Yolngu had lost.

The front-end loaders continued scooping, the ore crushers continued crushing, the landscape was turned inside out. In the years since, many millions of tons of bauxite have been extracted. A twelve-mile-long conveyor belt, one of the longest in the world, now carries ore from the open-pit mine to a stockpile near the export wharf on Melville Bay, one of the sacred Yolngu areas.

But in defeat, seeds of victory—if not total, then partial, and if not for themselves, then for others. When the Aboriginal Land Rights Act was eventually passed, in 1976, it was the Yolngu who deserved credit (and to some degree got it) for having awakened Australia to the realization that Aboriginal-occupied land was not empty land. That act, plus several later legal cases (*Mabo and Others* v. *the State of Queensland* and *The Wik Peoples* v. *the State of Queensland*), were major stages in the process, still continuing, by which modern Australia has tried to find reconciliation with its own forty-thousand-year human history. The Bark Petition had said to Parliament roughly what our own Declaration of Independence said less politely, less suppliantly, to King George: *We hold these Truths to be self-evident, that . . . Dhuwala wanga Arnhem Land yurru djaw'yunna naburrungala.*

Galarrwuy Yunupingu, having begun his public life as an interpreter for his father and the other elders, matured into a weighty political figure. He became chairman of the Northern Land Council, one of the oversight bodies established under the 1976 act. In that role he dealt constantly with issues of land rights, land use, land access, and native title, within both the balanda legal context and the Aboriginal one. As a member of the Gumatj clan, though, he had an additional perspective. "We believe that we came as a crocodile," he told one interviewer. "When our body's dead, gone, our spirit becomes crocodile." The landscape stood as voucher to this special relationship embraced by the Gumatj, in that "every one of the tribal lands that we own from our forefathers were created and given to us once by a crocodile." As for himself, Galarrwuy added, the big reptile is part of his identity. He belongs to crocodile, and crocodile belongs to him.

There's a defining intensity in this connection that can be heard from other Gumatj members too. There's a reverence of the most earthbound sort. But even a fully observant Gumatj might be willing to admit that Bäru isn't infallible. If the Ancestral Crocodile had been acting presciently when he created and bestowed those lands, he might not have burdened the Yolngu world with an underground lode of bauxite.

33

My time in and around Yirrkala resolves into a series of voices. One voice belongs to an elder named Djalalingba Yunupingu, who was himself among the signers of the Bark Petition. He tells me, "Bäru is very important part of the land. We act by Bäru."

Djalalingba is Gumatj, an uncle of Galarrwuy Yunupingu. Big-bellied and bare-chested, with a neat gray beard and gray hair encircling his round face with its sage, squinty eyes, he sits cross-legged on the sand beside his beachfront house, near the easternmost point of Arnhem Land, while several younger men open fire-roasted oysters with hammers. Having welcomed me to sit beside him, having heard what brings me here—that is, my curiosity about the Gumatj association with crocodiles—Djalalingba waves dismissively toward the small bay in front of us, where a single croc hangs afloat on the water's surface. "We don't worry about them," he says. "Only balanda people worry about them. We've lived with them forty thousand years. They're like pets to us. We just leave them in peace." Among the Yolngu, he reminds me, several clans have their special bonds to Bäru—not just the Gumatj but also another, the Madarrpa. Even within the Gumatj clan, there are variant kinds and degrees of attachment, specific to families. Bäru lies upon the coastal landscape like a river, gigantic and immanent, his head at the inland headwaters, his tail extending into the sea, and different families have their home countries in different sectors of that expanse. Djalalingba's own family, the Yunupingu, are associated with the head end, the freshwater end. Another family, the Munungurritj, occupy the tail end, down in saltwater country. Can Djalalingba tell me something about how those two family attachments differ? No, he can't. Or he won't. As with Jackie Adjarral, so with him: Some matters are too private, too sacred for bantering about with every poke-nose white man who stumbles up to the house. Balanda people don't understand the Yolngu connection to Bäru, he says. "They only think: only animal."

Noticing me scribbling notes, Djalalingba shrugs impatiently. No no no, foolish man, you can't learn anything from a short visit, you can't grab these truths and carry them off. Have you come a long way? Yes, I say, from America. He nods. Well, the only way you could grasp Yolngu beliefs, he says, would be by spending the years necessary to enter Yolngu culture through Yolngu language. Undeniably true, I think—and sadly impossible, within the scope of what I'm trying to do. (Before tackling Yolngu, I would need also to learn Gujarati, in order to see the lion through Maldhari eyes, and then there would be Romanian, for the brown bear, and then Udege, for the tiger, and then . . .) But I don't offer this lame excuse to Djalalingba. I merely set my notebook aside, to avoid causing him further aggravation.

The conversation has chilled. Before ending it entirely, he remarks that certain sites have profound significance within the life story of Bäru. There's a place down near Grindall Bay, for instance, known as

Bäniyala. It's the homeland for a group of Madarrpa people. Bäru is everywhere, of course, but at Bäniyala . . . Down there, Bäru is present in a special way.

Bäniyala? I've heard the name mentioned before, and I'm destined to hear it again. *Bäniyala*. Yes, all right, that much I can remember without notes.

I meet a man named Dula, of the Munyuku clan, who has recently done battle with a crocodile. The Munyuku, unlike the Gumatj and the Madarrpa, have no totemic bond with Bäru. Their pieties and obligations toward crocodiles are therefore different. Dula himself is just old enough to recall the times, before *Crocodylus porosus* was legally protected, when he and his brother used to hunt crocodiles for their skins. Working at night, they used hooks, ropes, and flashlights, and then sold the harvested skins to balanda traders for payment in tobacco, flour, tea, and sugar. The meat they kept for themselves. They would build an earth oven, cover the pearly flesh with bark and then sand, cook it up. But that was the bygone era, before balanda laws put crocodiles off-limits. As for this latest incident, Dula makes clear: It wasn't hunting but self-defense.

He and his brother had stopped to swim in a little creek, a tributary of the Dhuruputjpi River, down near Grindall Bay. The crocodile appeared suddenly—a very large beast for such small water—and slashed at Dula with its tail, trying to drown him. Dula's brother came quickly with a spear and stabbed it, but the croc was too strong. It broke the spear (Job 41:29, "he laugheth at the shaking of a spear") and climbed out onto land, making a guilty retreat. Dula fetched a shotgun and blasted it, hit it right here—he makes a gesture—right behind the ear. Shot it twice. They butchered the animal where it lay, and then built their earth oven. The meat was fatty, Dula recalls, but this was good fat, natural fat, not like what you find in a feedlot cow. "Yolngu people, as in past, we don't like wasting anything. If we kill something, we eat it."

Crocodile meat in particular, he reminds me, has been an important resource to the Yolngu. From the old times, forever, they killed and ate crocs. But then came that period, after World War II, when rifle-equipped hunters slaughtered crocodiles, so many crocodiles, solely for the cash value of their skins. No one knows how much of that meat went to waste. There was the population decline, the protective laws, and the years of scarcity, during which even a Yolngu in Yolngu country could be punished for killing a crocodile. Finally came the rebound, with a looser regime of regulations. Are there more crocodiles in the rivers now, I ask, than when Dula was young?

Oh yes, he answers. "Big more."

And is that good or bad?

"Really bad," he says. "Because the population is growing, and it's getting more dangerous. Same like tiger in India. The man-eaters. If the population is growing, they eat more people." The correlation, at least where he stands within the landscape, is axiomatic. Where he stands is among the unprivileged muskrats.

Dula has already told me (confiding a delicate story, which for spiritual reasons may not be repeated) about losing one of his very close relations to a lurking crocodile. He knows the costs of unpeaceful coexistence. He exemplifies Alistair Graham's conviction, reached during that year among the Turkana at Lake Rudolf, that "there is in man a cultural instinct to separate himself from and destroy wild beasts such as crocodiles." But I'm left wondering how it jibes with Djalalingba's attitude, equally grounded in ancient experience: We don't worry about them, they're like pets. Or with Galarrwuy's: They *are* us.

At a Yolngu arts festival I speak with Mandawuy Yunupingu, Galarrwuy's younger brother, who happens to be a national celebrity on his own terms—as lead singer and motive force of an Aboriginal rock-and-roll band, Yothu Yindi. Mandawuy, formerly a teacher, is a gentle-spirited fellow with a driving, galvanic baritone voice. In the group's most successful songs, such as "Treaty," "Mainstream," and the aptly titled "Tribal Voice," he has channeled the anger and dignity of the Yolngu into popular art. For the Gumatj clan particularly, Mandawuy tells me, that sort of calm ferocity has its source and its model in Bäru. "Crocodile is the basis of where we started from," he says.

Mandawuy spends a few minutes trying to educate me on the relationships among clans, animal totems, stories, and areas of landscape—arcane stuff that can barely be spoken about, let alone reduced to a crude summary. There are the shark people, the crocodile people, the stingray people, each clan having its special connection to this or that animal. During a primordial era these creatures—Shark, Stingray, Bäru himself, and other iconic fauna—existed in individual human form as the founding deities of Yolngu landscape and law. One ancient story recounts how Bäru and Stingray disputed the reach of their respective domains. Stingray speared Bäru in the leg, Bäru accepted the wound without retaliation, and there it was finished, settled, in an understanding that Bäru's country ended at the salty estuarine zone, where Stingray's began. Another story describes a domestic tiff between Bäru and his wife, Blue-Tongue Lizard. She had collected some snails, and while Bäru was sleeping she roasted them in a fire and set them on Bäru's skin, causing it to blister. Bäru woke, hollering, "What have you done to me?" He shoved Blue-Tongue into the fire, giving her the scaly-skinned disfigurement that she has carried, in her lizard form, ever since. This little tale of marital strife isn't so solemn as the primal creation narratives, Mandawuy says. It's just a legend of the sort Yolngu mothers tell their children, as his mother told it to him.

Speaking of children, I say, what about the danger to them from crocodiles? Is that a problem for Yolngu mothers, concerned as they must be about kids playing in the water? No, Mandawuy says, not if the children are taught the animal's nature and its place. "As long as you respect it, it will respect you. And that means, when you're in the bush or near the water, you always keep a close watch. If you see a

Bäru you leave it, let it be, let it have your space—and you go the other way." Over thousands of years, he adds, the Yolngu have learned to live with the crocodile rather than eradicating it.

When I mention Bäniyala, the little homeland near Grindall Bay of which Djalalingba spoke so reverentially, he volunteers, "That's where Bäru started from." And there's a nesting ground? Somewhere nearby? Where a river disappears into brackish wetlands just north of the bay? "That's Garrangali," says Mandawuy. "That's number one." It's as though, in my ignorance, I've said to a Muslim, *Have you ever heard of a place called Mecca?*

34

Even before the testimonial from Mandawuy, I've begun to feel Bäniyala beckoning. Could I go there myself? Yes, but not easily. Visitation requires a permit from the Northern Land Council, issued only after consultation with the local landowners. That much, thanks to generous help from my various contacts, is arranged. Logistics is another matter. The best way to reach Bäniyala, I'm told, is by small plane. A person can also make the trip overland—at least in the dry season, sometimes, maybe, if the person has a four-by-four vehicle that's both sturdy and expendable. I don't. So I charter a Cessna, which flies south for an hour and then lands on a grass airstrip at mid-morning. Climbing out, I walk past a sign reading WELCOME TO BANIYALA, neatly hand-lettered and decorated with an image of spears.

It's a well-kept little settlement consisting of ten metal buildings and a banana grove. A fire smoulders in an oil drum. A single tall pole, painted white, stands ready for use in a coming-of-age ceremony for adolescent boys. I'm met cordially by two sturdy men of early middle age, Djambawa Marawili and his brother Nuwandjali. They're both artists, their bark paintings galleried through the arts center at Yirrkala and widely admired across Australia. One of Djambawa's pieces, in fact, was chosen best bark in a national competition for 1996. Their elderly father, Wakuthi Marawili, is a revered clan leader as well as a painter himself. Within the formalized tradition of Yolngu art, in which families and clans hold proprietorship over decorative motifs as well as areas of landscape, some of the barks painted by Djambawa and Nuwandjali echo work done by Wakuthi decades ago. One subject of abiding importance to the Marawili family is Bäru, the Ancestral Crocodile. Bäru as bringer of fire. Bäru as link between saltwater and freshwater domains. Bäru's sacred nesting ground, Garrangali, as progenitive birthplace of the Madarrpa clan.

Today, by happenstance, Djambawa is at work on a Bäru painting. He has finished only a cartoon outline, but I can already see the emergent figures of Bäru and Blue-Tongue Lizard, enacting their legendary scuffle at the fire. It's just the sort of representation I've come looking for. (Although at that moment I have no inkling of its availability, this particular painting will eventually go to the Yirrkala gallery and, still later, hang in my home.) For now, easygoing and hospitable, Djambawa sets his brushes aside so we can chat.

After we've seated ourselves on a tarp, and I've offered my visitation gift (another carton of Winfield Reds—shamefully unwholesome, but concordant with local etiquette), I explain my interest in crocodiles. Nuwandjali says, "If they are hungry, they go for kill. A bit like lion. They are dangerous." But only dangerous to humans, he adds, if you tease them. Djambawa has meanwhile walked off without explanation, and returns carrying a present for me: a commemorative booklet, with text in a Yolngu language and also in English, plus photographs and maps, from a workshop for Yolngu children on the subject of Bäru. Held here in Bäniyala several years ago, it was one of a series of such events staged throughout eastern Arnhem Land, all with the dual purpose of transmitting Yolngu culture to youngsters and advancing their multi-lingual education. Bäniyala's workshop took for its focus the famed crocodile nesting ground, Garrangali.

I page through the booklet, seeing photos of Garrangali and cryptic diagrams placing it within the universe of Yolngu belief. "It's a natural crocodile farm," Nuwandjali says. "We never made it. Our grandfathers never made it. Our fathers, even. No. Crocodile made it. It's been there all the time."

In its reality on the landscape, Garrangali is a hummocky glade of mangroves, pandanus, and paperbark trees standing silhouetted above a low, reedy wetland. It lies ten miles north of Bäniyala, near where the Dhuruputjpi River performs a mystifying trick—vanishing underground, like a snake into a hole, just short of what would be its mouth on Grindall Bay. Conditions there are ideal for nesting by *Crocodylus porosus*. That alone would give the spot importance. But what makes Garrangali extraordinary, apart from the concentration of nests and the river's disappearance, is a network of narrow channels, kept open by flowing water and uncounted generations of crocodiles. The channels, each barely wider than the shoulders of a large croc, wind through the soil and the root network. In the grainy black-and-white photos of the workshop booklet, they resemble ditches for irrigating a hay field. And the tunnels, Nuwandjali tells me, allow crocodiles to travel underground beneath the banks.

Tunnels? He did say *tunnels*? Not just open-air ditches, then, but subterranean tubes? With crocodiles traveling through them, like those legendary alligators in the sewers of New York? It takes me a few moments to absorb this. The tunnels of Garrangali—though they're physical actualities, I gather—seem surreal and irresistibly symbolic. But I can't put my finger on what they might symbolize.

The children who came for the workshop have left behind dutiful little reports, collected in the booklet, about their excursion to this mystical site. "Yesterday we went to Garrangali looking for crocodiles and went to the jungle. It is a big jungle and different trees and big tunnels. The Garrangali was beautiful." Another: "We saw three Bäru and we saw the tunnels for the Bäru." Besides glimpsing crocodiles and peering into tunnels, the kids were evidently given another signal experience: sampling two different flavors of water from closely adjacent sources. "We tasted salty water and fresh water." Here, it was fresh and sweet; just over there, salty. Somewhere in the vicinity of Garrangali, discreetly concealed amid vegetation, channels, underground seepage and conduits, the Dhuruputjpi River meets the tidal reach. Exactly where? Well, dip from various water holes and taste. Fresh, *umm*. Salty, *yuk*. No doubt it's a good lesson for children of any culture—that the world's crucial boundaries are not always discernible to the eye.

From the contemplation of such briny ineffables, Djambawa brings me back to solid ground. In past years, he explains, the residents of Bäniyala collected eggs from Garrangali. They shipped them, for hatching, to a Yolngu-run crocodile farm near Yirrkala. It was a carefully limited harvest, leaving some nests untouched. Gumatj people operated the farm; Madarrpa people, his people, controlled the nesting ground and contributed the eggs. This yielded a modest trickle of income for both clans, which are linked in close kinship. But the Madarrpa have discontinued that. Suspended it, anyway. They have decreed a moratorium on gathering eggs. Something apparently happened at Garrangali—something ugly, something offensive, though Djambawa doesn't say what— to provoke the change. (Later I'll learn that a poacher's fishing camp was discovered there, littered with fuel drums, bedding, garbage, and, worst of all, a burlap bag containing the severed head of a crocodile. Although crocodile killing isn't unknown in Madarrpa country, the means and the manner are crucial, and this bagged head represented an appalling sacrilege.) So the wild eggs are presently left untouched. Garrangali is now an undisturbed sanctuary. It may be one of the most productive natural crocodile hatcheries in Australia.

The ban on egg gathering is intended to be temporary, Djambawa says. They'll let the spiritual wound heal and the crocodile population breed itself up. After that, the Madarrpa may choose to establish a croc farm of their own. They could harvest a sustainable fraction of the wild eggs and raise the hatchlings themselves. They might even try a little tourist operation, allowing outsiders to visit this crocodile refuge. (When he mentions tourism, I cringe. It's easy to envision, but not my place to say, that a handful of barramundi poachers might do far less damage to Madarrpa culture than a stampede of "ecotourists" booked in through Darwin.) Of course hunting isn't allowed at Garrangali, Nuwandjali says, adding two words to summarize its management status: "Just wild."

Then we climb into their Land Cruiser and drive north on a dirt track, winding among sand palms, eucalyptus, and orange termite mounds toward Garrangali itself. We don't approach close. It's the wrong time of year—or I'm the wrong kind of visitor—to be clomping around such a sanctorum. Across the floodplain I see its profile, standing up like a tussock on a short-grass meadow. The tunnels aren't visible. The crocodiles must be imagined. But the power of the place upon Djambawa and Nuwandjali, even at this distance, is unmistakable. They've escorted me out here and shared a glimpse of their spiritual omphalos because—well, I don't *know* why, except that they're gracious and hospitable men; also savvy. They understand that although their traditional culture may feel self-contained and complete (even as it embraces Winfield Red cigarettes, art galleries, and Land Cruisers), it can't escape ever-widening interactions with the bigger world beyond, from which I'm merely the most recent intermediary.

Back at the airstrip, my Cessna is waiting. Nuwandjali, whose cheeks protrude roundly below gentle, smart eyes, shakes my hand. "Thank you for coming to Bäniyala," he says. Djambawa's final words are less personal, more lapidary, a statement of fundamentals spoken by a senior clan leader: "Madarrpa means crocodile."

35

But before flying back to Yirrkala, I have another little mission. After barely more than a parabolic hop, the plane puts me down at a homeland called Dhuruputjpi, beside the Dhuruputjpi River, just a few miles upstream from where its flow disappears beneath Garrangali. I'm hoping to find MänMan Wirrpanda, a man of the Djapu clan who has complained to authorities that crocodiles are terrorizing his community.

A Cessna can't land unobtrusively at such a place, and by the time I step out there's a crowd. Having stated my business, I'm delivered to the presence of MänMan, a small, sixtyish fellow whose head and chest are smeared with white clay. He holds a decorated digging stick like a scepter. His adornment and equipage reflect important business I've interrupted, a circumcision ceremony for young boys, including his son. Maybe it's the same coming-of-age ritual that, back in Bäniyala, will involve the mysterious white pole. Anyway, only the preliminaries have commenced here in Dhuruputjpi, not the actual knife work, and MänMan can tolerate a pause. He welcomes my visit because, poor man, he misunderstands it as a chance to request help from someone official. We sit on the ground, beneath an awning, and he explains.

He's concerned about three very large crocodiles, fifteen-footers, that have been lurking in waters nearby. Haunting the community. He wants them captured and removed, or at least restrained by a fence. His people, he tells me, are afraid to go near the water. They're afraid to hunt wild geese. Himself, he lost a dog. The children aren't safe. "Not only kids, but adults too," MänMan says. No one feels safe, since no one knows just where these crocodiles may turn up at any given moment. Somebody's gonna get killed.

His own clan, the Djapu, have their primary totemic bond with

Mäna, the Ancestral Shark, not with Bäru. Shark is their governing presence, Shark made this area, he tells me. The community's hand-painted welcoming sign, I notice, bears the cheery image of a lavender shark. But the matter of crocodile control is made tricky, as MänMan well knows, by the fact that his neighbors, the Madarrpa, are devoutly affiliated with Bäru. "We have to go through channels," he says. Political ones, he means, not subterranean tunnels full of water and crocodiles. "And do the right way." He doesn't dare shoot the crocs, or spear them, for fear of causing grievous inter-clan offense. So he's hoping that rangers from the Parks and Wildlife Commission might swoop in and police up the menacing animals before a disaster hits. Or maybe cordon them behind a fence? Or in a cage? A cage, yes—and MänMan's people would promise to feed them. But he needs some form of separation, please, between his people and those crocodiles.

What strikes me most about MänMan's plea, besides its sheer urgency, is that it contrasts starkly with the insouciance of Djalalingba, the proud affinity of Mandawuy, the devout protectiveness of Djambawa and Nuwandjali toward crocodiles. Totemic affiliation offers its transcendent form of truce between beast and human, as manifest in the attitudes of the Gumatj and the Madarrpa. But if you don't happen to belong to one of those clans and you've got children in the water, then a crocodile seems to present itself quite differently. There's just no single answer to the question of how Yolngu people regard *Crocodylus porosus*. One man's monster is another man's god.

"It's only for safety," says MänMan. No, he's not looking for meat or for hides. He bears no malice against crocodiles, no disrespect toward Bäru. "Safety for kids. And for the community. Otherwise, if something happen—it's too late, you know? Just only for safety, David."

Although his anxiety is understandable, his appeal for outside assistance bears a certain freight of irony; he's asking the agents of alien governance, the same governance that effaced Yolngu sovereignty and denied Yolngu proprietorship over the landscape, to mediate a primordial conflict. And his misapprehension about me—that I carry some influence in the matter—is embarrassing but not easily corrected. After all, he radioed his concern to Nhulunbuy, and then here came this balanda out of the sky, asking about crocodiles, taking notes. "It's up to you," he says, trusting and wrong.

Okay, I say. I'll tell them. Only for safety. Got it.

I close my notebook. As he walks me back toward the Cessna, our business completed, MänMan lets drop a surprising fact. That fellow Dhakiyarr, long time ago? he says. That one who speared a policeman, got taken to Darwin for trial? That one who disappeared completely, never came home, never was heard from again? That one who, seems like he been killed and dumped away by the Darwin police? That fellow Dhakiyarr, says MänMan, that was my father.

A coincidence? Yes, but a weirdly befuddling one. I've taken a lot of testimony out here in eastern Arnhem Land, and much of it seems magnetized with strong antinomies—Yolngu way versus balanda way, native autonomy versus imperialism, indigenous predators versus a colonial power that would pacify not just the people but the landscape itself. Crocodile veneration versus crocodile fear and loathing and exploitation. Despite all that, MänMan has just reminded me that mingled destinies can't be sorted neatly into two piles. His close connection to Dhakiyarr Wirrpanda seems to coil the whole Yolngu story into an improbable circle, like a crocodile biting its tail. Bäru, sacred monster, is no simple animal and no simple idea. They never are.

36

Eight months after my first trip to Arnhem Land, I'm back in Maningrida, on the north coast, just in time for the crocodile-egg harvest. In the course of an afternoon I help the Djelk Rangers and their mentor, Jackie Adjarral, "the Professor," open four nests along the banks of the Liverpool River. We guard one another's backs against angry maternal crocodiles. We excavate the nest mounds by hand. We load dozens of eggs into Rubbermaid coolers.

It's delicate work—getting the eggs wiped clean, labeling them with location codes in felt-tip marker, taking care that each egg stays right side up (so as not to tear loose an embryo inside), stacking them gingerly in rows. The nest mounds, which are piles of rotting vegetation heaped up by the females after laying, have provided warmth by decay, a nice reptilian alternative to the body warmth of an incubating hen. The eggs, cozily buried, have come to term. Left untouched, they would hatch in the next few days. The little crocodiles would struggle out through the compost, drop into the river or be carried there by mama, and then suffer the natural rate of mortality, which is high. Most of them would be dead within a year. We're preempting all that.

Each egg we touch feels like a ticking time bomb of life, nearly ready to blow. The fourth clutch is so ready, in fact, that the eggs begin pipping as we handle them. A patch of shell fragments erupts at one end, behind which comes a small olive nose, and behind that the exigent force of a miniature crocodile impatient to be born. Quickly but carefully we finish loading the unhatched eggs, and then we head downriver, racing a rainstorm. But the dinghy's engine combines with the jostle of emerging siblings to create an infectious excitement that travels from egg to egg, and before we're back in Maningrida, the entire cooler load has gone off like a pan of popcorn. A neat pile of eggs has turned into a wriggling scrum of baby crocodiles. Several of the emergers, tangled and stuck, need assistance. Offering some, I thereby confirm that the first instinct of a hatchling crocodile is to bite the finger that has helped free it from its shell. But the teeth of these tiny beasts are like toothbrush bristles, and their jaws deliver barely a pinch.

Freshly hatched, albumen-smeared, the squirming green bodies are as slimy and fragile as salamanders.

At the Djelk work shed, we unload them into trays. The unhatched eggs go into separate trays. Among the four nests (coded as L-9, L-10, L-11, and L-12, for their numerical order along the Liverpool River, each egg marked with the code of its nest), we've collected ninety-nine viable eggs and forty-five hatchlings. By the time we've tucked them all away in the moist warmth of the incubator and sealed its big door, I feel the tender pride of a midwife. Baby crocodiles are cute.

But it's no justice to sentimentalize them, or their place within Aboriginal cultures. Just as well, then, that on the following night I receive a strong tonic against any such temptation.

On this outing, we're after adults. We carry an electric beam, two twelve-volt batteries, several shanks of rope, a wide roll of cellophane tape, an old rifle, assorted dull knives, two spools of parachute cord, and a harpoon. We're also equipped with a box of brass harpoon points, each with an eyelet for attaching cord. For dinner we've got hot dogs, which we eat—cold—at a quiet anchorage along the lower river while waiting for darkness. The moon is a day short of full, rising dimly behind a phlegmy layer of clouds. Before proceeding, we connect the beam to a battery and test it. Jackie directs us to put all the gear in order—gas carboys to the rear, spare battery to the rear, box of ropes in front, rifle to the rear, everything else stowed neatly, so that the floor of the dinghy is clear. Things may get hectic. Then we motor upstream to the Liverpool's juncture with the Tomkinson.

A young Ranger named Randy Yibarbuk sits cross-legged on the bow deck. Alister is at the tiller. Jackie stands forward, feet placed wide, a commanding figure though barely larger than a child—and as ready as Queequeg to engage the prey. Randy sweeps the banks ahead with his beam, scanning for crocodile eye-shine. The long shaft of light appears smoky blue against the green wall of mangroves and the brown water. Dodging here, dodging there, it probes the shoreline, lingering briefly in coves, behind logs, amid the cover of drowned snags, anywhere a crocodile might hang for a night's feeding. Jackie watches. We all watch. We see many sets of paired orange dots, but in most cases, as we glide closer, engine throttled low, they prove to be juveniles, only three or four feet in length. "Small," Jackie says, and waves Alister onward.

The river is lit with eyes lining each bank like highway reflectors. We investigate them one by one: a small crocodile, another small one, small again. But now Jackie rigs the harpoon, fitting a brass point onto the socket, and holds it ready with the cord pinched along the shaft. We spotlight several lunkers, which sink away as we come near. It's spooky that such a big animal can vanish so quietly, so effortlessly, while still lingering somewhere beneath the boat. But crocodiles don't survive to adulthood by being stupid—at least, not as stupidity is measured within their dim, perilous world. Just when I'm beginning to envision a long evening of futile approaches, Jackie slams his harpoon into the water.

Nothing much happens. He draws back the shaft. The brass point is gone. The cord has paid out a few yards, but it seems to be tangled among the roots of a mangrove. My impression is that Jackie has missed.

He knows better. He probes with the shaft toward the apparent site of the snarl, as though trying to retrieve a snagged lure. He shifts the angle of tension on the cord while Alister steadies the boat. Then he pulls firmly, easily, drawing in the cord hand over hand until, to my amazement, up comes the head of a large, angry crocodile. The harpoon point is buried in neck muscle just behind its brain.

For twenty minutes Jackie plays it like a five-hundred-pound catfish. When the croc insists violently, he lets it run. When the croc rests, Jackie reclaims the initiative, slowly cranking in line. Meanwhile we drift silently on the Tomkinson's current. The night is warm. The mosquitoes know we're here. Jackie speaks in a whisper, either from reverence or, more likely, to avoid exacerbating the crocodile's sense of alarm. At an opportune moment—has the crocodile given up? paused to gather its wind? the signals aren't obvious—Jackie asks me to pass

him another harpoon point. Randy rigs the cord and, with Jackie raising the croc's head back into view, rams home the second strike. The
crocodile is now double-tethered and nearly hopeless. Within a few
minutes we've got it beside the boat.

The voice of sobriety in my brain says, *Now what?*

Jackie lassoes its upper jaw with a loop-ended rope. The loop tight-
ens down at that characteristic narrow point behind the bulbous nose.
With the upper jaw held, the lower jaw remains agape, exposing rows
of spiky teeth and the gurgly pink maw. At first I mistake this for a
failed try, thinking that the intent was to catch both jaws together. But
no, it's exactly what Jackie wants. Cinching the upper jaw gives a tight
grip, snagged among teeth; cinching both would allow the rope to slip
off the snout. Still, the mouth remains free to slam shut on anything
within reach. The crocodile hisses, a scary but piteous sound expressing
both rage and exhaustion. Alister takes up the rifle, an old bolt-action
.22 from the Arms Corporation of the Philippines. I hand him a single
bullet from the pocketload that, for some reason, I've been assigned to
carry. Jackie and Randy haul on the noose rope, hoisting the crocodile's
head against the gunnel. The boat has become a crowded place. Alister
puts the muzzle to the croc's brain. The gun jams.

Alister works the bolt to clear it. He tries again. The single small *pop*
sounds almost trivial, a shallow slap against the deep quiet of the night;
but to the crocodile, it is not trivial. The animal flinches, releasing
another guttural hiss, like the air coming out of a punctured truck tire.
Now, only now, Jackie says, "Seven foot."

He's close, but too conservative. The eventual measurement will be
238 centimeters, more like seven foot ten. It's a female. The belly skin
looks good, though not premium—probably grade two. The meat will
go to the people of Maningrida.

37

It proves to be a long, harsh, instructive evening on the river. We harpoon two more crocodiles of roughly the same size. One is released alive—no simple task, believe me—because Jackie has noticed a flaw in its tail. The mild harpoon wound in the thick of its neck isn't life-threatening, and Jackie sees no reason to kill a good, healthy animal with a commercially negligible skin. The other is an eight-footer, a strong, tenacious animal that requires two harpoon lines, a snout noose, a length of parachute cord holding the mouth closed, two shots from the rifle, pithing the brain with my Leatherman blade, and then pithing again with a longer knife. Throughout all this, I do my best to be helpful. I do my best to stay out of reach of the jaws and out of line of the rusty, unpredictable rifle. I do my best to reconcile sympathy for the crocodiles with sympathy for the people whose tradition, and needs, dictate hunting them.

I arrive late and exhausted back at my lodging room outside Maningrida, peel off my filthy clothes, let them drop on the floor, and sleep. It all seems like a lurid dream. But when I wake the next morning, I find splatters of dried crocodile blood on the inside surfaces of the lenses of my glasses.

SHADOW OF
THE NINE-TOED BEAR

38

Totemic affiliation comes in many forms, not all of them severely reverential. Various peoples find strength, affirm values, and sometimes polish their own vanities by associating themselves with particular animal species—especially with alpha predators. Besides the Gumatj and the Madarrpa of eastern Arnhem Land, for instance, there's another group of Australians who link their collective tribal identity with the figure of *Crocodylus porosus*: Darwin's chapter of the Hell's Angels motorcycle club. My source for this bit of anthropological data is an exuberant taxidermist named Andrew Cappo, formerly of the Darwin Museum, whose portfolio of freelance assignments includes pickling gigantic crocodile heads to serve as decor for Hell's Angels clubhouses.

Cappo lives alone in a bush compound, miles beyond the outskirts of Darwin, near a crossroads called Humpty Doo. His workshop, which is also his home, consists of a metal roof standing over a concrete slab, open to the tropical weather on all sides. There's a bed, with a mosquito net, and a hotplate and a refrigerator, together constituting almost the sole domestic felicities. There's also a band saw, a drill press, a large freezer filled with dead animals and animal parts, a few tables, an unpainted cast of a crocodile head on a stand, and a metal lathe. In the years before he moved up to Darwin and learned taxidermy, Cappo worked as a machinist for General Motors, down in Adelaide. "Fitting and turning," he explains, with no trace of nostalgia. One day he was laid off, along with thirty-five hundred other workers, and soon afterward he caught a plane for the Northern Territory. That was sixteen years ago. He spent five years learning taxidermy on a training contract at the museum but didn't see eye to eye with the boss, and eventually he went independent, finding his own modest niche in the trade. Seven years ago he moved out here to Humpty Doo. "If you're not in a city, living in some sort of shit-box little cubicle, you're just not in the game anymore," Andrew says, and yet not being "in the game," whichever game that is, seems to suit him fine. Even here, amid what until a decade ago was virgin bush, he's troubled by suburban sprawl. "Too many people on the planet." Too many people and not enough wild animals, for his taste.

Andrew is a tall, tanned, vehemently congenial fellow with ringlets of sun-yellowed hair, a broad smile, and a mad enthusiasm for wildlife, outdoor sport, open spaces, liberty, bamboo horticulture, firearms and the right to bear them, and—above all—barramundi, the most tooth-some of northern Australia's native fish. He lives for fishing, he admits. That's what brought him north. Eats barramundi almost every day of the year. Fishes when he wants, works when he must. Was married at one time, but not lately. "Taxidermists don't have wives, mate," he tells me. "They don't go together. Mine bolted years ago." Want some barramundi? he asks generously, and proceeds to cook us lunch on the hotplate. It turns out to be one of the best meals I've had in Australia—a cold compliment, considering Aussie cuisine, but meant warmly.

Upon one of Andrew's work tables stands a cast of a leaping barra-mundi, representing the staple of his taxidermy practice as well as his diet. He does trophy mounts of dead fish and faithful effigies (based on precise measurements, plus expert extrapolation from his collection of fiberglass molds) of lunkers that fisherman have chosen to release. "What keeps me alive are barramundi," he says. "That's my job. Crocodiles, I don't rely on them."

But he does crocodiles when asked. The "bikies" (his casual label for such leather-clad, Harley-Davidson-riding gentlemen as we would call "bikers") have asked more than once, and they seem to be well satisfied with his museum-quality work. In addition to pickling crocodile heads, he also produces tanned skins and bleached skulls. The choice of preservation method in a given case depends on customer preference and the condition of the carcass when Andrew gets it. A tanned crocodile skin, if not sold wholesale for leather-goods fabrication, can be mounted flat for its sheer decorative appeal, like a brass rubbing or an heirloom quilt. A bleached crocodile skull makes a nice, sanitary, smell-free artifact, suitable for display in the open air of a rec room; though perhaps not so vivid as a pickled head, it's likely to have a longer shelf life. Very large heads lend themselves better to the boil-and-bleach process than to pickling. Given that saltwater crocodiles are the signature animal of the Northern Territory and that Darwin itself is the capital of croc schlock, Darwin's Angels have shown local pride (as well as a certain ghoulish élan) in their choice of totemic artifacts—and not just by accenting their own clubhouse with crocodile trophies, but by sending them as gifts to chapters elsewhere in Australia. The chapter in Adelaide owns two Cappo skulls. There are chapters across the country, clubhouses in every major city. A fellow could stay busy. And the Angels, Cappo tells me, they're good customers and good blokes. "You be careful using their name in print, too," he says. Then he laughs nervously. Together we laugh nervously. "Is it going to be flattering?" he asks. Of course, I say. Who would disparage a roisterous band of motorcyclists for their lighthearted appreciation of pickled crocodile heads?

The tanning of crocodile skins, although it's an older and more refined craft than head pickling, is no easier. In fact, Andrew says, it's "hard, dirty work." It's also hazardous to the health of the practitioner, because of the toxic chemicals used. In bygone days, when arsenic and cyanide were part of the standard recipe, taxidermists died young. Nowadays it's done with formaldehyde, formic acid, kerosene, sodium bicarbonate, salt, a tanning agent called Lutan F, and carbolic acid, otherwise known as phenol, which may be as menacing as the old poisons. "That's the one I don't like. Cancer in a bottle. Kills everything," Andrew says about phenol. And it doesn't break down, it lingers. Up there in Maningrida, he says (being familiar, through a mutual acquaintance of ours, with the Djelk Rangers' wild-harvest efforts), the matter of toxicity will be important when those lads begin their own tanning operation. Andrew has agreed to go up, teach them the skills, help them establish a tannery. From what he has heard, they've already got twenty skins (two of which would be from the animals I helped harpoon), salted and rolled, ready to go. But how will they dispose of their toxic wastes? And will they be conscientious about minimizing their own exposure? Skin contact with these chemicals can be bad, inhaling phenol fumes can be worse. Andrew is concerned. He doesn't care to do much tanning himself, and it's certainly no task for slapdash amateurs. But he seems sympathetic with the Maningrida community's desire for a little vertical integration in the skin trade, and he'll do what he can to instill good precautionary habits.

Tell me about technique, I say. Details. How do you tan a crocodile skin?

Well, it's a seven-stage process, he tells me. There's salting, soaking, scouring, pickling, shaving, tanning, and oiling. Oh yes, and dyeing, if you want your leather in rich shades of brown or black, which are more popular than that bleary, crocodilian olive-gray. Okay. So you've got a dead croc. You skin the animal, lay out the skin raw side up, water-blast it with a power nozzle to remove the last remnants of meat, and then salt it.

Salt it, I echo. You dust it with salt.

Dust it? No, he corrects me, you *pour* the stuff on, pack it into the crannies, moosh it in, covering every speck of skin. Don't be stingy. You can't have too much salt. And the salt has to penetrate. Then you roll up the skin and store it in a cooler until you're ready for step two. Step two is the lime bath. Soak it in lime to remove the scales, which will peel off as thin, translucent shucks. Then scour the skin in an agitator tub with water and detergent, which removes all the dirt, blood, and muck. Take it out, rinse it, and pickle it in a formic-acid solution. But be careful of your solution, make it strong but not too strong, because if the acid is overly concentrated (below a pH of, say, 2.5) you'll get an ugly phenomenon called acid swell, where the skin goes all slimy and spongy. "Acid swell is the nightmare of tanners."

Acid swell, I repeat. Too much acid. The skin swells.

"Yep."

"The nightmare of tanners?"

"Yep."

"Formic acid. Do you get it from ants?"

"I'm not sure where they get it. Smells like ants."

"Sounds like ants, too."

But at proper concentration, formic acid will loosen the collagen fibers, make the skin pliable, even take some of the calcium out of the bony internal scutes. It's wizard stuff for softening bone. Soak a human femur in formic acid, Andrew has heard, and you can lift the thing out and tie it in a knot. Because there's virtually no calcium left. With crocodiles, therefore, whose very hide contains those calcareous scutes—the osteoderms—you don't want to omit a good formic-acid pickle bath.

After pickling comes shaving. A crocodile skin is thin and irregular and relatively fragile, not uniformly thick like a buffalo hide, so you need to be delicate about shaving. Don't run it through a shaving machine, Andrew warns. Use a circular wire brush on an electric grinder. That's delicate enough.

And this grinding, you do it on the *inside* of the skin? I ask. It's a risibly ignorant question, but I'm groping for clarity. Me, I've never flayed any animal larger than a roasted turkey. My taxidermic experience was limited to butterflies.

"Yeah, the flesh side," says Andrew. Of course, you dolt. "Always the flesh side. You don't touch the other side."

"That's your cosmetic side."

"Yep. Handbag side. Shoe side."

Then the tanning. You stir up a solution of salt, formaldehyde, formic acid, sodium bicarbonate, Lutan F, and a synthetic oil known as "fur liquor," which replaces the natural oils you've leached out. If you're having trouble with bacterial growth, here's where you add the phenol. Don't breath those fumes. Don't be soaking your hands in this potion. And don't let it get strongly acidic, not like the pickle bath; use your sodium bicarb to bring its pH up, somewhere between 3.0 and 4.0. Give the skin at least twenty-four hours of tanning at thirty degrees Celsius. After tanning, you've got to oil it again. And now with heat. Either you soak it in a fur-liquor solution warmed to thirty-eight degrees, or you just swab the stuff on hot. Let it penetrate. Also, by the way, this is when you do your dyeing. Finally, after tanning and oiling and dyeing, you've got to break the skin. You can do that by hand—working it back and forth over the blade of a shovel, say, as you might with a buffalo hide. But again, no, that's probably not how you want to go, because a croc skin is too lumpy. Better to toss it into a big revolving drum, with sawdust. Flopping around in there will loosen the fibers. Just let it flop for ages and ages.

Revolving drum, I say. No shovels.

"You can also pound it with a rubber mallet," he suggests. "Very hard."

Mallet pounding. That's labor-intensive, right? And flopping it in a drum is . . . what, capital-intensive? Drum-intensive? I'm trying to follow closely. I'm interested in the nitty and the gritty. So far I've got seven simple stages: washing, salting, soaking, scouring, pickling, shaving, tanning, oiling, dyeing, and breaking.

Yeah, that's basically it, Andrew says. Your skin is finished. Doing a head, though, is a different story. Doing the head of a big animal, with the skin fused to the skull (as it is in crocs, unlike in mammals) and unremovable in any salvageable form, is trickier still. Different processes for different anatomical parts. You don't want to immerse a crocodile head in your tanning bath, because formic acid attacks the teeth. They go all bendy. Or they might flake away. That's the acid, extracting calcium from the dentin. Not good. Nobody wants a pickled crocodile head with bendy teeth. So you've got to protect them. You can submerge the head just partially—this way and then that, bottom jaw, top jaw, always keeping the teeth out of the bath. Which is a slow, clumsy hassle. Meanwhile you attack the head with a big syringe full of formaldehyde. Your job is made more difficult by urgency, since it doesn't take long, not in this climate, for a head to go off. The rot sets in fast. A matter of hours. Because they're such filthy animals to begin with, Andrew says. Covered in bacteria. They don't brush their teeth, you know. Nasty microbes lurk in the mouth—a fact that's reflected, he notes, in the incidence of nasty infections among taxidermists. Last time he did a head, he scratched his finger on a tooth. He was unwrapping the thing from plastic swaddling, and he nicked himself—just a little slice, like a paper cut. Once the infection set in, it seemed unstoppable. A creeping sore, with pus. "Darn near lost my finger. Bacteria ulcerated all the way around the knuckle." Wear gloves, wash frequently, use heaps of detergent, and nuke the head with salt, is Andrew's advice. Get that bacteria in you and you've had it.

On a medium-sized head, say from a nine-foot animal, you can put a syringe into the nostrils, into the eyes, up through the lower jaw. But a large head from one of those big old crocs, you just can't penetrate it with a syringe. Too solid, too tough. So for those jobs Andrew resorts to a hand chisel. "I chisel the whole top palate out. Inside the lower jaw,

I chisel out where you can't see." He extracts a sizable quantity of flesh and bone, leaving the basic structure and, unavoidably, a fair bit of material still susceptible to rot. Into the chiseled cavities he pumps formaldehyde, salt, alum, borax, anything he can think of that might impede bacterial growth. Then he seals the entire mess with an acrylic finish. If he has squirted in enough formaldehyde, splashed on enough acrylic, maybe the locked-away traces of living bacteria won't find conditions hospitable, won't multiply and thrive, won't expand into seething, pustular abscesses—at least for a while. "I'm not real happy about that," Andrew confides. "It's a very hard thing, heads taxidermy."

How long is it good for? I ask.

Behind a glass case, with bug proofing, maybe ten years or more, he figures. But there's no warranty on pickled crocodile heads.

Who's the typical customer?

Hard to put a label on them, he says. Museums, wildlife exhibitors, government institutions, private collectors, motorcycle clubs. That sort of folk. A big head is worth thousands, so you can't sell something like that to the man off the street. Reptile enthusiasts.

Considerations of scale and putrescence, not to mention cost, are exactly what make Andrew uneasy about his current commission from Hell's Angels. This crocodile was a seventeen-footer, a huge beast with a huge, fleshy head. It was trapped by a bloke who does commercial harvesting, Andrew says, one of the few white fellas venturing into the trade since restrictions eased. He's not allowed to do harpooning (that's still reserved to Aboriginal groups); trapping only, and in this case he left his catch in the trap too long. The animal struggled, went crazy, scraped away the end of its nose and broke some big teeth. Further deterioration occurred afterward, because it takes a team of men just to skin a seventeen-foot croc, and while they were busy at that, the head got no attention. No syringe-loads of formaldehyde to hold off decay. It was just lying there in the heat, going high. Still, the bloke sold this thing to the Angels, and by the time Andrew got it, the condition was . . . uh, suboptimal. "It's stinking rotten!" he told them plaintively. Stinking rotten but frozen solid, which minimized the smell but didn't change the fact that, as Andrew could see, its eyes were bulged out from the pressure of internal gases. Now that we've finished our barramundi lunch, he offers to show me.

Digging into his freezer, he unloads the stiff carcasses of bonefish, barracuda, barramundi, and other fish destined for trophy preparation, as well as a buffalo hide, folded and jammed into a cardboard beer crate. From the freezer's bottom, he lifts out a crocodile head—a sizable armload, big as a duffel bag. Its left eye protrudes like a cup of raspberry sorbet. The front of the snout, from that last crazed struggle, is abraded down to bone. A majestic and dangerous beast has died an unpretty death, and now someone hopes that its majesty can be recaptured, represented, owned, and possessed (minus the danger) in the form of an artifact derived from this hunk of frozen, putrid meat. Andrew Cappo, a stalwart professional, will do what he can.

But there's not much to work with. The bare minimum. It wasn't skinned properly for a full-head preparation, he tells me. The trapper bloke took away all the skin he could get with the belly hide, including a crucial flap underneath the jaw. And the tongue, that's gone too. There's all these bits missing. Nose rubbed off. Teeth shattered. I'll have to fake it, Andrew says. Don't want it to look second-rate. I'll defrost it, he says, and cut away some of the meat, while watching to see how fast the decay comes on. But once I start to chisel, he says, there's no turning back. The sensible thing would be to forget about the head as a head and go straight to the fallback option, a boiled-and-bleached skull. Better a good, dignified skull than a half-rotten chop job of bone and meat, yeah? That's Andrew's considered view. He's got misgivings about this project.

When it comes to taxidermy, which is an arcane craft as well as a business, the customer is not always right. Then again, he implies, try telling that to Hell's Angels.

39

Half a world away, in the highlands of Romania, other fruits of taxidermy are displayed at an institution called Muzeul Cinegetic al
Carpaților—that is, the Carpathian Hunting Museum. No crocodiles
inhabit the Carpathian Mountains, but there are bears, many bears.
And the primitive human impulse to convert formidable beasts into
inert, decorative trophies, vouching for machismo and supremacy
within a tribalistic culture, has been played out again here, amid the
boreal forests. In place of a motorcycle club there was the Romanian
Communist Party and, risen from within it, one very odd little man.

The museum stands in a town called Posada, along a scary two-lane
highway winding north toward the city of Brașov. The highway is a
gauntlet of soot-farting trucks, lane closures for repair, horse carts
impeding flow, and impatient drivers who pass blindly on switchback
turns, as though the possibility of a head-on crash were some esoteric
abstraction, like relativity theory, that they've never troubled themselves to consider. In other words, it's an average Romanian road. If
you attend to your driving as carefully as you should, you might easily
miss the museum's sign. Stay alert, plunge out of traffic into curbside
parking, and sigh with relief when you're there.

Cinegetic is an important term in Romania, especially among those
charged with managing wildlife. *Vânătoare* is the straightforward
word for hunting, whereas *cinegetic* (possibly derived from the French
adjective *cynégétique*) carries high-flown, scientific nuances that suggest the use of hunting as a tool in sophisticated game management.
The traditions and protocols associated with hunting in this country
are ancient, deeply embraced, and peculiarly European, though the
European patterns have their Romanian variants, which are more
peculiar still. The Romanian Forest Department (Regia Națională a

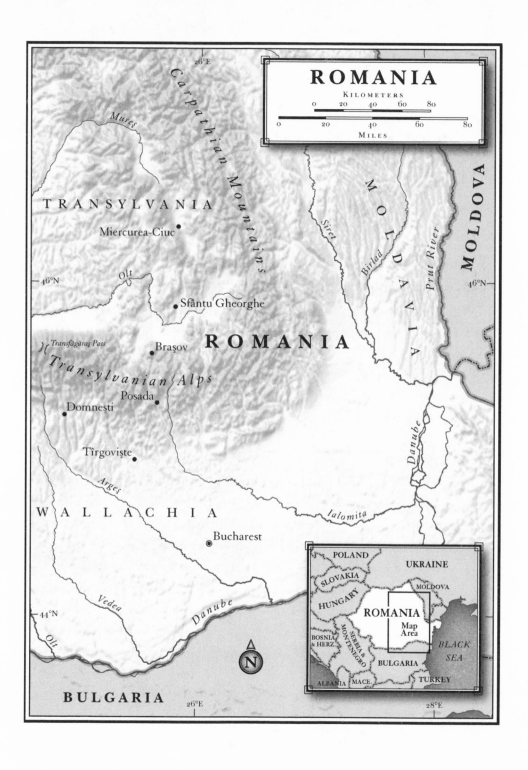

Pădurilor, or RNP) prides itself on a high degree of cinegetical expertise. And the woods are full of creatures considered game. Besides bears, in extraordinary number, there are wild boar, red deer, roe deer, lynx, and wolves. The native bear species is *Ursus arctos*, conspecific with the grizzlies of North America and the brown bears of Scandinavia, northern Asia, and Hokkaido. It's known here informally as *ursul brun*. For a combination of reasons, some natural, some strange, Romania harbors a far larger population of brown bears than any other European country west of Russia. That unusual abundance is reflected at Muzeul Cinegetic.

The museum is a boxy steel-and-glass edifice, a modern building that looks anomalous on the grounds of the old mountain estate of the Bibescos, one of the princely families prominent during the late phase of the Romanian monarchy, which ended amid the upheavals of World War II. The Bibesco palace was destroyed by fire, but the servants' quarters, a hulking stone manor in its own right, has survived to suggest its flavor. Within the museum itself, one section features ornate hunting gear and opulent rustic decor from that milieu, including tapestries, antique firearms, signal horns, powder horns, a pearl-handled sword, a set of hunting knives with deer-foot handles, a heavy walnut table and chairs that once graced the royal family's own mountain palace (just up the road in Sinaia, now surrounded by a tinselly ski resort), a silver jewel box in the shape of an ostrich, and a bronze sculpture of Saint George killing his dragon, which in this rendering closely resembles the dragonoid monitor lizards of Komodo. Elsewhere the museum is filled with mounted heads, skulls, stuffed animals, antlers, horns, boar tusks, and big furry pelts spread-eagled on the walls. The stolid silence of the trophies, and the dearth of explanatory legends in Romanian or any other language, are offset by a young woman in a cream-colored sweater and scarf who, presiding as guide on the day of my visit, pours forth unbidden (but welcome) commentary in singsong, schoolroom English.

Although her outfit seems casual for a museum functionary, her manner is quite official. The place was created in 1996, she says. (She doesn't mention that an earlier National Hunting Museum existed, elsewhere and under far different political circumstances, in the 1930s.) This is a black goat, she says, gesturing at a chamois. And these Carpathian stags, she says—motioning toward a wall festooned with forty-some skull-and-rack mounts of red deer—all of them, they rated gold medals. To earn a gold medal, she explains, requires greater than 220 CIC points. (She offers no explanation of CIC points, but with them I'm already familiar. CIC is the acronym for Conseil International de la Chasse et de la Conservation du Gibier, a Paris-based body that maintains a record book on trophy animals, roughly equivalent to the Boone and Crockett Club in North America. Its point system, based on a series of measurements and judgments applied to trophy heads and skins, is the means by which competitive European hunters keep score.) Here's a stag that received 261 CIC points, she says, a world record when it was killed in 1980. We step past a lynx pelt on the floor. We admire an array of fallow-deer antlers. We move from the mouflon (*Ovis musimon*, an exotic bighorn sheep), to the dainty roe deer (*Capreolus capreolus*), to the stocky and pugnacious wild boar (*Sus scrofa*), this last represented in various forms: a boar head, a full body, three pelts on the wall, and two dozen sets of tusks mounted on wooden plaques. A large boar, says the woman, can attain 250 kilograms. Most of the record-book animals here, she says in a tone of ambivalent pride, were shot by Nicolae Ceauşescu.

She doesn't need to tell me who he was. And yet Ceauşescu's name is conspicuously absent from this national shrine of blood-sport mementos. CIC scores are posted beside many of the trophies, but there's no information as to who killed what. That part comes only orally, from the young woman in the sweater-and-scarf uniform. Yes, Ceauşescu. Yes, Ceauşescu again. This one too. And we have many more Ceauşescu trophies in storage, she brags or (it's hard to tell which) admits. I keep hearing his name, I can almost feel his presence, but there are no photos, no printed attributions. Ceauşescu, he's everywhere and he's nowhere. It's a post-Communist museum filled with discomfiting evidence of, on the one hand, the country's wildlife treasures and, on the other hand, the degree to which those treasures were pillaged for decades, during the Communist era, by a pipsqueak autocrat who fancied himself a great hunter. No wonder the young woman sounds conflicted. The whole place is uneasily balanced between remembering and forgetting.

Nicolae Ceauşescu, of course, was the tin-pot dictator who ruled Romania for twenty-five years with ever-increasing harshness and megalomania, eventually treating it as his personal kingdom. Like others of his class, he was a small, boring man whose chief talent was machination and whose life story ascends to drama only by way of woeful consequentiality and evil. He came from humble origins (the son of a drunken farmer, he was a shoemaker's apprentice in Bucharest at age eleven), showed no early promise or flair, and made his political contacts in prison during the years when Communists were persecuted as criminal agitators. He managed eventually to get hold of power, then gradually tighten his grip, because he was a deft manipulator of people and situations. One of his former minions, a director of foreign intelligence who defected to the West, considered him a man of "native intelligence, phenomenal memory, and iron will," although other portraits are less flattering.

During his early years as chief of state, Ceauşescu seemed progressive, at least as compared with most Communist bosses of the time. He was always more nationalist than Marxist, a Romanian leader in the homegrown style that unites a whole rogues' gallery of fascists, Christian vanguardists, and anti-Semites. He distanced himself from the USSR, attenuated his participation in the Warsaw Pact, and criticized the Soviet seizure of Czechoslovakia in 1968. Moves like that made him, for a while, the favorite Commie potentate of Western democracies. Richard Nixon came calling in 1969—it was his first visit as president to any Communist state—and posed with his arm thrown cordially over Ceauşescu's shoulder. George McGovern in 1978 rated Ceauşescu "among the world's leading proponents of arms control." George Bush the senior, while vice president in the 1980s, called Ceauşescu "one of Europe's good Communists," although by that time his brutal side had become manifest. He was never a democrat.

From the hopeful beginning to the grim end, Ceauşescu's regime became increasingly more personal, more ruinous, more desperate, and more vicious. In the 1970s, dissatisfied with the vision of Romania as an agrarian breadbasket for the Eastern Bloc, he put the country on its own crash program of industrialization, from which the main results were heavy-equipment factories building inferior Romanian-brand trucks and tractors, petrochemical factories dependent on imported oil, forced urbanization, polluted air, polluted rivers, and the dissipation of a huge sum in foreign loans. Products rolling out of the factories were generally too shabby for export. Craving a larger national workforce, he prohibited not just abortion but also contraception for any married woman under forty with fewer than four children, and he enforced that rule by way of compulsory monthly medical examinations. With illegal abortion the only available form of birth control, Romania's death rate among women from complications of abortion became the highest in Europe. The infant mortality rate was high also, so high that births weren't recorded until after the child had survived into its fourth week. Collectivized factory farms were set up, though many small farmers in the mountains managed to preserve their independence. Another program, called *Sistematizare* (Systematization), entailed the willful destruction of villages to force people into towns, and the razing of old neighborhoods in towns and cities to replace them with high-rise, concrete apartment blocks. Once peasants had been transformed into industrial laborers and packed into state-controlled urban hives, they were at the mercy of his government in ways that they hadn't been while living on the land. For instance, their heat and their electricity could be cut off. Their habits, even their thoughts, could more easily be monitored. All these ugly measures, as well as Ceauşescu's hold on power itself, were supported by a large security-police agency—Departmentul Securităţii Statului, known colloquially as the Securitate—and its vast network of paid or coerced informants. By one knowledgeable estimate, the Securitate apparatus included three million people among the national population of twenty-three million. Typewriters, by law, had to be registered. Telephones were tapped, mail was scrutinized. Ethnic minorities, notably the citizens of Hungarian and German descent who had lived for many generations in western Romania, lost some of their rights. Arrest and torture enforced compliance and silence. The Securitate as Ceauşescu used it, combined with a certain historically based stoicism among the populace, prevented the growth of any Romanian dissident movement equivalent to those in the Soviet Union, Czechoslovakia, and elsewhere within the Eastern Bloc.

Then, during the 1980s, while Systematization and other costly programs were grinding along, Ceauşescu became obsessively determined, as a point of pride and independence, to pay off the foreign debts. He succeeded in doing it—by rerouting the modest consumer economy into exports and bleeding his own people pale. Food went abroad while Romanians suffered rationing and hunger. Energy went abroad while Romanians in chilly apartments lived by the light of forty-watt bulbs. Gasoline grew scarce, and horse carts, which had always been part of the rural scene, became vehicles of national purpose. The country's single television channel carried hour after hour of fulsome attention to Ceauşescu, his wife (a harpy named Elena, his chief adviser and full partner in bad governance), and their supposed achievements. But within the cool blue glare on the other side of the TV screens, people hated as well as feared the reigning couple. When the wave of revolutions broke across eastern Europe in 1989, Nicolae

Ceauşescu achieved distinction as the sole Communist leader who was not only deposed but, along with Elena, promptly executed.

That occurred on December 25, 1989, at an army barracks in the city of Tîrgovişte, to which the Ceauşescus were brought after their helicopter escape from Bucharest went awry. The fatal slip of his political grip had happened abruptly, just four days earlier, during a speech to a mass gathering in Bucharest that, after so many years of oppression and docility, all at once became weirdly rebellious. To their own surprise, and everyone else's, the Ceauşescus were now on the run. Then caught. They remained under guard at Tîrgovişte for a few days, while dangerous decisions were pondered in the capital by a hastily formed junta, the National Salvation Front. Then came a kangaroo court, lasting fifty-five minutes and recorded on videotape, during which Ceauşescu irascibly denied the legitimacy of the men (from the junta, which included some of his prominent underlings, now turncoats) who had come out to enact this pro-forma tribunal and then kill him. Power relations in Romania had changed suddenly, like a phase shift in a beaker of fluid, and that change wasn't precisely a triumph of good over evil. The army had lost patience with Ceauşescu and thrown its fealty to the people; the most adroit of the Communist apparatchiks were now reinventing themselves quickly as populist opportunists. To fit the purposes of those opportunists, and to guarantee their safety against the possibility of a restoration, Ceauşescu had to go—not just from command, but all the way down. Then again, it could hardly have happened to a worthier fellow.

After five minutes of what passed for deliberation, the death sentence was read, and the couple were led outside. Ceauşescu, possibly still assuming he'd be helicoptered back to Bucharest, began humming the "Internationale." Instead of a helicopter there was just a blank wall and four soldiers with rifles. "Stop it, Nicu," Elena snapped. "Look, they're going to shoot us like dogs." A moment later she spoke what may have been her last words: "I can't believe this. Is the death penalty still in force in Romania?" For her and him it was. The backstop wall later showed scars from more than a hundred bullets.

During his heyday Ceauşescu had styled himself the Conducǎtor, a highly resonant Romanian word suggesting supreme leader, boss, master, which linked him with an earlier *conducǎtor* (the military strongman Marshal Ion Antonescu, Hitler's Romanian accomplice in the Holocaust) who had led the country during World War II. After 1989 Ceauşescu was snidely remembered by some citizens under a different nickname: Împuşcatul, meaning "the Shot One."

But the heyday lasted many years, as reflected in this building, with all its skinned and stuffed witnesses vouching for the Shot One as Romania's preeminent and most privileged shooter. Passing beyond the boar, the museum guide leads onward, and I follow her into an innermost chamber appointed to serve as the resplendent culmination of a visitor's experience. She calls it the Carnivore Animal Room. There's a wildcat, mounted in a tree, and below it three wolves chorusing their voiceless howls. Elsewhere around the room, in poses intended to seem lifelike, are a dozen bears, mostly cubs and yearlings. Another dozen, each in the form of a pelt with head and claws attached, decorate the walls. And alone on the west wall is the centerpiece of the collection, the room's focal specimen—a vast bear skin spread flat, like the largest flying squirrel ever imagined. Its great head hangs downward, nose to the floor, eyes empty. Its front claws are almost three inches long. Its fur is umber, with blond highlights across the shoulder hump and along the forelegs, beautifully catching the afternoon light. The mount is backed with green felt and punctuated by four florets of green bunting, one at each hip and shoulder, as though the bear itself were adorned to march in a Saint Patrick's Day parade. Beside it is a laconic plate reporting only the CIC score, which is 640.46. This trophy—so the young woman claims, anyway—is the biggest bear fur in the world. The animal who wore it was shot by Nicolae Ceauşescu.

Whether she's correct about its surpassing size (is it *really* larger than any grizzly from our lower forty-eight, any Alaskan brown bear, any Kodiak from Kodiak Island, or any polar bear?) is an issue I have no heart to dispute with her. Without question it was a gigantic individual. Yes, she adds, 650 kilograms when killed. That was 1984, in the Mureş district, on the west slope of the Carpathians. It was recognized as a world-record trophy one year later, she says, when the CIC met in Leipzig. The size of the skin, the density of the hair, the coloration—all these considerations contributed to its high score.

Then the woman shifts from her vaunting tone to a confessional one, as abruptly as those underlings shifted in December of 1989. "This bear, I must tell you sincerely," she says, "was artificially feeded for Ceauşescu to kill." I haven't badgered her. Apparently it's just something she needed to say. Carrying that piece of information as a lead worth exploring, I make my way out past Saint George and the Komodo, into daylight.

40

The Carpathian Mountains form a large, L-shaped divide across central Romania, with the letter turned backward, its open side facing west toward Hungary and Serbia, its vertical stem arcing northward into Ukraine and Poland. The crest of the Carpathians within Romania itself serves as the boundary between the country's western province, Transylvania, and the provinces of Moldavia (to the east) and Wallachia (to the south and southeast, sharing a border with Bulgaria). The east-west stem, the base of the L, is sometimes called the Transylvanian Alps. Both stems of the range are steep, serious mountains that have helped delineate and enforce political divisions in the region for centuries. It seems improbable that a single country, Romania, could have coalesced with the Carpathian partition as its internal frame. Then again, much about Romania is improbable.

The province of Transylvania was traditionally part of the Hungarian ambit, dominated by ethnic Hungarians and ruled as part of the Hapsburg empire from the late seventeenth century until World War I. Moldavia and Wallachia endured several centuries of subservience to the Ottoman empire, centered in Constantinople (which was closer than Hapsburg Vienna), and then came under the dubious protection of czarist Russia during the mid-nineteenth century, when Ottoman power started to fade. Romanian was an ancient language and an ethnic identity, but in those years there was no Romanian state. Foreign armies sloshed back and forth across the landscape like waves on a sand spit, and would continue to do so, though in 1859 Moldavia and Wallachia found an opportune moment to ally with each other, under a jointly elected leader and the label United Principalities. In 1862 they took the name Romania. In 1877, seizing their chance during still another Russo-Turkish war, they declared independence. Transylvania was added in 1920, as a reward to Romania for having fought (though not very well, judging from the fact that both Moldavia and Wallachia stood occupied by German and Bulgarian troops when the Axis collapsed) in World War I. The treaty makers at Versailles chose to recognize friendly intentions, not results. This left Romania as one of the largest nations of southeastern Europe.

The lowlands were flat, fertile plains, draining gently south to the Danube or (in Moldavia) east to the Prut River. Those plains supported most of the country's human population and were largely settled, dissected, plowed, and otherwise committed to old-fashioned, almost medieval agriculture. The mountains were still covered with hardwood forests of oak and beech, giving way at higher elevations to spruce and fir and, above that, rocky spires. The hardwood forests, with their mast crops of acorn and beechnut, their mushrooms and berries, their abundance of roe deer and other ungulates representing potential prey, were excellent habitat for brown bear.

Records suggest that in 1940 the country harbored about a thousand bears. By 1950, after the disruptions of war and the austerities of the early postwar years, during which people were desperate and poaching went mostly unpoliced, the number had fallen to around 860. Given the circumstances, such population estimates shouldn't be taken as precise, but the general range and the slight downward trend are plausible. In coming years, both the size and the precision of the numbers would increase.

Although *Homo sapiens* and *Ursus arctos* had coexisted in the Carpathians for thousands of years, interacting enough to acquire some degree of mutual wariness, the tradition of hunting bears wasn't widespread or deeply grounded. Most hunters were peasant men, and most peasant men were more interested in game animals that delivered better meat for less risk, such as red deer, roe deer, and wild boar. Under a hunting law promulgated in 1891, the bear was considered a pest, subject to extermination without license or limit. The royal family did some bear hunting, and probably also the lesser aristocrats, no doubt in a spirit of noblesse oblige as well as rustic adventure, but to the people who made their living in the mountains—the poor side-hill farmers, the shepherds, the woodcutters—bears seem to have been a menacing nuisance more often than a desired form of game. During the 1920s, according to one authority, "the bear was considered a bad and very dangerous animal, and all the bears that were found were shot."

In 1927 some protective restrictions were introduced—against shooting bears without a license, against shooting sows with cubs, against killing bears in their dens. Implicit in these restrictions were two premises: first, that bear hunting was a form of sport, to be governed by orderly and ethical principles; second, that the bear population was a valuable and finite resource. Recognizing that value, that finitude, was not the same as requiring that bears be managed sustain-

ably, but it was a step. What proved helpful for Romania's bear population was not so much the lofty ideals of sustainable management as
the realities of Communist autocracy. After the war, things were different in the mountains. Common people had no guns. Common people were afraid of the central government, its regulations, and its
means of enforcement. Bear hunting became a prestigious privilege
reserved mainly to the nomenclatura, the Party elite.

The last Romanian king was forced to abdicate in 1947, after which
the country became a People's Republic. For a handful of years, while
consolidating its power, the Communist regime endured an internal
struggle between two factions, one of which had spent time in Moscow,
the other in Romanian prisons, until the prison-annealed group
(including Nicolae Ceauşescu, then a minor functionary) succeeded in
overpowering and purging the others. In 1952, a former railway-
workers organizer named Gheorghe Gheorghiu-Dej, who led the
prison faction (although he was more Stalinist in style and ideology
than most of the Moscow group), emerged as premier of the country as
well as secretary-general of the Party. By that time, according to
records from Romania's Forest Research and Management Institute
(Institutul de Cercetărişi Amenajări Silvice, or ICAS), the bear population had grown to 1,500. In 1953, *Ursus arctos* became a protected
species—which is to say, managed carefully for hunting.

Two years later the population reached 2,400, a brisk increase that
reflected the good habitat, the natural rebound of a depleted population, and the new regime of protective sanctions. The numbers continued rising—to 3,300 bears in 1960 and then 4,014 in 1965, the year
Gheorghiu-Dej died and was succeeded by Ceauşescu. The history
books don't say whether Gheorghiu-Dej, an urban agitator in the pro-
letarian vein, fancied bear hunting personally, although there is a
record of his hosting Nikita Khrushchev to a hard-drinking, bear-
killing junket up in the Harghita district. Nicolae Ceauşescu, similarly,
had shown no interest in woodland shooting sports during his earlier

years. But in the late 1960s, while Ceauşescu solidified his position as supreme leader of both the Party and the country, he did discover a zeal for hunting—or, more accurately, for the sort of pampered travesty of hunting that only a despot gets to experience and only a delusional egoist would enjoy.

To understand Ceauşescu's relationship with bears, it's necessary to consider a few facts about the bureaucratic arrangements that govern *Ursus arctos* in Romania. Within the country's forty-some administrative districts (*judeţi*, roughly equivalent to counties) are 2,226 game-management units, known as *fonduri de vânătoare*, or hunting areas. Many of those areas are in the Carpathians, and more than four hundred contain bears. The average size of one hunting area is less than ten thousand hectares (about thirty-nine square miles), small enough that a vigorous person could walk its perimeter in a day. To each area is assigned a professional gamekeeper, a hardy field man who comes to know its terrain and its animals closely. Some of the hunting areas, notably in the lowlands, support only pheasant, roe deer, waterfowl, and other small game. If the hunting area contains bear habitat, as the best of the mountainous areas do, then an important part of the gamekeeper's job is to nurture the resident bears and to familiarize himself with them as individuals. How does he nurture them? By putting out supplemental food—apples and pears and plums by the bushel, cobs of corn by the cartload, granular pellets of a specially blended bear chow, and occasionally the carcass of an old horse. Spring, summer, and fall, the gamekeeper and his helpers deliver vast quantities of such stuff for the delectation of bears, serving it at feeding stations sited strategically throughout the forest. A typical station includes a feeding trough— just like one you'd see in a barnyard, except with bear scat as well as old corncobs strewn about—and a tall iron frame for hanging large items of meat. How does the gamekeeper familiarize himself with local bears? By walking the forest, studying their sign, spending long days and nights observing them at the feeding stations.

Another feature of the typical station is an elevated blind, within eyeshot of the trough, from which the gamekeeper does his observing. In some cases, this is a simple platform of planks about ten feet off the ground, like a child's tree house; in others, it's an enclosed structure on sturdy pilings. If the blind is used as a shooting position as well as an observation post, it's known as a "high seat." A high seat may be spartan or comfortable. At the comfortable extreme, it's essentially a two-room cabin, furnished with cots, a woodstove, a window overlooking the target area about fifty yards away, a firewood bin, a toilet, and maybe a bottle or two of vodka. Under such circumstances a bear hunter is not put to great inconvenience or challenge, let alone danger. Although a recent change in regulations prohibits shooting bears from enclosed high seats, the prohibition reportedly isn't always enforced. In any case, the real work of understanding the quarry has already been done by the gamekeeper, and the necessity of stalking is obviated by baiting the bears to the trough.

Gamekeepers of the old school, diligent and devoted, keep their bear-watching data in notebook diaries. They give the bears names and record their activities, minute by minute, like a field biologist studying animal behavior. From such a diary the gamekeeper can report, months or years later, say, that a certain mature sow came to the feeding station at dusk on October 4 and then was driven away after an hour by a notoriously ill-tempered male, or that a younger female arrived at midnight the same evening with only one of her cubs—where was the other? A further responsibility of the gamekeeper is to estimate, in advance, the trophy quality of each bear in CIC points. That task is sometimes aided by hoisting part of a horse carcass on the iron frame so that a bear will stand erect and reach for the meat, showing its full size against calibrations on the frame. The gamekeeper's role, vis-à-vis his resident bears, combines the services and attitudes of a nanny, a zoo attendant, a field naturalist, a sniper's spotter, and a pimp. To call it an ambivalent relationship would be understatement.

On a mild summer evening I visit one of these gamekeepers at his home in the mountains, near the Transylvania-Wallachia border. His name is Ion Moşu and his position is *maistru de vânătoare*, master of hunting, for an area of some thirteen thousand hectares. He's a slender, youngish man with hazel eyes and a stubble beard, wearing a jaunty warm-up jacket, an olive uniform shirt beneath the jacket, and an olive fedora. Having just returned from hoeing potatoes, he invites me into his house but then suggests we'll have a pleasanter ambience out in the twilight. Settled outside on plastic chairs, we talk for two hours about bear management and bear behavior while his wife and his mother-in-law, wordlessly, bring us coffee and sliced cake.

Moşu's area at present contains twenty-five bears, including cubs of the year, he tells me. During hunting season he feeds them five kilos of special pellets per bear per day, and in summer (when wild foods are more available) somewhat less. Occasionally he also treats them to a horse carcass from an abattoir, or to a load of rotten apples. In autumn and spring he concentrates this bait at the three high seats within his area. There are several other feeding stations more remotely sited in the forest, at which no one is permitted to hunt. Those he uses in summer, placing food there to entice his bears away from mountain meadows where they might otherwise make trouble—and find it—among shepherds' flocks.

Moşu can recognize each of his bears individually from its fur color, its footprints, and its behavior. And the bears recognize him in return—by his smell. They're accustomed to him, Moşu tells me. They trust him. They remain calm in his presence but grow consternated when they catch the stink of a stranger. Sometimes he toys with a bear, sidling close, closer, seeing how near he can get before the animal demands a little more distance. Moşu himself is always listening, watching, perceiving smells. He's attuned to the dynamics of the forest, to the interplay of creature upon creature, to the balances and the disruptions, in a way that his Forest Department bosses are not. He's a field man—minimal education, lots of experience. He can sense the presence of a bear from the way birds respond. He knows how a bear reacts to the sight of a human. A wolf reacts differently. A fox or a lynx, differently. Ever since he was a small boy, Moşu says, as though this were an explanation, hunting has been his passion.

When I ask about his current crop of bears, he dashes for the house, returning proudly with a small notebook. It's his field diary of observations, page after page of dates, times, terse behavioral notes on animals that he knows by the pet names he has given them. Fricosu ("the Shy One") is a big fellow, worth about 430 CIC points as of April. And here's Furiosu ("the Angry One"), a beast with an attitude. Frumosu ("the Beautiful One") was shot by a hunter several months ago, earning the department a good fee. Yes, of course he was present for that hunt, Moşu says. The gamekeeper must always be present, in case the animal is wounded (he'll track it and kill it) or something else goes wrong.

Does it ever make you sad, I ask, to see one of your bears shot? Waiting for my translator to work, I realize that I don't really expect this question to engage Moşu. Sadness over one dead bear? Probably that's an alien and irrelevant vein of sentiment. But I'm wrong. "Sure, it makes me sad," he says. "I play with them. I *know* them." It's a job, never easy, not always agreeable, raising bears for other people to kill. But he's a professional.

As for himself, Moşu no longer cares to hunt. He'd rather watch animals, he has discovered, than shoot them.

41

The close attention of gamekeepers like Ion Moşu to their respective hunting areas, each with its few dozen bears, is what allows the Romanian Forest Department to offer precise (and maybe even accurate) annual counts of the total bear population. Biologists at ICAS, the forest research institute, do the arithmetic based on numbers that come in from the field men. A glance at the tables and maps prepared by ICAS will tell anyone which districts are richest in bears—Harghita, Bistriţa, Argeş, and Covasna, among others—and which hunting areas within a given district are particular hotspots. Of course, Nicolae Ceauşescu didn't need to consult tables and maps, not personally, because he had toadying bureaucrats whispering in his ears.

Beginning in the late 1960s, Ceauşescu made himself the hunter in chief of Romanian forests as well as the commander in chief of the military. He arrogated hundreds of hunting areas—the best of them, so far as large game was concerned—to his personal use. Forest managers at the district level, and the hunting wardens who worked for them, and the gamekeepers who reported to the wardens, came to realize that any estimable animal emerging within their purview was an animal that the Conducător might want to kill. They recognized that pandering to his blood lust, to his lazy greed for trophies, was good professional politics. One district competed against another for his visits, offering big bears and rack-heavy stags as easy targets for his pricey imported rifles. For a typical hunt, Ceauşescu would fly in by helicopter and land on a cleared pad within the hunting area itself. From there he'd be taken by rough-terrain vehicle (in earlier years he favored Jeeps, then a Russian make, the Gaz, and still later a rattletrap Romanian imitation, the Aro) along forest roads, to a point very near the spot where hungry bears or rutting red deer were expected to appear. He would walk the short distance to a strategically placed high seat—in a tight little draw that served as a game corridor, say, or along a stream, where the gurgling water would cover noises made by a hunter. Usually he was accompanied by at least one security officer, who would carry his weapons and ammunition, and a forestry official from the district office. Many other Forest Department personnel would have been involved in preparing for his visit, but they were kept at a distance during the actual hunt. In the high seat, he had little patience for waiting and watching. His attention span, according to a witness who worked with him often, was five minutes. But for this brand of hunting, patience wasn't necessary. Bears came to the feeding troughs; red deer stags congregated in response to hormonal imperatives and the attraction of hinds; or, in some cases, both bears and wild boars were pushed toward a high seat in organized drives involving dozens of beaters. Ceauşescu took his shots, admired his kills, posed for photographs, and then departed.

The report of his short attention span comes from Vasile Crişan, a forestry official who later published a memoir, in German, the title of which translates as *Ceauşescu: Hunter or Butcher?* The gist of the book is that his boss was the latter—a *schlächter*, not a *jäger*. For instance, Ceauşescu would continue firing at an animal until it collapsed or ran away. If he wounded a stag, he'd command Crişan and other attendants to find it and bring him the trophy. If he missed altogether, they would tell him the stag was wounded and that they'd find it; that stag or a similar one would then be killed and delivered. "Sometimes more stags were 'found' than were shot," Crişan testifies. "Once, after a hunt, a party secretary called him the next day and told him that all the six stags were found. 'The hell,' Ceauşescu said, 'how can you find six stags, if I only shot four?'"

Crişan reveals the tricky processing method that helped augment the size, and therefore the CIC scores, of Ceauşescu's bear skins. Once cleaned of meat and fat, a pelt was nailed onto a specially designed table with movable panels that could be cranked outward, like a medieval rack. Treated with certain oils to prevent tearing, the skin was thus stretched "to the limit of its resistance." Taxidermists who showed their adeptness at such stretching became Ceauşescu favorites. "It was an honor for them, but not for us, the forestry staff," Crişan writes. "Many pelts were totally deformed through this method. Every expert could tell that these were not the natural proportions of a bear, and this was awkward for us." Furthermore, the bear skins didn't need such abuse, he adds, "as they were already big enough."

During the twenty-five years of his reign, according to Crişan's tally, Nicolae Ceauşescu shot about four hundred bears. In the earlier years, he sometimes hosted shooting parties at which guests were welcome to kill game—deer, boar, even some of those precious bears. On a day's hunt in 1974, Ceauşescu himself shot twenty-two bears and his guests another eleven. In later years he more jealously kept the bears for himself. Between 1983 and his death in 1989, Crişan reports, Ceauşescu bagged 130. His most notable fit of excess occurred in the autumn of 1983 when, during a single day, aided by four separate game drives toward his position, Ceauşescu personally shot twenty-four bears.

That slaughter occurred in a hunting area called Cuşma, within the Bistriţa district, not far from a luxurious hunting lodge known as Dealul Negru (the Black Hill), which had been built expressly for Ceauşescu and his wife. Informed that the 1983 bear crop was bounteous at Cuşma, Ceauşescu announced his intention to visit. This triggered a scramble of kowtowing preparations. The high seats were repaired. The forest roads were improved. The bears were fed—generously, with two tons of fruit and two hundred kilograms of bear chow poured into the area each day for six weeks. The hunting lodge, Dealul Negru, was made spiffy. The local Party office recruited four hundred citizens to serve as beaters, and from among the local police and the Securitate came a hundred more. Ceauşescu arrived by helicopter on the morning of the hunt, October 15. The plan was to split the beaters into three groups, for three separate drives, and then marshal them all into a giant sweep of the forest for a climactic fourth. Crişan describes how the day unfolded, with Ceauşescu blasting at bears, killing bears, wounding bears as they fled toward his position in one high seat and then another. After the first drive, in which he killed three medium-sized animals and injured two but missed two others that ran back into the forest, Ceauşescu complained petulantly about the arrangements. God forbid that two bears out of seven should escape—or, if God wouldn't forbid it, the Conducător would. Next year, he commanded, there should be a fence along here, dammit, to channel the animals inexorably toward the high seat. Yes yes, the district director promised, next year there would be a fence.

After the second and third drives, having killed seven more bears, Ceauşescu was still unsatisfied. The fourth drive began, the big one, with hundreds of beaters moving down brushy hillsides toward a valley. The security men carried semi-automatic rifles; the foresters had small-gauge shotguns; they all shouted, fired into the air, setting up a din. Vasile Crişan took refuge on a high seat, from where he could watch without too much danger of being mistaken by Ceauşescu for a bear. As the beaters pushed within a couple of hundred yards of the firing line, they came virtually shoulder to shoulder. "The bears were running in every direction, trying to escape," Crişan writes. "But it was useless, it was impossible." Bears fell dead, bears were wounded, and amid the chaos Crişan couldn't tell just how many; but few if any seemed to be escaping. Ceauşescu blazed away with a pair of Holland & Holland .375s, a minion beside him reloading one rifle while he fired the other. When the shooting and the hollering stopped, the forest workers started dragging in carcasses. Twenty-four dead bears were lugged back to the hunting lodge (where Elena could admire them) and laid out in two rows, framed with freshly cut brush, like trout on a platter garnished with parsley. Ceauşescu posed for photos. "We, the foresters, gathered at a certain distance," Crişan recalls, adding the tight-lipped understatement "Contrasting feelings governed us." He had devoted much of his life to hunting, but he labels this sorry episode the Massacre of Bistriţa.

Vasile Crişan was just one of many such facilitators. Forest Department field men in various parts of Romania still tell tales, some proudly, some loathingly, of the hunts they helped arrange for Ceauşescu. Down in the Argeş district there's Viciu Buceloiu, a *tehnician de vânătoare*, supervisory warden of three hunting areas, who remembers Ceauşescu not just for his imperious temper but also for the grandfatherly sweetness he often showed to Buceloiu's little daughter, Petra. In Valea Bogaţii, a valley not far from Braşov, there's a gamekeeper named Gheorghe Bumbu, a thin man who wears an olive Forest Department uniform, a Tyrolean-style felt hat from which his ears protrude widely, and a sad, sheepish half-smile like Stan Laurel's. Bumbu assisted a hunt there in the spring of 1989, during which Ceauşescu took a bear worth 616 CIC points, one of the biggest—and among the last—that the Shot One ever killed. Near the headwaters of the Mureş River, working out of a local office in a village called Izvorul Mureşului ("Source of the Mureş"), is a short, red-faced man, Laszlo Kedves, whose head tops his wide shoulders and squared-off body like the cap on a pint of whisky. Kedves, another supervisory warden, is laconic at first acquaintance but gradually, over coffee, can be enticed to share some of his own rec-ollections of the many occasions when Ceauşescu came hunting in Kedves's areas. There was the visit in April 1989, for instance, during which Ceauşescu bounced from one high seat to another by helicopter in order to kill ten bears in a day; and the time Ceauşescu hosted Muammar Khadafy, and the two of them shot every animal within sight. Up in Bistriţa, that famously bear-endowed district, is a former warden named Tudor Tofan, a husky and plainspoken fellow who rose through the ranks to a point where, at retirement, his duties cov-ered three hunting areas, including Cuşma, as well as the Dealul Negru lodge. Back in the autumn of 1983, the time of Ceauşescu's twenty-four-bear slaughter, Tofan had been a mere gamekeeper. Probably he knew at least some of those dead animals by names he had given them himself.

Tudor Tofan consents to meet me for a chat. In a borrowed office at the district headquarters, he rests his thick, brown forearms on a bureaucrat's desk and responds to my questions with answers so terse and direct that they seem like karate chops. He resembles the actor Ed Asner—or rather, what Asner might look like if he'd spent much of his life outdoors, feeding rotten apples to bears. Yes, Tofan says, Cuşma and two other hunting areas were his responsibility. Together they encompassed forty thousand hectares of premium bear habitat, attended by four gamekeepers. Within the three areas, in any given year, lived about forty bears. Of those forty, as things are now managed, maybe twelve to fifteen would be hunted (by which he means killed) each year. The "harvest" quota for each hunting area is flexible, to be adjusted annually by upper Forest Department officials based on recommendations from the warden's level.

But wait—*twelve to fifteen*, from a resident pool of just forty? To me that sounds far too high. It sounds unsustainable. Offering numbers and memories, not justifications, Tudor says yes, and it was higher still in the time of Ceauşescu. More like twenty-five bears, he says, all of which were killed by Ceauşescu himself, nobody else. Next question?

I'm not sure what has compelled Tudor Tofan's receptiveness to this interview—a lingering sense of duty? the summons of a district chief who may somehow still control his pension?—but certainly it wasn't the sheer joy of reminiscing to a snoopy foreign writer. Nor did he come to bellyache or emote. We discuss the stages of his career. We discuss the schedule and amounts of supplementary feeding within his three hunting areas. We discuss the contents of those special pellets— oats, corn, bone meal? He doesn't seem to know or care. He can't say where they are made. We touch upon the problem of bears in orchards during autumn when the fruit is ripe, and on the conflicts between bears and shepherds. He mentions several cases of bear attack on humans—a wounded bear, in each case, lashing back at a hunter or a gamekeeper. He describes the arrangements under which foreign hunters pay big money to the Forest Department for the privilege of shooting a bear. Usually it's a cash transaction, he says, the money being split between ministry and departmental coffers in Bucharest and the district office. What reaches the district goes back into food and other expenses for supporting the bear population. A large, handsome bear might carry a price tag of $15,000 (usually paid in deutschemarks or, more recently, euros), not inclusive of what the client pays to a foreign hunting-tour company for packaging the trip. Bear hunting amounts to big business in an economy as decrepit as Romania's. *Ursus arctos* is an export product, yielding significant sums of hard currency. Although I'd like to know, for comparison, what Tudor was earning by the time he retired—sixty dollars a month? a hundred? two hundred?—I'm not cruel enough to ask.

I do ask whether he's a hunter. "*Da*," he replies. Mostly for wild boar, red deer, roe deer. And bears? No, not bears, Tudor says. Bears, they're too expensive. That's a rich man's game. And the department favors foreigners for the permits, he says, because law requires them to pay higher fees. Germans, Austrians, Spaniards, Mexicans, Belgians, Frenchmen—they're the ones who kill Romanian bears. All foreigners, nowadays. Back before 1990, it was only Ceauşescu himself hunting bears. No one else could, no one else dared. Tofan Tudor remembers it.

He remembers October 15, 1983. The big boss, the little man, came in his helicopter to Dealul Negru. We had prepared for him, Tudor says, by closing off other hunting areas and putting out extra food. Us, the gamekeepers, we organized those drives. Ceauşescu stood in his high seat with two attendants, one of them always reloading the second gun. He took what was handed to him, and he fired. That day he killed more than twenty bears. Was he a good shot? I ask. "*Forte bun*," very good, Tudor concedes. Anyway, good enough for the circumstances. But there's more to a real sportsman, Tudor implies, than good marksmanship. After swallowing a gulp of anger, and in response to no particular question, this stolid, dedicated ex-professional volunteers, "He hunt too much bears. Too much, for a single hunter, to shot so much bears. He was not allow somebody else to shot bears—only him."

And the skins? They were sent to Bucharest for processing. Most of them, Tudor guesses, are probably now at that museum in Posada.

42

"The cult of the dictator," says Ion Micu, a well-known Romanian bear biologist and the author of a book titled *Ursul Brun*, "is not only one personality, but the whole institution." Dr. Micu knows something about dictatorial prerogatives and compromised institutions, having been part of the Romanian Forest Department throughout Ceauşescu's rule.

The corruptions result not just from a potentate's personal character, Micu explains, but from the force field of incentives and fears surrounding him, within which people will go to lengths—sometimes ugly or foolish lengths—in hopes of pleasing the leader. That effect was well evidenced by a misguided gambit in bear management down at Râuşor, in the Argeş district, south of the Carpathians. It was a captive-rearing facility—or, in plainer language, a concentration camp for cubs. The idea arose not from Ceauşescu himself, Micu says, but from Argeş politicians who hoped to butter him up. They imagined they might accomplish that by kidnapping bear cubs from elsewhere in Romania and raising them at Râuşor for release into the Argeş forests, creating an inordinate abundance that would entice Ceauşescu to come there and hunt. They wanted to curry his special favor. Of course, every district wanted that. This man controlled the siting of factories, of electrical plants, any number of national projects, even the flow of food. He already possessed all the money he needed, and he was too old to be tempted by women, Micu says, so these politicians thought they might pander to his lust for killing bears.

Micu speaks with the calibrated detachment of a semi-independent scientific observer. He has studied Romanian bears for decades, though not always with as much freedom as he'd wish. During the regime, he directed the Forest Department's district office for

Harghita, in the northern Carpathians, and filling that bureaucratic role allowed him little time for research. He also spent two years down in Bucharest as head of the hunting division. Now he's glad to be back in the mountains, close to the bears, serving merely as Harghita's chief of wildlife. Much of his work has been biometrical—that is, he specializes in the sedulous measuring of bears, both live ones (insofar as possible) and dead ones killed by hunters. Ceauşescu's visits helped provide him with data. In fact, Micu notes with a complacent smile, he himself managed to persuade Ceauşescu to shoot *small* bears when he hunted in Harghita, not just the record-book animals that Ceauşescu targeted everywhere. It was important for scientific research, Micu argued (at several removes, through party bosses in the Conducător's entourage), to have little dead bears available for measuring, not just big ones. Diversity of specimens would allow better profiling of the population. Ceauşescu took the point and was glad to oblige.

Micu is a robust, dark-haired man in his late fifties, with sparkling brown eyes and a goatee going from grizzled to white. He wears a nicely pressed uniform of Forest Department olive and a watchband of camo green. His second research interest, besides biometrics, is ethology (the study of animal behavior), as reflected in his bear book's subtitle, *Aspecte Eco-Etologice*. Most of his behavioral observations come from high seats that overlook feeding stations, but he has also drawn on data gathered at the Râuşor rearing compound and on the published literature—or at least on those select bits of it that come to hand in Romania. He knows Marcel Couturier's book, *L'Ours Brun*, published in 1954, and Frank Craighead's 1979 *Track of the Grizzly*. Micu's office, on the second floor of a Forest Department headquarters in the city of Miercurea Ciuc, mixes trophy-den motifs with research resources. The shelves are filled with books and journals dealing with big game generally and bears in particular, and some videos, including one titled *Bear Attacks*. There's a map of the district, showing the high-mountain topography that provides good summer habitat and winter den sites, and a caption in English: *Harghita—the Country of Bears.* Probably that's a misprint for "county," though Harghita is indeed bear country. Micu's walls are decorated with skulls of red deer and bear, boar tusks on plaques, and two framed portraits above the desk: the crucified Christ and the Austrian ethologist Konrad Lorenz. On a shelf in the corner sits a clear tank, like the largest pickle jar in an Italian delicatessen, holding a newborn bear cub preserved in alcohol. The cub, pale as a baby, mouth agape, seems to paw the glass as though drowning.

I stare at the cub while Micu's attention is elsewhere. If you were writing a novel about a Romanian bear biologist, you'd be embarrassed to invent such a thing. Too much symbolic freight. But in its physical reality, it's just a pickled bear.

The cubs sequestered at Râuşor would have been larger, too large for a deli jar. They were six or seven months old at the time of capture, which was typically accomplished by tracking and treeing them with dogs. The sow, their mother, had to be driven off or killed before the cubs could be wrestled down alive. The cub collectors—enterprising peasants or hunters—earned a bounty. A man named Mitica Georgescu ran the rearing compound. During its seven years of operation, from 1974 to 1981, it took in 227 such kidnapped cubs. Of that total, three died in captivity, eight were given to circuses, and 216, after having been fed up into hefty yearlings, were released in Argeş forests that were already quite full of native bears. The cubs' diet at Râuşor consisted of bread, corn porridge, potatoes, carrots, and fruit. During an average year the compound held a cohort of thirty-two animals, all confined within two high-fenced pens, each pen barely larger than a tennis court. The pens contained flowing water and some sort of artificial dens. As part of their processing, the cubs were marked, so that they'd be forever identifiable as Râuşor bears. The marking was done by cutting off toes.

There was a numeration system that, as Micu describes it, sounds not very different from the one used by crocodile researchers when they mark captured animals for release. With crocodiles the method entails slicing off dorsal scutes from four different areas along the lower back and tail. Since the dorsal scutes don't grow back, removing one from each area codes the animal with a four-digit number, adequate to distinguish it permanently from hundreds of other crocodiles. The scutes are little served by blood vessels or nerves, and the marking process (I've assisted in it) doesn't seem to cause much discomfort to the crocodiles. Bear toes don't grow back either, but they're definitely sensitive to pain. The system at Râuşor involved amputating a toe from one of the paws. Because bears are plantigrade (meaning that, unlike cats or dogs, they place five toes from each paw on the ground, not just four), all twenty of their toes normally show in a set of prints. The first twenty Râuşor cubs could therefore be enumerated by removing a single toe from one of the paws—the pinkie from the left front paw for cub number one, say; the fourth toe on the same paw for cub number two; for cub number three the middle toe, and so forth, until each of those twenty animals had been reduced to its own nine-toed (among the front paws or among the rear) pattern. After that, two toes were removed in various combinations, from dozens more victims, until the system reached its mathematical limit and had to be started over. The cubs, given no anesthetic, were left to howl away their hurt and outrage. That took about two days, according to Micu. Once the toe stumps had healed, and the grown cubs were released, every set of footprints represented a coded signature that would distinguish a given bear from several hundred others, and would announce to any hunter (that is, to Ceauşescu, in those years the only hunter who mattered) that this sporting target had been supplied through the generous diligence of the managers of Râuşor and their sponsors within the Argeş party and government.

But the bear farm was a failure. The cubs, separated from their mothers prematurely and raised in the squalor of crowded pens, didn't acquire the skills and the habits of wild ursines. Bear ecology involves a crucial component of learned behavior, and that learning comes primarily from maternal example during the first two years—a realm of experience that these cubs had largely missed. At the end of their year's incarceration, taken to release sites by helicopter, they acted scared and confused. Some immediately climbed trees, replaying their last instinctive defense before they'd been captured. Others bumbled around, looking for something recognizable to eat. Over time, they tended to pal up in small clusters, consoled by the company of familiar penmates. During their first winter back in the wild, they generally didn't hibernate, never having learned the advantages or methods of that. Not illogically, they associated human presence with food, and so even the feeding stations placed deep in the forest—filled with apples and bear chow, and meant to draw bears away from orchards and sheep—didn't keep these animals from straying into trouble. Some of the Râuşor cubs showed up in railway stations, expecting who knows what. Others attacked livestock, or raided orchards and crops, and got themselves killed. A group of fifteen became scavengers at a military camp in the mountains, feeding on garbage. When the soldiers broke camp and rode away on their trucks, the bears chased them forlornly for hundreds of yards. One bear from Râuşor was a familiar character along the Transfăgăraş Highway—a steep road that winds to a pass across the Transylvanian Alps—where it cadged food from passing cars and acquired the nickname Milogul ("the Beggar"). This animal did well for a time, growing to almost four hundred pounds before being hit by a truck.

Among the bears' varied fates, a single category of misadventure is notably missing. Not one mangle-toed Râuşor cub (according to what I've heard from another source, independent of Micu) ever grew into an adult bear eventually shot by Ceauşescu. They all either leaked away into oblivion or, as luck and wit blessed a few of them, found ways to survive.

In 1981 the program ceased. The last cubs were released. The empty compound was converted into a breeding facility, in which two imported Kodiak bears were encouraged to contribute their genes (for large size? for ferocity? for ever-greater CIC scores?) to the Carpathian brown bear population. But the Kodiaks found no atmosphere of romance in Romanian cages and were consigned to a zoo when Râuşor closed for good.

Why was the cub-rearing operation shut down? Why did those Argeş politicians, and their busy Mr. Georgescu, abandon the dream of beguiling Ceauşescu with hothouse bears? "Because they realized that it was stupid," Ion Micu tells me. "They realized that, from the ecological point of view, it was stupid to make something like this." Stupid and triply stupid, because it was wasteful, unnecessary, and damaging to the native bears of the region. Any well-informed fisheries biologist could have warned them of that. Planting hatchery-raised animals into habitat already occupied by a wild population is nowadays recognized to be futile at best, and more likely counterproductive. But politics was still politics in the Romania of 1981, and Ceauşescu was still in charge. Micu doesn't explain how "the ecological point of view" ever happened to enter into it at all.

43

The growth curve of the Romanian bear population over the past sixty years, as charted by ICAS, shows a peculiar correlation between ursine abundance and autocratic oppression: positive. Bad governance, of a very particular sort, has been good for *Ursus arctos* in the Carpathians.

At the start of the Communist era, in 1945, following decades of parliamentary turmoil and disorderly constitutional monarchy, there

were fewer than a thousand bears. By the time Gheorghiu-Dej established his chieftaincy of the regime, in 1952, the population had
increased by fifty percent, thanks to good habitat and a postwar decline
in subsistence-driven poaching. In 1965, when Gheorghiu-Dej's death
opened the way for Nicolae Ceauşescu, as I've mentioned, the bear
count was just above four thousand. Over the next decade, the early
phase of Ceauşescu's bear-killing mania, during which he shared his
hunts with select foreign dignitaries and a limited circle of Romanian
courtiers, the population dipped slightly and then rose back. By 1978,
it stood at 5,204. The Forest Department was now essentially farming bears, through the program of supplemental feeding, which
probably helped increase both the growth rate of individual animals
and the survival rate of cubs. Meanwhile, the department's conscientious forestry management (devoted to timber extraction, but on a
long-rotation schedule) had the effect of preserving large expanses of
excellent habitat. It was a miracle of good fortune that, as Ceauşescu
drove the country toward clumsy industrialization, then squeezed its
economy flat to pay foreign debts, he never launched a program of liq-
uidating Romania's forests for exportable timber. So the bear population flourished, and the Forest Department reserved it for Ceauşescu's
personal amusement.

By 1984, despite the Massacre of Bistriţa a year earlier, the tally was
6,713. Although this figure and the others are misleadingly exact (the
real population may have been lower or higher by hundreds), unques-
tionably there were many, many bears. The precise numbers are suspect, but the directional trend can be trusted, since the methods of
counting didn't change much from year to year.

The trend continued upward. In 1988, the last full year of Ceauşescu's
dictatorship, the Forest Department calculated a bear population of
7,780. That's amazingly high for the area of habitat in question, which
totals somewhere in the range of 3.1 to 7.7 million hectares, depending
on which Romanian source you accept. Back in America, the Greater

Yellowstone Ecosystem encompasses roughly the same total area (5.6 million hectares) but supports only about five hundred grizzly bears. In other words, at the peak of the curve, the density of the species *Ursus arctos* in the Romanian Carpathians was *fifteen times* the present density in Yellowstone National Park and its contiguous forests, which constitute one of the world's most celebrated bear sanctuaries. This discrepancy says as much about political arrangements, and about thresholds of social tolerance, as it does about the sustenance value of acorns, bear chow, horsemeat, and deliquescent fruit.

But what exactly does it say? That autocracy is beneficial to bears? That sport hunting, confined to a small circle of self-pampering politicos, tends to be far easier on predator populations than more democratic systems of resource allocation? That dictatorship yields better conservation results than private land ownership, open hunting, and cultural priorities that value livestock above native carnivores? These questions point us back toward the Nawab of Junagadh and his last, precious Asiatic lions.

When the Nawab (with goading from Lord Curzon, who as Viceroy of British India was a neighboring potentate) issued his ban on lion shooting within the Gir forest at the start of the twentieth century, he accomplished something that couldn't and wouldn't have been achieved—not at that point in history—by democratic methods. He saved the subspecies from extinction. By protecting the lions, he also kept open the chance of conserving their habitat, which remained less hospitable to invasion by farmers and settlers so long as it harbored those dangerous cats. Only the Maldharis, whose economy walked on buffalo hooves, whose lives were grounded in the forest as they found it, and whose expectations of the world did not include immunity from danger, were really comfortable within the Gir landscape. As decades passed, as the successive nawabs and the British disappeared, the Maldharis remained; they became on-the-ground guarantors of the protection originally declared from above. To live as their culture and their inclinations decreed, they needed that little island of forest as much as the lions did. They had no political power to defend it, but simply by occupying the habitat they helped give it tangibility and status. Conservation, having begun as autocratic fiat, continued as an issue of tribal integrity—until, still later, came another fallout of protective decrees from above, one of which evicted the Maldharis themselves from the innermost zone of the forest.

In northern Australia, on the other hand, it was a colonizing invasion that overrode tribal autonomy. James Cook made his first Australian landfall in 1770, Matthew Flinders explored the north coast in 1803, J. L. Stokes and others followed, after which an alien and centralized authority took hold, represented by British administrators in Darwin, Melbourne, eventually Canberra. But there was never a single autocrat comparable to the Nawab, the Viceroy, or the Conducător. Arguably the difference is moot, just a decorative matter of Anglo-Saxon pieties, since colonial Australian policies and practices toward the various Aboriginal peoples (not unlike colonial American policies and practices toward the Mohegans, the Pequots, the Seminoles) were often brutal, hypocritical, opportunistic, and even genocidal in the fullest sense of the word. If in Australia those severities didn't happen to issue from a single autocrat, it's hard to imagine how one perverse, autocratic governor could have made them much worse. Obscure administrators of the colonial authority sent out decrees—telling the Yolngu, for instance, that they could no longer hunt crocodiles in their traditional way, during the years when *Crocodylus porosus* was fully protected, and that they were obliged to accept aluminum mining on their sacred lands. Such decrees superseded ancient tribal protocols and played a role in disrupting the equilibrium between people and landscape. But that wasn't the whole problem, or even the worst part of it.

The other disruptive vector of colonial occupation was the wave of venturesome white individuals—miners, hide hunters, cattle-station pioneers, wool growers, settlers—who arrived in the outlying areas and made their force felt even more potently than the distantly issued decrees. One result among many was that the saltwater crocodile, having enjoyed millions of years of adaptive success, having achieved forty thousand years of ambivalent coexistence with Aboriginal people, suddenly went to the brink of extinction, just a couple of centuries after Cook's landing and a couple of decades after the species was first attacked with motorboats, spotlights, and good rifles. Nearly exterminated by white hunters (helped, in some cases, by paid Aboriginal collaborators), the crocodile received last-ditch statutory protection that put it off-limits, at least for a stretch of years, to black hunters. For the Yolngu and those other groups, it was a loss that went far beyond diet.

I have a small theory about all this, applicable to the lions (and the tigers) of British India as well as to the crocodiles of colonial Australia, though admittedly inapplicable to the brown bears in Romania. My theory—really only a notion—is that the extermination of alpha predators is fundamental to the colonial enterprise, wherever that enterprise occurs. It's a crucial part of the process whereby an invading people, with their alien forms of weaponry and organized power, their estrangement from both the homeland they've left and the place where they've fetched up, their detachment and ignorance and fear and (in compensation for those sources of anxiety) their sense of cultural superiority, seize hold of an already occupied landscape and presume to make it their own. Killing lions from horseback or in shikari-assisted hunts, shooting crocodiles for the skin trade from a dinghy, is more than sport and more than commerce. It's more than bloody-minded adventure. It's one aspect of a campaign by which the interlopers, the stealers of landscape, try to make themselves comfortable, safe, and supreme in unfamiliar surroundings.

Bring this notion back to America, and it casts some light on the murderous loathing that many ranchers (of European extraction) in Montana, Wyoming, and Idaho continue to harbor for the grizzly bear. At some subliminal level, the grizzly is perceived as a guerrilla warrior, fighting the final noisome skirmishes in a war of territorial seizure that began with Lewis and Clark, continued with the great cattle drives up the Bozeman Trail, and reached its provisional culmination with the surrender of Chief Joseph and his harried remnant of Nez Perce in the Bearpaw Mountains. That war won't be over, not quite, until the last individuals of the animal once known as *Ursus arctos horribilis* have been eradicated from the northern Rockies and the forests (on public land as well as private) are safe for white people and their cows.

Achieving military victory over the indigenous tribes, whoever they are, is sometimes the easiest part of the whole process. The land itself, the ecosystem, must be defeated too—or so the invaders think. The foreign wilderness must be mastered, made tractable, if not utterly subdued and transformed. That requires, at the lower end of the size scale, coping with pestiferous local microbes and parasites, which sometimes present the fiercest resistance of all. Malaria certainly slowed the white conquest of Africa. At the upper end of the scale, it means rooting out those big flesh-eating beasts that rule the woods and the rivers and the swamps, that offer mortal peril to the unwary, and that hold pivotal significance within the belief systems of the natives. Kill off the sacred bear. Kill off the ancestral crocodile. Kill off the myth-wrapped tiger. Kill off the lion. You haven't conquered a people, and their place, until you've exterminated their resident monsters.

44

But the role that Nicolae Ceaușescu claimed for himself, in such a personal and exclusive way, was different. It was nationalistic, not colonial. It was a caricature of the ancient notion that a strong leader must be also a great hunter, a great slayer of dangerous animals—like the

Assyrian king Ashurnazirpal II, who bragged of having taken fifteen lions, or the Egyptian pharaoh Amenophis III, who reportedly killed 102. Implicit in these boastful records, as in Ceauşescu's collection of bear skins at Posada, is the suggestion that a king serves the same function for his people that a shepherd (such as the biblical David, in his obscure years before Goliath) serves for his flock: ridding them of the menace of predators. What distinguishes this tradition from mere livestock guarding is that the slaughter is preemptive, not reactive. The good ruler, like the bold shepherd, devotes himself to exterminating predators whenever and wherever they can be found. It's a paradigm of valorous leadership that traces back through some of the earliest masterworks of literature and some of the most durably resonant myths. Killing monsters, on one pretext or another, is something that has always allowed heroes to seem heroic. Take *The Epic of Gilgamesh*, for instance. You find the battle between a king and a monster at the core of the story.

Gilgamesh is most famously known as a Babylonian poem composed in Akkadian, the imperial language of early Mesopotamia, presumably for oral recitation, and inscribed in cuneiform script on a series of clay tablets around 1100 B.C. There were antecedent versions, including some in Sumerian a thousand years earlier, and multiple copies made of the Babylonian text by apprentice scribes practicing their cuneiform hand. Many tablets have been recovered from archaeological digs since the first rediscovery of *Gilgamesh* in 1850, but not a complete set for the full poem in any of its earlier or later forms. What's called the "standard version" comes from seventy-three tablet manuscripts, including thirty-five that survived in the ruins of Ashurbanipal's library at Nineveh. The traditional title of that version was *Sha naqba īmuru*, or "He Who Saw the Deep," referring to Gilgamesh as a man who plumbed for life's fundamental meaning. Among the seventy-three tablets are gaps, missing sections, only some of which can be provisionally filled from earlier fragments in Sumerian or Old Babylonian.

So the scholarly study of *Gilgamesh* has necessarily been an exercise in patchwork and extrapolation, and to read the results as a living poem is alternately thrilling and frustrating, like an important conversation on a cell phone while driving the freeway.

Most of it comes through clearly, even after three thousand years. Gilgamesh, tall and magnificent, is king of the city-state of Uruk, on the lower Euphrates. (There was a historical Gilgamesh who ruled Uruk around 2800 B.C., but he seems to have been just a dim template for the bardic poets who cooked up the literary character.) His monstrous adversary is Humbaba, a gigantic ogre who guards the Forest of Cedar. In the poem's early lines, Gilgamesh himself is portrayed unflatteringly; he has trials to endure, lessons to learn, before showing the depth of purpose that will make him, by the end, a transcendent figure. Son of a mortal man and a goddess, he's a paragon of strength but not of virtue—an imperious, randy ruler who tyrannizes his own people by such habits as harassing the young men, molesting the young women, and asserting *droit de seigneur* over brides on their wedding night. So robustly barbaric is Gilgamesh, in fact, that the gods feel obliged to offset him with a contrary figure. They therefore create a wild man of the forest named Enkidu, hairy-bodied and feral and very unlordly, content to graze among gazelles as though he were one of them. Maybe this rowdy bumpkin, the gods hope, will somehow damper or deflect Gilgamesh's misbehavior. The two men meet, they fight, and after whomping each other nearly senseless, they become great friends—just a pair of compatible, knockabout guys. The plan of the gods backfires. Having found his ideal running buddy in Enkidu, Gilgamesh rises to a new level of reckless adventurism.

What he wants most are fame and glory, the only forms of immortality available to a half-mortal being in the earthly realm. Wouldn't it be glorious, wouldn't it be a hoot, Gilgamesh suggests, if we marched off to the Forest of Cedar and kicked old Humbaba's ass? Enkidu, who has spent time enough in the woods to be acquainted with Humbaba, thinks it's not such a brilliant idea:

I knew him, my friend, in the uplands,
when I roamed here and there with the herd.
For sixty leagues the forest is a wilderness,
who is there would venture inside it?

Humbaba, his voice is the Deluge,
his speech is fire, and his breath is death!
Why do you desire to do this thing?
An unwinnable battle is Humbaba's ambush!

That cautionary statement by Enkidu, known from a clay tablet in Old Babylonian, is spliced into what was otherwise a gap, in the well-edited new translation by a scholar named Andrew George. One of the earlier Sumerian fragments offers an even more lively version of the same warning:

My lord, you have not set eyes on that fellow, your heart
    is not stricken,
But I have set eyes on him, my heart is stricken:
a warrior he is, his teeth the teeth of a dragon,
his eye the eye of a lion!
His chest is a torrent in spate,
his brow devours the canebrake, none can go near him,
like a man-eating lion, on his tongue the blood never
    pales.
My lord, you go on up the mountain, but let me go
    home to the city.

Enkidu is no fool. Yet the dragon teeth and the leonine eye mean nothing to pumped-up Gilgamesh. What kind of weenie are you? he demands. "With your spineless words you make me despondent." But he doesn't actually despond; he's too keen for action. And his friend seems quickly, even abruptly, to catch the spirit; either some transitional moment has been lost in another textual lacuna, or being called cowardly by Gilgamesh is all the persuasion Enkidu needs. Off they

go, armed with freshly forged war axes and daggers, to beard
Humbaba in the Forest of Cedar.

One question that arises from reading the various Gilgamesh versions, in various translations, is whether Humbaba should be seen as a
gargantuan humanoid enemy or in some other bestial or chimerical
form. Andrew George calls him a "fearsome ogre" cloaked in "seven
numinous auras, radiant and deadly." Stephanie Dalley, another translator, describes him as a "monster" with a face like coiled intestines,
prefiguring the later Greek gorgons such as Medusa, who were snake-
haired and sharp-clawed composites. N. K. Sandars labels him a giant
but also explains: "Humbaba is 'Evil.' The first time he is referred to it
is simply this, 'Because of the evil that is in the land, we will go to the
forest and destroy the evil'; so Gilgamesh plays the part of the knight
who kills the dragon." And Humbaba, whatever his shape, fills the
role of that dragon.

A scholar of Mesopotamian art, Wilfred G. Lambert, has offered his
own cautious hypothesis based on a terra-cotta plaque, datable to the
Old Babylonian kingdom before 1600 B.C., that resides now in the
Berlin Museum. The plaque shows a monstrous creature with "a
strange face and long hair, feline paws and a bird's talons for feet." The
talon-footed creature is held pinned by two human figures, one of
them poised to drive a sword into its neck. Lambert suspects that this
tableau, seen also on cylindrical seals from a later period, depicts
Gilgamesh and Enkidu in the act of killing Humbaba. Whether or not
Lambert's guess is correct, it does suggest the range of Humbaba's possible anatomical forms. Clearly he's not just a gigantic human lout like
Goliath, formidable in the style of any murderous bully, because at one
point Humbaba boasts:

> So I shall bite through your windpipe and neck,
>   Gilgamesh,
> And leave your body for birds of the forest,
> Roaring lions, birds of prey and scavengers.

In the era before Dracula and Hannibal Lecter, that represents non-human menace. The lethal bite never happens, since Gilgamesh gets the better of Humbaba with a stab in the neck (as in Lambert's Old Babylonian plaque), after which Enkidu joins the fracas, ripping out Humbaba's lungs. In death, the monster also reveals some distinctly nonhuman dentition, when Gilgamesh cuts off his head and then takes from it "the tusks" as booty.

Rain begins to fall. The seven auras have dissipated, Humbaba is vanquished, the mountainside has turned quiet. All this occurs within a small number of lines that, following such a big buildup, feel anticlimactic. The most interesting thing about the fight scene, so terse, so one-sided, is what happens immediately afterward: Gilgamesh goes down "to trample the forest." He and Enkidu set to felling the sacred cedars, of which Humbaba had been god-appointed protector. "Gilgamesh was cutting down the trees; Enkidu kept tugging at the stumps." Come to that point, an ancient and complex poem seems just another story about humanity's compulsive eradication of predators and habitat.

Gilgamesh and Enkidu build a raft. They float triumphantly down-river, taking with them the head of Humbaba. They also bring a single great cedar log, stripped from the largest tree they could find, to be carved into a door for the temple of Enlil, "Lord Wind," the Earth-ruling god. Enkidu, the accomplice, the former wild man, will later regret his involvement with the tree. On his deathbed, as he succumbs to a fatal curse sent by Enlil, Enkidu whines that he should have given that "door of the woodland" to a different god, who might have received it more gratefully. And the real killer of Humbaba, Gilgamesh himself, escapes punishment but not his own eventual death. By the poem's end, the hero is gone, the monster is gone, the forest is gone; the city of Uruk remains. The march of civilization has begun.

45

The Gilgamesh-Humbaba contest is no anomaly. The list of legendary battles between heroes and monsters is long. There's Marduk versus Tiamat, Theseus versus the Minotaur, Perseus versus the sea monster, Ninurta versus Anzu, Sigurd versus Fafnir, Rama versus Ravana, Saint George versus the dragon, Odysseus versus Polyphemus, Nayenezgani versus Teelget, Bellerophon versus the Chimera, Oedipus versus the Sphinx, Minamoto-no-Yorimasa versus Nue, Tishpak versus the Labbu, Heracles versus the Nemean lion, and of course Beowulf versus Grendel, followed by Beowulf versus Grendel's mom and then finally against another dragon. Even these represent just a sample. Before considering the ecological realities to which such stories might be partly traceable, and the psychological verities that they might signal, it's worth reviewing a few details.

Teelget was a huge man-eating beast with the antlers of a deer, familiar in Navajo lore as one of the dangerous creatures killed by the hero Nayenezgani. Ravana was the ten-headed king of the Rakshasas, a race of grotesque demons, eventually conquered by the blue-skinned god Rama, as told in India's great Sanskrit epic, *The Ramayana*. Sigurd the dragon slayer dealt with Fafnir, "the most evil serpent," by hiding in a ditch until Fafnir slithered over him, then ramming his sword up into the creature's left shoulder (yes, this "serpent," also referred to as "the worm," did have shoulders) until it was buried to the hilt, which left Sigurd's own arms soaked with blood. Sigurd cut out the dragon's heart, roasted it on a spit, poked it with his finger to test the meat, and put the finger in his mouth. As soon as the sizzling blood touched his tongue, he graduated into a new consciousness, a new intimacy with nature that allowed him, like a gore-smeared Dr. Dolittle, to understand the language of birds. That's recounted in *The Saga of the*

Volsungs, a thirteenth-century Icelandic prose work based on earlier Norse poems. In the later German variant of the same story, the *Nibelungenlied* (which became the source for Richard Wagner's *Ring* cycle), Sigurd is known as Siegfried. Upon eating the dragon's heart, he becomes not just sapient but invulnerable, except (like Achilles, with his unprotected heel) in a fateful spot that the magical galvanizing didn't cover. One of Siegfried's enemies puts a dagger into the very spot, proving again that, in the legendary past, killing ferocious beasts was not enough to win a man immortality. It still isn't, as Nicolae Ceauşescu more recently learned.

Saint George was another venerated dragon slayer, in his case within the Christian vein of European folklore. He's linked tenuously to a real person who may have come out of Syria during the Dark Ages. By some accounts, the saintly George vanquished his dragon in Libya, where it lurked in a lake or a marsh, emerging furtively (as a real crocodile might have) to prey on sheep. The Minotaur was half bull and half human but, unlike your standard bull, anthropophagous. Held captive in its Cretan labyrinth, it ate seven maidens and seven boys in a blood-feast every nine years, until Theseus killed it. Nue was the monkey-headed monster—also tiger-footed, serpent-tailed, and badger-bodied, a raucous mishmash—that Minamoto slew because its infernal racket was disturbing the Japanese emperor. The lion of Nemea was, thanks to its impenetrable hide, "a beast no weapons could wound." Heracles solved that problem as the first of his famous twelve labors (set to him as penance for having murdered his family during in a fit of insanity). He *strangled* the lion no weapons could wound. Earlier, he had already killed a different lion in the woods of Cithaeron. Staying very busy with monsters throughout his career, Heracles would also exterminate the nine-headed Hydra, steal cattle from the three-bodied Geryon, kidnap the three-headed dog Cerberus from Hades without getting bitten or pooped on, and drive off the man-eating mares that belonged to King Diomedes. It seems a shame to imagine him, after such feats, succumbing to a toxic shirt.

Anzu, as known from Babylonian poetry, was a furious lion-headed eagle. Polyphemus, son of the sea god Poseidon, was the cyclops who ate several of Odysseus' men, scarfing them like shucked crawfish, before sneaky Odysseus paid him back with that archetypal affliction, a poke in the eye with a sharp stick. The Chimera was a fire-snorting goat-lion-snake. The Sphinx was a sadistic woman-faced lion, who devoured people after teasing them with her stupid riddle. The Labbu, another formidable Babylonian monster, was 630 miles long, with huge eyelids. Its high-protein diet included fish, wild asses, birds, and people, until Tishpak or some other heroic intervener (the sources are patchy) vanquished it. The original meaning of the word *labbu*, by the way, was "lion."

All these stories took shape within what was, at the time, alpha-predator habitat. They arose not purely from human imagination but also from the materials and concerns at hand—from life on the landscape. The Babylonians, in the fertile river valley they shared with a healthy population of lions, worked the hero-against-monster motif again and again. Marduk's battle with Tiamat appears in a creation epic known as *Enuma Elish*, sometimes called "the Babylonian Genesis." Tiamat herself, "the glistening one," was not a mere monster but a queen of monsters, who spawned eleven subsidiary "monster-serpents," each one sharp of tooth and merciless of fang:

> With poison instead of blood she filled their bodies.
> Fierce monster-vipers she clothed with terror,
> With splendour she decked them, she made them of
>   lofty stature.
> Whoever beheld them, terror overcame him,
> Their bodies reared up and none could withstand their
>   attack.

The conquest of Tiamat and her minions is placed in a time before history, preliminary to the formation of the world as we know it, and scholars say she represents watery chaos. Marduk, the "wisest of the gods" and eventually the patron deity of the city of Babylon, draws the unenviable chore of killing her, which he accomplishes with the aid of a net, an evil wind, and a spear, bursting her belly and piercing her heart. Then he splits her carcass out "like a flat fish into two halves," stretching one half overhead to shelter the sky, leaving the other half below to constitute Earth. Having subdued the chaos and sorted wet from dry, Marduk creates mankind, and nothing afterward is ever the same.

46

Vivid as they may be, none of these heroic tales is more mysterious, complicated, and compelling than *Beowulf*, the Old English poem about a brave Scandinavian thane who conquers a pair of monsters and then dies fighting a dragon.

Why does *Beowulf* hit harder? I can think of three reasons. First, because we have the complete work, not just fragments, preserved as a single manuscript from roughly A.D. 1000 and now lodged in the British Museum. Second, because (at least for Western readers) the world of early-medieval Germanic tribes, with their mead halls and swords and political marriages and drunken banquets, seems just that much closer than ancient Babylon, ancient India, or even pre-Homeric Greece. Third, because the hero is so poignantly human that he grows old, even wise, but not wise enough to admit the waning of his own powers or resist the temptation of one final contest. After traveling overseas as a young man on a public-service adventure—to kill

Grendel, and then also Grendel's mother as an unavoidable sequel—
Beowulf returns to his homeland, rules it for fifty years, and finally
takes up his sword again to fight the local dragon. This trajectory has
been described by one scholar, Michael Alexander, in his introduction
to a translation published in 1973, with a sports metaphor: "Beowulf's
life, though full of away wins, ends in a home defeat." Nowadays we
could go a step further, viewing Beowulf as the Michael Jordan of
monster-extermination quests: He retires with unsurpassed glory but,
alas, just can't tolerate staying retired.

Grendel is the most engaging of epic monsters (as the late novelist
John Gardner recognized, re-imagining him as the protagonist in
Gardner's own little novel, *Grendel*). He appears first as "a powerful
demon, a prowler through the dark," in the boggy terrain surrounding
a great hall that belongs to Hrothgar, king of the Danes. By lineage
Grendel is among "the banished monsters, / Cain's clan, whom the
Creator had outlawed / and condemned as outcasts." A sad, grumpy
character, annoyed by the music and rowdy laughter blaring out from
Hrothgar's banquets, Grendel turns into a man-eating marauder, raiding the hall repeatedly (with the preternatural stealthiness of one of
Jim Corbett's leopards) and preying on Hrothgar's people. The Danish
warriors are vulnerable because, after each evening's festivities, they
sleep drunk on the benches of the big hall. Grendel skulks in after
lights-out, kills thirty or so men, and drags their bodies away (in a big
pouch made of patched-together dragon skins) to his lair, where he can
devour them at leisure. Time after time, the liquored-up Danes finish
their evening by swearing to stay vigilant, swords ready, and to deal
with this gruesome visitor. But every time, as described in a vibrant
new *Beowulf* translation by Seamus Heaney,

> . . . when dawn broke and day crept in
> over each empty, blood-spattered bench,
> the floor of the mead-hall where they had feasted
> would be slick with slaughter. And so they died.

Beowulf's arrival makes all the difference. Apart from his credentials
as a mighty young warrior, he has apprenticed for this task by killing
nine sea monsters. He plants himself on a bench among the boozy
Danes, inviting Grendel to make the first move.

> Nor did the creature keep him waiting
> but struck suddenly and started in;
> he grabbed and mauled a man on his bench,
> bit into his bone-lappings, bolted down his blood
> and gorged on him in lumps, leaving the body
> utterly lifeless, eaten up
> hand and foot. . . .

As that passage shows, Grendel is not just murderous but also hungry,
a carnivore with a lusty appetite for humans. Is he a cannibalistic ogre,
humanoid in form like Polyphemus, or is he some other sort of hulking, toothy beast? Grendel's descent from the race of Cain suggests
humanoid, and his anatomy is left mainly undescribed. But a few
strokes of chilling detail imply a shape other than human:

> . . . Venturing closer,
> his talon was raised to attack Beowulf
> where he lay on the bed; he was bearing in
> with open claw when the alert hero's
> comeback and armlock forestalled him utterly.

The ensuing fight scene offers what Gilgamesh-versus-Humbaba
doesn't: knock-down, drag-out action. They wrestle. Straining fingers
press into monster flesh. Grendel stumbles backward. They waltz
violently through the hall, smashing benches, breaking gold fittings,
wrecking the joint, making the very timbers shudder and squeak. The
Danes set up a chorus of terrified hollering. Beowulf's own entourage
of warriors try to help, slashing wildly with their swords and accomplishing nothing, since Grendel is impervious to conventional blades.

By sheer strength, Beowulf gets the monster in a hammerlock so forceful that he succeeds in ripping off Grendel's right arm. Grendel flees, charging back to his den in the depths of a swamp, where he'll languish and die. Hrothgar's mead hall has been relieved of depredation, at least temporarily. As evidence, Beowulf hangs a memento near the roof beam: Grendel's shoulder and arm, at the end of which is that single claw.

But facing Grendel has been just the first of Beowulf's three major trials. A night later, after a round of premature celebration, the Danes again go to bed groggy, and so are caught off-guard when Grendel's mother, "grief-racked and ravenous, desperate for revenge," breaks into the hall. She snatches a victim—only one, but it's Aeschere, an elder warrior who happens to be Hrothgar's senior adviser. Called back into action, Beowulf tracks her to a lake in the boggy outlands, dives down, and almost loses his life in a harrowing underwater battle with this *brim-wylf* ("wolf of the deep," as Heaney translates) near the bottom. She catches him in her "savage talons" and drags him to her lair, where she commands support:

> . . . and a bewildering horde
> came at him from the depths, droves of sea-beasts
> who attacked with tusks and tore at his chain-mail
> in a ghastly onslaught. The gallant man
> could see he had entered some hellish turn-hole.

Beowulf fights, but his sword is useless, glancing harmlessly off Grendel's mother's head. So he grabs a huge, ancient blade from her armory and, with one blow, chops through her neck. Then, spotting her son's soggy corpse lying nearby, he whacks off Grendel's head, takes it as a trophy, and swims back to the surface.

This time the celebration at the mead hall, with its reward giving and speechifying, isn't premature. Hrothgar's domain is now free of monsters, with only the more routine troubles of feuding, political treachery, and intertribal strife to darken its future. In addition to warm Danish thanks, Beowulf receives seven horses, a precious sword, assorted other battle gear, and, from Hrothgar's daughter, a fancy necklace. Everybody is happy, but happiness can't last, since this is only the midpoint of the poem. The evening's most serious moment comes when Hrothgar himself, the old king, offers the young hero Beowulf a bit of parting advice:

> . . . Do not give way to pride.
> For a brief while your strength is in bloom
> but it fades quickly; and soon there will follow
> illness or the sword to lay you low,
> or a sudden fire or surge of water
> or jabbing blade or javelin from the air
> or repellent age. Your piercing eye
> will dim and darken; and death will arrive,
> dear warrior, to sweep you away.

That's pretty downbeat for a thank-you toast. However ungracious it may sound, though, Hrothgar is right. His gloomy warning is central to *Beowulf*, not just by position, but because it states the main theme of the poem and delineates the trajectory of its second half.

Beowulf returns home to Geatland (in southern Sweden) and serves under two kings before reluctantly inheriting the throne himself. He rules sagely for five decades, then finds his kingdom terrorized by "sudden fire," as Hrothgar predicted, in the form of an incendiary dragon on a rampage of flame-breathing attacks, which reduce Geatish homesteads to ashes. The dragon, a gold-guarding creature like many other dragons of legend, has been provoked by a human robber who broke into its den and made off with a gem-studded goblet. Instead of delegating this problem to a young warrior, as Hrothgar once delegated fighting Grendel to him, old Beowulf himself totters out to battle the dragon. Deserted by his comrades (all except one, brave young Wiglaf), he shatters his ancient sword on the dragon's skull but still manages, with a little help from Wiglaf, to kill the thing. But it's not a clean victory like the ones in the old days; the dragon, before succumbing, wounds Beowulf mortally. When the craven comrades return, they find both victims laid flat:

> . . . The great man
> had breathed his last. Beowulf the king
> had indeed met with a marvelous death.
>
> But what they saw first was far stranger:
> the serpent on the ground, gruesome and vile,
> lying facing him. The fire-dragon
> was scaresomely burnt, scorched all colours.
> From head to tail, his entire length
> was fifty feet. He had shimmered forth
> on the night air once, then winged back
> down to his den; but death owned him now,
> he would never enter his earth-gallery again.

Gruesome and vile as he may be, the dragon in death is also pathetic. Vanquished, immense, roasted like a rotisserie chicken, he receives a faint nod of sympathy from the poet, as Grendel and Grendel's mother did earlier.

Grendel was a "banished" monster, spurned and joyless, "God-cursed," who fought a "lonely war." His mother, who'd been condemned to lurk in cold, deep waters for her association with Cain, mourned her son with maternal anguish, and "brooded on her wrongs." The dragon had been content to den impassively in the treasure-filled barrow until the goblet thief, "the trespasser who had troubled his sleep," caused him to perk up, "writhing in anger" and working himself into a vengeful tizzy. They're all awesome creatures, hideous yet formidable—fated to kill, fated to die, and piteous in final defeat. Among the realistic touches that make them memorable and compelling as opponents to Beowulf is that, once killed, these monsters stay dead. That's not myth. That's biology.

47

What's it all about? What are the sources, and what are the meanings, of this archetypal encounter between the hero and the monster, repeated so often throughout early literature and legend? Nobody really knows, of course, but some thinkers have made interesting efforts to answer.

Joseph Campbell, the famed scholar of mythology, saw the conflict as subliminal allegory. "Psychologically, the dragon is one's own binding to one's ego," he told an interviewer. "The ultimate dragon is within you, it is your ego clamping you down." Campbell discerned two general patterns of hero-monster adventure. In the first, the hero himself is swallowed—as Jonah was, in the Old Testament, by a whale. The swallowing proves to be a transformative ordeal, and the swallowee is later resurrected. In the second pattern, the hero battles a beast and defeats it—as Siegfried defeated his dragon in the *Nibelungenlied*. The actual killing is only one step in overthrowing the dragon's dark power; Siegfried was required also to ingest some bit of the creature (reversing the fate of swallowed Jonah) in order to subsume that power within himself. "When Siegfried has killed the dragon and tasted the blood," Campbell said, "he hears the song of nature. He has transcended his humanity and reassociated himself with the powers of nature, which are the powers of our life, and from which our minds remove us." Our minds imagine that they control our destiny,

Campbell added, but of course that's just the illusion of a secondary organ perched atop a human body that's more profoundly animated by, in one sense or another, sipping dragon's blood.

Another view comes from the medievalist J. R. R. Tolkien, an erudite scholar of early Anglo-Saxon texts long before he found himself an acclaimed author of mossy fantasy novels. Tolkien studied *Beowulf* closely, and unlike most of his scholarly contemporaries, he genuinely relished it as a work of literature. In the early decades of the twentieth century, that grizzliest and most-analyzed Old English epic was regarded primarily as an archeological artifact, filled with evidence about language and culture in Britain during the eighth century (the context of its composition) or, beyond that, in southern Scandinavia during the sixth century (the setting of its events). It was also valued— or at least used—in the academic world as an appropriately challenging text for literature students, who were obliged to swallow a dose of Old English like cod liver oil before proceeding to the tangier delights of Shakespeare, Dr. Johnson, and Oscar Wilde. In 1936 Tolkien delivered a lecture to the British Academy, later published as "Beowulf: The Monsters and the Critics," which did much to supplant those perspectives with a fresher one. He saw the great poem not as a pile of philological potsherds, not as a schoolbook exercise, but . . . well, as a great poem. A lifetime later, his lecture remains the single most influential treatment of *Beowulf* on its literary merits.

Among Tolkien's big contributions is the recognition that its two monsters and one dragon, far from being cheesy distractions from some weightier essence, as earlier critics had suggested, are central to the poem's meaning. They came from somewhere, those figures, and they went somewhere. "A dragon is no idle fancy," Tolkien wrote. "Whatever may be his origins, in fact or invention, the dragon in legend is a potent creation of men's imagination, richer in significance than his barrow is in gold." Beowulf's dragon signified malice, greed, destruction, and, most pointedly, "the undiscriminating cruelty of for-

SHADOW OF THE NINE-TOED BEAR   271

tune," delivering death as the inescapable fate of all mortals, without exemption for their virtues or deeds.

Grendel and Grendel's mother are no idle fancies either. Although Beowulf defeats them and lives to celebrate, they support the poem's main theme, encapsulated by Tolkien as "man at war with the hostile world, and his inevitable overthrow in Time." They do it by allowing the poem to build suspense, offering the youthful hero challenges against which to achieve heroic stature, toward the point fifty-some years later when even he, mighty Beowulf, the wise ruler and monster slayer, succumbs. What he succumbs to is simply the same scary fate that faces every human, the same one from which Gilgamesh futilely sought immunity: death. Tolkien located that theme in a single Old English line: "*lif is læne: eal scæceð leoht and lif somod.*" Life is temporary: everything perishes, light and life together.

Another point Tolkien made was that the horrific creatures in this poem, written by an anonymous Christian in eighth-century Britain about Germanic pagan cultures in sixth-century Scandinavia, were not monsters created and sponsored by various gods (as Polyphemus had been by Poseidon, and Humbaba by Enlil, and Leviathan by Yahweh) but monsters set *against* the single beneficent God of Christian belief. Grendel, notwithstanding the dabs of sympathetic portraiture, is a "hell-serf" (in Heaney's rich but conscientious translation), reflecting his descent from Cain. In the original, Tolkien noted, Grendel carried the onus "*Godes yrre bær.*" He bore God's wrath. Grendel's mother is a "monstrous hell-bride," later a "hell-dam," who comes as a "force for evil" to avenge her son. And the dragon, called "the evil one," is likewise on the dark side of a dark-against-light Christian dualism. So in fighting his own monstrous foes, as they pop up from the landscape, Beowulf prosecutes a holy war against the adversaries of God. Within the Christian framing of these ancient pagan stories, "the old monsters became images of the evil spirit or spirits," Tolkien wrote, "or rather the evil spirits entered into the monsters and took visible shape in the hideous bodies," such as Grendel's.

This is very different from the battles, equally heroic but morally less stark, fought by Greek and Mesopotamian mortals against monsters set up by one or another petulant god. It's different from Sigurd's struggle against serpentine Fafnir in the *Volsungs* saga, derived from pre-Christian Norse legends. As a pagan hero contending with monstrous enemies that are demonized in Christian terms, Beowulf stands somewhere in an unmapped moral zone between Sigurd and Saint George. Maybe that strengthens our empathy with him. It's a dilemma that seems modern.

What gives *Beowulf* its final, eerie tension is that despite the explicitly Christian tone of the poet's voice, with its characterization of "Cain's clan" and its pious nods to "the Lord of Life," there is no Christian ending. There is no eternal salvation for the protagonist, only heroism, fame, and death. The beautiful harshness of the old legendary material hasn't been fully digested by the new credo. This poem couldn't have been written as it was, Tolkien judged, "but for the nearness of a pagan time. The shadow of its despair, if only as a mood, as an intense emotion of regret, is still there." Beowulf, having died righteously, doesn't ascend to a celestial reward. Instead his death opens the way for renewed feuding with, and eventual invasion by, his kingdom's blood enemies, the Swedes. His body is burnt on a big pyre, yielding billows of wood smoke and smudgy fumes of incinerated flesh that rise darkly, while a Geat woman sings out, in grief,

> . . . a wild litany
> of nightmare and lament: her nation invaded,
> enemies on the rampage, bodies in piles,
> slavery and abasement. Heaven swallowed the smoke.

The apex of that rising smoke is as close to heaven as Beowulf gets.

Meanwhile the dragon's carcass has been shoved off a cliff into the sea, a mode of disposal not much drearier than cremation. The point I take is that this hero, and these monsters, live and die in a single universe of concrete reality. They are all memorable, and they are all mortal. Their molecules reenter the landscape. They are parts of a food chain.

The ecologist and philosopher Paul Shepard touched on hero-monster contests in his last book, *The Others*, published just before his death in 1996. Subtitled *How Animals Made Us Human*, it's a sweeping survey of the ecological underpinnings of culture, taking its epigram from Joseph Campbell and its scope from Shepard's own lifetime of research and wild-hearted speculation. Its chapter on chimerical beasts begins:

> Culture teems with animals who have no exact equivalent in nature. A huge, make-believe fauna of monsters, prodigies, and wonders slithers and swarms and storms through all of the arts, as though the natural world were somehow deficient. It is necessary to ask: To what end are these creatures of the imagination? Are they really substitutes for ordinary animals or do they have their own purposes? What are they, where do they come from, and what are they doing here?

Discussing dragons in particular, Shepard notes that they've carried one set of meanings and nuances in Europe, and a much different set in the Orient. This is a broader difference than the Christian-pagan opposition within Europe. It's more than scary godly monsters versus scary ungodly ones. Oriental dragons were generally benevolent, graceful creatures, often connected with ideas of renewal, wisdom, enlightenment, blessed birth, and cyclical weather. A typical dragon of the medieval West, in contrast, was "a fire-breathing old grump guarding his gold beneath a mountain," as the dragon of *Beowulf* did. "Since gold was a patriarchal symbol of virginity," Shepard continues, "the dragon was therefore associated with the violent, virile control of women that dominated Europe for three thousand years. Its tenacity and ferocity against masculine challengers and its ownership of the cave in the earth are suggestive of that association."

So all those dragons stand for draconian sexism? Maybe so, maybe not. In any case this assertion, unsupported by evidence or argument, is less persuasive to me than a point Shepard makes, earlier in the book, about wild predators and their transmogrification in the human psyche. He writes that lions and wolves, toothy and carnivorous, have helped shape our worried imaginings about the jaws of Hell, and about other agents or avenues of supernatural menace. "Our fear of monsters in the night probably has its origins far back in the evolution of our primate ancestors, whose tribes were pruned by horrors whose shadows continue to elicit our monkey screams in dark theaters." Such fear arises from signals beyond memory, beyond oral tradition, beyond ancient poetry and cave art, Shepard argues; it's programmed into human DNA. "As surely as we hear the blood in our ears, the echoes of a million midnight shrieks of monkeys, whose last sight of the world was the eyes of a panther, have their traces in our nervous system." Evolution doesn't quickly erase important genetic memories, and the memory of being a barefoot anthropoid on the East African savanna is relatively recent. Isn't it likely that the monkey shrieks and the panther eyes have their traces in our epic literature too?

Call that the ecological reading of monster and dragon mythology. There's also a paleontological reading. One famous case derives from Klagenfurt, in southern Austria, a riverside town whose name translates as "ford of lament." Ancient legend held that floods and drownings suffered thereabouts were the work of a dragon that denned in a nearby swamp. As the story went, a benign duke had set some brave knights to fish for the dragon, using a bull tied as bait on a barbed chain, like a dough ball on a treble hook. The dragon bit; the knights pulled it in and slaughtered it. Who says you *can't* draw up Leviathan with a hook? Klagenfurt, liberated from the terror, eventually embraced a dragon image on its city arms.

Physical evidence became conjoined to that story in the fourteenth century, when a large, weird hunk of fossil material was unearthed from a local gravel pit. The fossil, clearly some sort of cranial fragment, was construed to be a tangible remnant of the conquered dragon. It was put on display at the town hall and, in 1590, served as model for a nicely garish piece of civic statuary, sculpted by one Ulrich Vogelsang: a dragon-shaped fountain. The fountain, still standing in Klagenfurt, has a snarly head with small ears, plus wings and a long tail. The fossil itself remained in the Klagenfurt hall until 1840, when a visiting paleontologist named Franz Unger recognized it as the cranium of a wooly rhinoceros. Soon afterward it was transferred to a natural history museum, while the statue back in Klagenfurt came to be considered, by some scholars, the earliest known attempt (however inaccurate) at reconstructing an extinct species. Whether it deserves that distinction or not, it reflects the readiness of people to affirm a psychological connection— and to invent a physical one—between the monsters of febrile fantasy and the real beasts who prowl (or once prowled) familiar landscapes.

More recently, a classical folklorist named Adrienne Mayor has made a better-documented argument about the origin of griffins, those lion-bodied creatures with raptorial beaks, in her book, *The First Fossil Hunters: Paleontology in Greek and Roman Times*. She tracks the legend of gold-guarding griffins to its earliest known appearance—in an epic poem by a Greek traveler, Aristeas of Proconnesus, about his journey across Scythia in the seventh century B.C. Mayor's hypothesis about what came before Aristeas, based on geography and paleontology as well as textual scholarship, is nifty and persuasive.

Stretching north from the Black Sea and eastward beyond the Caspian, a region that nowadays encompasses Ukraine and southern Russia, Scythia was known to be rich in gold ore. Aeschylus later put Prometheus on a cliff there, in his play *Prometheus Bound*, evoking the

Scythian landscape as he imagined it from Aristeas' account and furnishing it with both gorgons and griffins. The latter were described as "silent hounds with cruel sharp beaks." In the third century A.D., the Roman scholar Aelian compiled hearsay reports that the griffin was "a quadruped like a lion; that it has claws of enormous strength," as well as white wings, black feathers along its back, variegated blue ones covering its neck, and "a beak like an eagle's." Aelian placed griffins farther southeast than Scythia (in Bactria and India) but repeated the rumor that they guarded gold. From these clues and many more, Mayor deduces that the griffin legend may have originated from ceratopsian dinosaurs as their fossilized skulls and other bones weathered out or were excavated at Scythian gold-mining sites. *Protoceratops* in particular, with its beaked face, its bony collar frill sweeping backward like wings, and its lack of conspicuous horns (such as those on *Triceratops* and other late ceratopsians), might have been combined with living eagles, lions known by reputation from Greece and Persia, Caspian tigers then surviving in Scythia, and a bit of nervous fantasy, to produce griffins.

Farther west, farther north, into the lands of Sigurd and Beowulf, there were no native lions during the early centuries of our era, when half-invented heroes battled monsters in the forests and fens of human imagination and the great Norse and Germanic epics were dreamed into verse. There were no crocodiles or giant snakes in those latitudes. There were no notable lodes of ceratopsians. But there were bears— both living and extinct, the latter sort represented by oversize skulls and other bony remnants left behind in a number of European caves. The extinct ursid, now known to paleontologists as *Ursus spelaeus* (the cave bear), seems from the evidence of its teeth to have been mainly if not completely vegetarian. But impressionable humans, when they began rediscovering those huge skulls, probably weren't put at ease by herbivorous dentition. The abundant cave-bear remains may have inspired even more nightmarish mythmaking than any stray rhino or ceratopsian fossils. An Austrian paleontologist named Othenio Abel, back in 1914, suggested exactly that connection between *Ursus spelaeus* and the old tales of cave-dwelling dragons.

Meanwhile the brown bear, *Ursus arctos*, more carnivorous and presumably more dangerous than *Ursus spelaeus* ever was, survived even on the island of Britain until about A.D. 1200. It loaned its burly aura to Beowulf himself, whose name derives from "Bee-Wolf"—that is, bear. It thrived across northern Europe and Scandinavia through the Middle Ages. It hung on still longer in a few enclaves of high, untamed terrain. The most hospitable of all those enclaves, for reasons good and bad, turned out to be the Carpathian Mountains in Romania.

No one can prove just how much Grendel and Fafnir and Grendel's mother (charging ferociously, enraged by the death of her cub) owe to the real physical presence of the brown bear, and I don't propose to try. But it's interesting to remember that the old dark woods weren't empty of inspiration. And it's dreary to reflect that the twentieth century's most notorious avatar of the Beowulf figure—chieftain, heroic slayer of monsters, wise king—was a megalomaniacal little Communist martinet who required underlings to load his rifles and stretch his pelts.

48

Given that humans and brown bears have shared habitat in the Carpathians for thousands of years, you might expect to find strong mythic traditions in Romania surrounding *Ursus arctos*. You don't. The poem *Beowulf* has no Romanian counterpart. Neither does Goldilocks.

Several folk tales of the storybook sort—one about how the bear lost his tail, one about a bear who wrestles the devil—suggest a fondness for the animal, a level of comfort with its presence, an appreciation for its greatness of body and heart. The devil-wrestling tale involves a man who meets Satan, dreads tangling with him, and weasels out by suggesting that the devil first try his strength against the man's old uncle, who lives in a cave. The man steers Satan to a bear's den and advises him to rouse the "old uncle" from sleep with a stick. Satan, gullible enough to wrestle the bear, takes a bad thrashing before he escapes. If your old uncle is *that* strong and fierce, Satan tells the man, then I don't believe I'll go at it with *you*. In this story and others, the bear is estimable and gruff but also beneficent, not cunning and menacing like the popular image of the wolf. An old belief, reported from northwestern Transylvania, holds that anointing a newborn baby with bear fat will endow the child with a shield of magical imperviousness, like Siegfried's. A mundane variant is the notion that bear fat rubbed on the scalp will cure baldness—bushy hair being a form of imperviousness, I suppose, to the harrowing realities of middle age.

According to one Romanian authority—a bear biologist named Ovidiu Ionescu, who holds a senior position at ICAS—his country's folkloric tradition casts the brown bear as "a kind of partner in the forest," powerful but cute. If there's a darker and more fearful perception of *Ursus arctos*, embedded deeply in any national epics or myths, neither Ionescu nor any other Romanian I've talked with considered it worth mentioning.

That's not to say the Romanian bear isn't dangerous. It can be. Ionescu and his co-workers have assembled statistics reflecting the scope and the nature of bear-human conflicts. Between 1990 and 1997, he says, there were 119 cases of bear trouble (or, from the bears' perspective, human trouble) serious enough to be recorded. In the course of those incidents, eighteen people died. Most of the altercations (fifty-seven percent) involved livestock. A much smaller portion (twelve percent) involved hunting or poaching. The rest were accountable to a variety of circumstances that either brought humans into the forest (gathering mushrooms, wild fruits, firewood) or drew bears out of it (for instance, to raid orchards). Among the livestock-related incidents, cows, horses, donkeys, pigs, and goats were minimally at issue. By far the largest share (seventy-one percent) of problematic bear-human encounters occurred in connection with sheep.

Romania contains millions of sheep—by one rough estimate, nine million head. Many of them spend the summer grazing the moist grass of high mountain meadows, surrounded by forest that harbors bears and attended by shepherds and guard dogs. The guard dogs are toothy and belligerent. They're a mongrelized army, conforming to no pure-bred type, but the most typical is a large-bodied, big-shouldered canine with medium-length fur, a short snout, and a furry face, like an Airedale wearing the head of Lon Chaney's Wolfman. Some of the dogs are fitted with spike collars as protection from neck bites in a fight against wolves. All are required to wear a sort of T-bar hobble dangling below the collar, a short piece of wood known as a *jujee*, which jounces enough to prevent the dog from charging off on rene-gade pursuits of roe deer or other game. The jujee also signals to hunters that the animal is a shepherd dog, not a feral pest to be killed. Puppies learn their sheep-guarding habits by galloping along with the adults. Nobody sends Romanian dogs to obedience school. The typical flock of three or four hundred sheep is attended by at least a half-dozen of these irascible curs. The shepherds, taciturn and self-pos-sessed men who can be tough when they need to, carry no weaponry other than stout staffs. If you as an interloper stroll into a Romanian shepherds' camp, you'll want a stout staff yourself—for defense against the dogs.

Guns are unthinkable or, if thought of, considered unattainable or unnecessary. So the nature of interactions among Romanian bears, Romanian sheep, and the guardians of those sheep probably hasn't changed much in the last thousand years. That may account for why the problem seems chronic but not acute, and why its resolution is usually nonlethal (except for the sheep). Bears, humans, and dogs were well acquainted with one another, in the Carpathians, long before the first rifle came along.

In many cases the sheep of a flock belong to several different owners, who contract with an *organizator de stîna* (camp organizer) to have their animals pastured in the mountains for a summer season. The organizer acquires grazing rights for a certain area, maintains a base camp centered at a simple cabin, hires a chief shepherd (*baci* is the Romanian term) or fills that role himself, assembles a crew of other shepherds, supplies guard dogs, adds a few dozen dairy cows and maybe some horses or donkeys for pulling a cart, takes his costs and his profit (if any) in the form of cheese made from fresh milk (usually a mixture of bovine and ovine) at the cabin, and returns the sheep to their owners in early autumn. The baci himself often functions as the expert cheese maker. The owners get a specified share of the cheese; the rest is left for the organizer to sell in local markets. If a ewe or a ram has been killed by a bear or a wolf, the organizer is absolved of responsibility on one condition—he must show evidence that the sheep wasn't simply allowed to stray off and disappear. Generally that means producing one of the sheep's ears, inside which is an identifying tattoo, like a brand on a cow. If the organizer can't deliver that tattooed ear, he may have to pay for the lost sheep himself (though the settlement terms are sometimes flexible). No effective system of government compensation exists, and insurance is impracticable for small operators. So the organizer has strong incentive to battle for the life, or at least the ears, of every sheep. He passes that imperative along to the shepherds, who know their duties, value their jobs, and may themselves be held financially accountable for any animal lost to a bear through their negligence. One result: Shepherds get injured, not just while defending sheep against attack but, even more perilous, while trying to take a fresh carcass away from a bear. "This is the situation,' says Ovidiu Ionescu, "in which the great majority of conflicts occur."

A colleague of Ionescu's, a Swiss-Italian scientist named Annette Mertens, studied the economics of this system. Visiting about twenty different shepherds' camps during each of two summers, for a total sample encompassing 17,449 sheep, she found that annual losses to predators averaged two percent. Most of those were inflicted by *Canis lupus,* the wolf, which is more dangerous to sheep (though less so to humans) than *Ursus arctos.* Bears accounted for 36.5 percent of the losses. Considering the market value of an adult ewe, plus the value of lost cheese, Mertens calculated an average loss at each camp, from bear predation, equivalent to $169. It's serious money to a camp organizer who earns only about a hundred bucks a month.

After learning this much from Mertens and accompanying her on some visits, I decide to make an informal survey of my own. With a venturous young translator named Ciprian ("Chip") Pavel, I set off on a series of hikes into the high country. We climb steep, wooded slopes above the mountain roads, above the trails, following hoofprints and sheep shit toward the alpine meadows where shepherds spend the summer. Sometimes we carry simple provisions (red-deer sausage, peanuts, raisins, chocolate) as well as sleeping bags and a tent. Sometimes we carry gifts (fresh oranges, cigarettes, pint bottles of a Romanian plum brandy called *palinca*) for the shepherds. Usually we carry walking sticks for defense against the dogs. We see some gorgeous landscape—mountain lakes, stony ridges and peaks, swales of verdant grass, fog-shrouded expanses of krummholz and heather, all of it chilly and wet even at the height of what, down in the lowlands, is a summer drought. We talk with dozens of nut-hard, gentle-spirited men. We're welcomed often with warm hospitality, candid comments, log seats beside a cook fire, hot *mamaliga* (Romanian polenta, grainy and yellow, served as a steaming paste) and freshly made cheese of the three traditional kinds: *urdă* (sweet and crumbly), *telemea* (salty, like feta, but easily sliced), and *brânză de burduf* (sweet and smooth, packed for market in cylinders sewn from fir bark). Rarely we face the cold truculence of a camp boss who wants no visitors and no palaver. Among the many things I discover is that this research regimen, so rich in exercise, cold mountain air, and cheese, may be bad for my cholesterol count but is highly agreeable to my disposition.

I also learn something of how Romanian shepherds feel toward bears. In the Târlung valley just east of Braşov, I meet a baci named Ion, a great-bellied man in a lavender flannel shirt, his head big as a pumpkin, who tells me that he has been lucky this year—no bear trouble yet, though a wolf recently ran through camp. Last year, on the other hand, Ion lost two sheep to a very large bear on the ridge behind us. The bear jumped his flock in broad daylight, and even his dogs (which seem more than adequately fierce) couldn't stop it. That bear had a maimed paw, maybe from a trap, Ion recalls. It devoured both sheep. Nothing left to recover, no ears—total loss.

As we chat, Ion sets out some fresh urdă, sliced, in a tin bowl. The cheese is luscious and creamy, like a guilty dessert. I ask about earlier years. The translation process allows me time to keep eating. Well, in prior summers Ion put his camp in the Bucegi Mountains (a small, majestic range within the Carpathians), thirty-some miles to the southwest. Many bears down there. He'd lose ten sheep and one or two cows in a season. And so, I presume, he moved up here to escape them? Escape, naw, he has no such delusion. "Where is *not* bear?" Ion says. "Where can you go, and *not* seeing bear?" But he declines to complain that there are too many bears in the mountains. No, too many is not problem. Problem, he says, is that Forest Department doesn't feed them sufficiently. When the feeding troughs are empty, *that's* when bears come after his sheep. Pondering this peculiar edifice of assumptions and expectations, I gobble another chunk of urdă.

On the next ridge over, across several meadows and a small creek, Chip and I walk into another camp and are met again by angry dogs and an amiable baci. This is Nicu, a compact man in a brown sweater-vest and a green felt hat. Bear trouble? No, not much. One *urs* did visit several nights ago, but Nicu's dogs—he has thirteen, and they're good, he says—ran it off. His camp, simple but tidy, is centered around a pair of plank huts, one devoted mainly to cheese making, the other to cheese storage. The shepherds sleep outside, wrapped in sheepskin robes, near the flock. Nicu gives me a tour of his operation, including the barrel of brine in which float big chunks of telemea and, bobbing among them, a single raw egg. When the egg sinks, he explains, it's time to add more salt. And over here, these are the tanks in which the boiled milk, activated by the culture, undergoes its magical change. As the cheese begins to set, he presses out the water with a wooden rake, or else through cheesecloth—one gauge of cloth for telemea, a different gauge for *caş* (the bland, intermediate stage toward brânză de burduf). In the second hut, under towels, rest many large rounds of caş, hefty and squat. They'll be held for a week and then put through a grinder to make the brânză, which is then packed into pig bladders for long-term storage or else tamped into those fir-bark tubes for market. Nicu serves us a generous spread of samples—telemea and brânză and urdă, along with a pile of salt for putting edge to the urdă's sweetness—plus slabs of cold mamaliga, sliced like meatloaf.

We're joined by another fellow and a handsome, weather-tanned woman, who seem to be here on a Saturday visit from the valley. Seeing my gusto for Nicu's product, the fellow jokes to me in Romanian: *Brânză de burduf,* that stuff is good for a man. He raises his forearm, big and stiff as a stallion's penis, and knocks on the door: Makes you *hard.* The woman rolls her eyes and swats him.

"As good as Viagra?" I ask in English.

Global citizen, he recognizes the word. "*Viagra? Nu. Brânză!* Forget your fancy American drugs; Romanian cheese will do fine.

We express our thanks and hike on. Later in the afternoon, on another stretch of ridge, we encounter Nicu and three of his shepherds, still agitated after chasing off some Gypsies who tried to steal a few horses. Watch out for those bastards as you hike down through the woods, one shepherd warns, they'll rob you too. Gypsies suffer from lingering prejudice in Romania, but today's episode, at least as Nicu recounts it, helps fuel the bias. His parting wisdom to me: Never mind bears and wolves—*Gypsy* predation is more troublesome to the shepherd.

In a remote drainage called Valea Rea ("Mean Valley") of the Argeş district, beyond the end of the road and then half a day's climb to the tarn-dotted alpine meadows, I speak with Nelu, a forceful young man with black bangs dangling out beneath a white sheepskin hat. He holds the authority, it seems, or at least the talking role, within a little squad of shepherds and dogs. Nelu leans on his staff, a wool blanket over one shoulder, as he hears why I've intruded onto his mountainside. He tells me that two or three sheep is his average annual loss to bears; that he, Nelu, has been *this* close to a bear, yes, four meters, but never injured; that some years, up here, they'll see a bear every day; and that dogs, good dogs, are the answer, though not an infallible one. When I ask my vaguest and most fundamental question—how does he *feel* about bears?—Nelu smiles coldly and answers bluntly. "The hell with them. Kill them."

That night, warming myself at fireside after a cold rain, I hear stories from an old baci, another Ion, who thinks that the current bears just aren't so fierce as their parents were. He recalls the time, fifteen years back, in this very cabin, when he himself nursed a young shepherd whose leg had been torn open by a bear. After a week, the shepherd's wound started looking ugly; he was sent down to a hospital, got zero care because he had zero money, and lost the leg, but survived. The baci also remembers a Moldavian shepherd who married in hereabouts (that is, among Wallachians) and, intemperate guy, took to trapping bears with a leg snare, then killing them with a long steel pike. Didn't own a gun. No, he never got caught at it, though that would've brought serious trouble. Nobody ever snitched, says the old baci, despite him being Moldavian.

Not many miles away, on another high slope above the Mean Valley, we rouse from his afternoon nap another elderly baci, nearly deaf, with a smile brightened by brown eyes and gold incisors. His name is Ion Petrică, and he's got four hundred sheep up here, tended by four shepherds. Where are *you* from?! he asks in a cordial shout. "Braşov," says Chip, and I say "America," both of which seem inconceivably distant to the man. Lost my hearing back when I worked for the postal service! he explains, hollering above the din of his deafness. From driving a truck! a drafty one! through the cold winters! Sheep losses?! Maybe two or three a year! that's to bears! and another ten to wolves! Last year there was a big red bear!, it took a couple of sheep! Three years ago, we had six bears up here! he recalls, but the dogs did their job! Five years ago, one shepherd got injured! and went to the hospital because he tried to stop a bear! The moral, as our host briskly takes it, is: Don't upset the bear! If your dogs can't stop him, let the beast have what he wants! Say, what's the current price of bread down below?! Having retired from the postal service seven years ago with a paltry pension and spent the subsequent summers up in shepherds' camps, he evidently gets hungry for news. Who's going to win the election?! If they privatize all the factories, what will be left for Romania?! This leads me to a question about Ceauşescu, toward whom the deaf baci harbors no post-Communist nostalgia. Ceauşescu?!! "He killed a lot of people," Chip translates, at normal volume. "Somebody else kill him. When you're making good, you find good. When you're making bad, you find bad." Want some cheese?!

He stirs up a fresh pot of mamaliga and then dumps it out, *whump*, a soft mound of corn mush on a bare wooden table. Beside it he sets a bowl of brânză, from which we each take a few cheesy nuggets, pressing them inside a handful of warm mamaliga and then shaping the whole thing like a snowball around a rock. We put our golden spheres into the fire's embers for ten minutes of slow roasting. This simple delicacy is known by Romanian shepherds as *bulz*. Toasted half-crisp on the outside, raked from the coals, broken open, each ball is a corn sandwich filled with molten cheese. You haven't tasted the Carpathian ether, nor begun to understand what might draw men to these high, severe places, until you've had bulz from a shepherd's fire, washed down with buttermilk and a nip of cheap palinca.

Ion ("John") is a common name in Romania. It seems to be commoner still among the aged shepherding class. In two weeks of walking the highlands I encounter more Ions than I can keep straight. One man who's unforgettable among them is Ion Dincă, sixty-seven years old and not a baci, not a boss, not a cheese maker, merely a shepherd, still climbing out of camp each day with his flock. I find him on a patch of meadow near the Transfăgăraş Highway, hundreds of yards above a switchback curve just south of the Transylvanian pass. He wears a lambskin hat, a wool jacket, another jacket atop that one, and rubber boots. He carries a staff, a small day-bag over his shoulder, and a silver pocket watch. Despite several missing teeth, he looks years younger than his age. Shepherding, it's like a virus, he tells me jovially. He's up here because he likes it, can't resist it, not because of need. After retirement from the Forest Department nine years ago, he returned to what he'd done as a boy. It's late season now, he says, the sheep are giving less, we're milking them only twice a day. We'll stay till mid-September, he says. Then back down to the village below, a two-day walk.

I ask about bear troubles. Oh yes, he says, down in the village a bear sometimes comes raiding with terrible force, breaks into corrals, takes pigs or chickens. And up here among the sheep? Occasionally here too, yes. But today Ion Dincă isn't concerned—no bears or wolves nearby, he declares, with the confidence of a man who knows intimately his own little patch of mountainside. Which predator is more troublesome, I ask, wolf or bear? "Urs," he answers unequivocally. Wolves are smarter, sure, they lurk, they move like thieves. But bears bring more danger and damage. One bear, a blackish-red animal with white neck markings, took a sheep and three donkeys in a single night. Would it be better, I ask, if there were no bears at all? Well, better for *him*, yes, it would be. But the bear, it's *podoaba pădurii*, the treasure of the forest. "If you lose this, you lose the treasure," he says. "A forest without bears—it's empty."

Not all shepherds see it that way, I tell him. He agrees, noting that most people who claim to like bears are gentlefolk. They live far away, he says, with no bear troubles of their own. Easy for them. Shepherds, plain men at work in the mountains, don't enjoy such distance. Their attitude tends to be *la naiba cu urşii*, to hell with the bears.

Here it is again, then: the Muskrat Conundrum. Ion Dincă doesn't call it by that label, of course, nor does he draw any parallels between the predator problems of Romanian shepherds and those of vulnerable, marginalized, rural people elsewhere. He doesn't even explain why, despite living the life of an unsheltered muskrat himself, he holds a more appreciative view of the creature that plays the role of the mink. He's simply a man of transcendent and generous spirit. As we sit talking, the neck bells on his leader sheep toll soothingly, *cloonka cloonka cloonka*, in the rhythm of their waddle across the slope. Life is hard, life is good, life is enriched by complications and—he seems to feel—so should it be. A forest without bears is empty.

49

But even a forest replete with *Ursus arctos* will not necessarily offer them to plain sight. You can walk for many days through the woodlands and high meadows of the Carpathians, as I've done, without ever chancing upon a bear.

This is normal, a matter of low statistical odds. It's inherent to the nature of large predators and the ways they inhabit landscape. Even when they're relatively common, they're rare, and their cagey furtiveness makes them seem rarer still. If you go searching for saltwater crocodiles at night, in good habitat, with a spotlight, you'll probably find some, thanks to the eye-shine from their retinal tapeta and to the fact that their distribution is essentially linear—crocodiles tend to align themselves along riverbanks. Lions aren't so obliging. Leopards are as elusive as final truth. You might live half your life in Montana or Wyoming, hiking the mountains, traipsing through cougar habitat, without ever glimpsing a cougar. In the Russian Far East, biologists who track Siberian tigers day after day, for years, seldom lay eyes on one. As for brown bears, they're evasive too, though not so fastidiously invisible as tigers or cougars. If you spend long, quiet evenings with a Romanian gamekeeper on the platform overlooking a feeding station, then yes, you will likely see a few dark, ursine shapes, moving to the trough, eating their fill, indistinct but recognizable under whatever dim moonlight seeps through the canopy. Alternatively, you can go to a place called Răcădău.

Răcădău is a housing project on the outskirts of Braşov. It was built in the mid-1980s, at the height of Ceauşescu's Systematization program of forced urban resettlement. It's a wedge-shaped zone of high-rise apartment blocks along the city's south boundary, the wedge defined by a narrow valley draining north between forested mountains. Twenty-five thousand people live there, low-income workers and their families, packed together at industrial density within a concrete peninsula flanked by a sea of greenery. Interspersed among the apartment buildings, which tower six and seven stories tall, are sidewalks, narrow cross streets, a few shops, the occasional derelict car, broken glass, several patches of open pavement on which children can kick a soccer ball, and about twenty garbage dumpsters. Each dumpster is roughly the size and shape of an ore car. They stand clustered, half a dozen here, half a dozen there, on concrete pads along a perimeter road that separates Răcădău from the forest.

Just beyond the dumpsters, rising steeply into slopes thick with oak and beech, is bear habitat. Since the dumpsters are constantly replenished with odiferous organic refuse, this juxtaposition hasn't been lost on the local bears. They come down off the wooded hillsides at night to eat Răcădău's garbage.

Being omnivorous, brown bears are content to forage among middens that include little delicacies of moldy sausage, rotten tomatoes, orange rinds, soured milk, potato peels, wasted cheese, and leftover mamaliga, all amid such less appetizing stuff as cardboard, discarded shoes, eggshells, ragged T-shirts, coffee grounds, broken lamps, and plastic trash. It's a situation reminiscent of Yellowstone Park in the slapdash old days, when garbage generated by tourists and employees was consigned to open-air landfills, which attracted grizzlies.

For decades, until a Park Service policy change in the late 1960s, bears were allowed to feed freely at the Yellowstone dumps. People gathered to watch them on tranquil summer evenings, and it could be quite a show. Near the Canyon dump was a special feeding area designated for bears—a sort of stage—above which sloped a natural bowl that had been furnished with bleacher seats. In 1937, according to one park historian, "there was not enough parking space for the 500 to 600 cars that appeared every evening" at Canyon, where the spectators' curiosity was gratified by, sometimes, as many as seventy bears. In 1966, on a single night, eighty-eight grizzlies visited the Trout Creek dump. By that time human garbage constituted an important nutritional resource for Yellowstone's grizzlies. The invidious term "garbage bears," occasionally applied to dump-visiting opportunists, reflected the dubious assumption (not shared by all biologists) that there existed two distinct groups within Yellowstone's population: bears that frequented the dumps and ate garbage, and bears that remained in the remote backcountry and did not. Even independent biologists such as Frank and John Craighead, who studied the Yellowstone grizzlies from 1959 to 1971 and doubted the two-groups assumption, recognized the crucial role of the dumps. In fact, the Craighead brothers used them as focal sites for censusing bears. During the 1965 season, according to Frank Craighead's *Track of the Grizzly*, Trout Creek alone attracted 132 different individuals.

Soon after that, in the late 1960s, several dramatic events reminded people (park managers in particular) that grizzly bears can be dangerous. On a single night in August 1967, up in Glacier National Park, two young women were killed by grizzlies in two separate incidents. The coincidental timing, after many years without a fatal mauling there, suggested a causal pattern and a crisis. Glacier Park is hundreds of miles from Yellowstone, part of a distinct ecosystem with some different management issues, but the ramifications of those deaths nonetheless traveled. In addition, there were further incidents. Two summers later, a five-year-old girl was mauled by a grizzly near Fishing Bridge in Yellowstone. A month after that, two more tourists were injured by bears, again near Fishing Bridge. The newly appointed superintendent of Yellowstone, along with his supervisory research biologist, began to rethink the old, easy tolerance of garbage-eating bears and their human audiences. Around 1970 the dumps were closed, with ill-advised abruptness (against which the Craigheads cautioned). Rather than solving the problem of bear-human conflicts, that move exacerbated it. Denied garbage from the familiar sites, hungry bears foraged elsewhere, sometimes recklessly, causing more concern to the Park Service and lethal trouble for themselves. In 1971, at least forty-three grizzlies died in the Greater Yellowstone Ecosystem.

At Răcădău, the numbers are smaller and the circumstances aren't complicated by the regulations and expectations associated with a world-famous national park. Still, there's a degree of hazard. How much? It's hard to gauge. *Ursus arctos* in Romania, with its own ancient history of experiences, is generally not so ferocious, impetuous, and defensively aggressive as the grizzlies of the northern Rockies. Bears come down brazenly to the Răcădău garbage feast, people behave stupidly, but so far no human has been mauled and no bear has been killed. Annette Mertens, monitoring the situation as a sidelight to her work on shepherd camps, counted twenty different bears making dumpster visits during a recent year. She worries that someone is going to get hurt—and, a broader concern, that one serious injury or death in Răcădău might turn public sympathy against Romania's bears. City officials are nervous too. Braşov's sanitation department has tried three different dumpster designs, none satisfactory, in an effort to exclude bears. The most recent design, the ore-car model, includes a bear-proof lid that can easily be slid shut after a person has tossed in an armload of garbage.

But the good folks of Răcădău leave the dumpster lids ajar. Why? Evidently because they enjoy watching bears. Amid all the concrete, amid the dreariness of their "systematized" environment and the austerity of their lives, here is *nature*. In Romanian, the dismissive "garbage bears" label would be rendered as *ursii gunoieri*, but many Răcădău residents are appreciative of this urban anomaly, not dismissive. A round-faced young woman named Daniela, who works as a chambermaid at the hotel where I sometimes stay, comes alight with pride when she happens to hear that bears are what keep bringing me to Romania. "*Ursii?*" She's delighted to inform the American mister that she has *ursii* in her very neighborhood. No kidding, I say, that wouldn't be Răcădău, would it? Yes, Răcădău! "We live . . . *across* the forest," she says, groping for the right English words; she means *next to*. "And every night they came. There are so many. They are so nice."

Unfortunately, she adds, the politicians want to give these lovely animals away—to other countries, to zoos. For the politicians, Răcădău's bears are a nuisance, or a threat against public safety, or some such. Daniela views them otherwise: "They are innocent. They don't do you anything." She's not atypical, it seems. The dumpster-diving ursines have become neighborhood mascots for people who don't realize how dangerous a brown bear can be, and who probably wouldn't behave much differently if they did. The nightly visitations bring a touch of magic to a place badly in need. "Every children, when is eleven, twelve," says Daniela, "they stay near the garbage to see the bears. And they are so excited."

Late on a June evening, Annette Mertens and I stay near the garbage ourselves. In the space of two hours, Annette taking photos while I handle a strong beam drawing juice off her car battery, we watch half a dozen different bears gourmandizing happily. We see three yearling cubs in a group, two subadults, a sizable adult (probably a female, as are most of the bears that forage in Răcădău) whose handsome brown fur glistens with silvertip highlights, and several others. They climb in and out of the dumpsters, push a lid open wider when necessary, paw through the refuse, feed upon choice morsels, gambol on the pavement, chase one another halfheartedly, glare back at my spotlight, lose interest, retreat up the hillside, and return again to feed. Meanwhile, Annette and I aren't alone. Taxicabs arrive, disgorging a few bearwatchers. Young men and women loiter beside cars parked haphazardly in the street, radios blaring, within thirty feet of foraging bears. Someone throws a wadded ball of paper toward a bear, and the bear turns curiously to inspect it. Dogs bark. A fox skulks in from the dark forest and moves toward the buffet until a jealous bear spots it. A glance, a subtle woof, are enough. The fox withdraws.

I take special notice of the silvertip female. Her head is wide, with big cheek muscles, and concave down the profile like a grizzly's. Her coat glitters frostily under the spotlight. Her eyes shine back orange, a rich orange, somehow warmer than a crocodile's. I admire her sadly. The *ursii gunoieri* of Răcădău—what a weird, ambivalent version of harmonious coexistence between people and bears. Is this the future, or is it the past? I watch her tug at, and then swallow, a plastic bag.

50

With the fall of Ceauşescu in December 1989, Romanian bear management became less political and more commercial. Soon afterward, a young ethnic-Hungarian forestry engineer, Arpad Sarkany, set himself up as a broker between the Romanian Forest Department and foreign hunters who might be tempted to come to the Carpathians. Sarkany wasn't the only new entrepreneur in this field, but he seems to have been among the most adept or well-connected, and his company, based in the town of Sfântu Gheorghe, amid the mountains north of Braşov, now claims a sizable share of the trade. He partners with hunting-package marketers in the countries of origin: They find the customers, he finds the animals (not personally, but through the Forest Department) and arranges the permits. Wild boar and red deer are popular among Sarkany's clients, who are mostly from western Europe and look to Romania for a mildly exotic but not too arduous game-shooting experience, with the chance to take home an impressive trophy. Bears are less abundant than boar and deer, less available, but bring higher fees per animal.

"We are the *organizator* for the bear hunt in Romania," Sarkany says categorically, if not quite grammatically, seated behind his fine wooden desk in an office decorated with skulls and tusks. A pudgy man in a bright sport shirt, cordial but indifferent, like an Ottoman sultan, he has summoned coffee for me but taken none himself. "Eighty percent of the bear hunting is coming through us." His pretty good English is seasoned with the Transylvanian variant of a Hungarian accent.

Eighty percent of how much, I wonder, amounting to how many bears? Like any businessman, Sarkany is not eager to open his books to a prying stranger, but he does mention some numbers, which can be put into context with information available elsewhere. Each spring the

Forest Department proposes a total kill quota, based on its area-by-area estimate of the entire bear population. Last year's estimate (according to ICAS) was 5,616, reflecting a decrease from the peak during Ceauşescu's final years but a level that's been stable now for half a decade. Last year's quota for hunting (according to Sarkany's casual recollection) was seventy bears, or maybe seventy-two. Either way, that's barely more than one percent of the total. (Other data I've seen, again from ICAS, indicate an "annual harvest" in recent years of about three hundred animals, and I've heard it said that the "optimal level" of bear density, considered socially acceptable to humans, would be at least ten percent lower than the present level, implying an argument for reducing the population. One way of doing that, of course, would be to reduce the supplemental feeding. Another way, quicker and more lucrative, would be to set hunting quotas higher.) If Sarkany's figures are correct—let's assume for a moment that they are, seventy bears taken from 5,616 overall—then the annual cull should be sustainable, provided that additional restrictions protect females with cubs.

But sustainability is only one issue. Whether commercialized hunting is the best way to conserve a population of large predators is another question. Arpad Sarkany argues that it is indeed the best way, his logic echoing what Grahame Webb has said about saltwater crocodiles and every other sort of dangerous, inconvenient beast: If hunting were prohibited, if no bears were shot, then the bear population would be doomed. It would soon disappear from Romania. Why? "Because is no interest, no *economical* interest, to this *especies*." Without those hunting fees flowing to the Forest Department and those payments (for food, lodging, travel, taxidermy) to local businesses, the incentive for close management would vanish, the supplemental feeding would cease, and the habitat would be lost to excessive timber harvest and other forms of land conversion in the new, desperate Romanian economy. To have bears we must kill bears, honoring each death with the electronic tweet of a cash register. So goes the argument. To me it's a tedious paradox, not a liberating insight, and no matter how often I hear it, applied to one or another magnificent species in their various corners of the world, each time I find it tedious afresh. But, beyond quibbling over details of linkage and enforcement, I can't rationally disagree.

On the wall behind Sarkany's desk hangs a sizable bear skin. Yes, he tells me, that one was worth 320 points. Did he shoot the animal himself? No, he replies, it was poisoned by a baci, up in the mountains. Years ago. Back in Ceauşescu's time. No hunter, no fees, no tourism— no economic benefit to anyone, Sarkany says. What a waste.

A waste? I'm not sure it's so simple. Shepherds, as I've learned, have a relationship with these animals that's more intimate, more mutual, than what you can get through the scope of a Holland & Holland .375 as you stand on a high seat, sighting down. They share habitat with bears. They have reason to fear them. To detest them. They have occasion to know bears the way King Hrothgar knew Grendel—as a grievous affliction, a reminder of personal mortality in a harsh, impermanent world. They have their own, old-fashioned means of coping. They measure bears in a dimension deeper than deutschemarks and CIC points. Maybe that relationship itself, not just Romania's population of *Ursus arctos*, is something too valuable to lose.

51

Viciu Buceloiu, tall and darkly handsome, like a soap-opera actor in his career's twilight, smokes too many cigarettes for his own good. As supervisory warden for three hunting areas near a town called Domneşti, in the Argeş district, he feels job pressure, though perhaps not so acutely as in years past. He has worked for the Forest Department since 1972. This is the fellow whose little daughter Petra, now a grown woman, brought out a certain doting tenderness in Nicolae Ceauşescu during the otherwise tense occasions when the Conducător descended into Domneşti for a hunt. Viciu, having assisted those hunts several times yearly for eighteen years, owns a considerable archive of mixed memories. A survivor, a dutiful functionary who balances professional discretion with instinctive generosity, he proves to be one of my most hospitable Romanian contacts.

He sends me into the forest with his gamekeepers. He hosts me for long evenings of bear watching from his high seats. He opens his house, his bottle of homemade *ţuica* (another endemic brandy), and his family snapshot albums to me. He provides me with photocopied excerpts from *Catalogul Trofeelor*, the record book of national hunting trophies (which tells a wet story dryly: Of the fifty highest-scoring bear skins ever recorded in Romania, forty-five came from kills credited to "N. Ceauşescu"). I impose upon Viciu my specific requests and my greedy, unfocused curiosity, to which he and his wife respond with graciousness, coffee, fresh strawberries, chocolate cake, corn on the cob, a deep-fried Romanian pastry called *gogoş*, and information. Then, one afternoon as we sit in his living room, which is spread with a half-dozen bear skins like so many throw rugs, Viciu puts a video-cassette into the player. He offers no preface or explanation. Music, credits, it's a little documentary about *Ursus arctos*.

The narration is in Romanian. The production quality is rough. There's some footage of a bear attacking sheep. Another bear, in the act of pilfering a chicken, is interrupted by a woman with a broom and a fierce little dog, who together drive it away, panicked and chicken-less. There's commentary on the anatomy and diet of the species. On the mating habits and reproductive schedule. Scenes of big yearling cubs being loaded into wire cages, and then onto a helicopter, for release from the rearing compound at Râuşor. Shots of smiling Forest Department men, in Tyrolean hats, opening the cages at a release site. Eleven cubs emerge simultaneously, looking confused. If they're missing toes, as they presumably are, the camera fails to show which and how many. Within days of release, Viciu comments caustically, the first of them would be blundering into fatal encounters with humans. Râuşor, it was futile, he says. The video ends. He pops in another. This one is more political.

It begins like a fulsome biographical documentary: clips of Ceauşescu in his public role, kissing babies, attending banquets, making state visits abroad. Ceauşescu the family man. Ceauşescu the sportsman, on a hunting excursion. He arrives by helicopter. Later he admires his day's kill, dozens of dead chamois laid side by side. In the background of one shot, fleetingly, appears Viciu Buceloiu himself, looking a dozen years younger. Then the scene shifts, the mood shifts, and I see Ceauşescu before a cluster of microphones on a balcony, addressing many thousands of Romanians in an urban square. Suddenly the camera captures something indecorous and unexpected—the crowd interrupting him. People start to jeer. Ceauşescu waves, as though to hush them. They refuse to be hushed. He looks flustered. More jeers. On the balcony, behind, alarmed minions slide into motion. Again he waves—it's a befuddled, ineffectual gesture. This is the very moment, on the morning of December 21, 1989, in Bucharest, as he stood before a populace so disgusted, weary, and angry that they would accept their own docility no longer, when he lost control. Recalcitrant hoots from the crowd. "Hallo," says Ceauşescu. "Hallo, hallo." Wake up, he seems to be trying to tell them, remember your manners, remember your fear; this is *me*. But they are awake, now more than ever. Elena, in the background, starts hollering at someone. She seems to have sensed, more quickly than her husband, that this break in the national trance could be catastrophic for both of them.

An abrupt cut. The next scene shows Ceauşescu stepping disconsolately from an armored truck. He wears a rumpled dress shirt, a tie, a dark topcoat. That coat, Viciu tells me—an idle aside, reflecting his own sense of intimate connection to the drama—that coat, it was lined with bear skin.

Inside a drab room, Ceauşescu sits at table. Elena is beside him. The room seems to be cold; anyway, they don't remove their coats. Bundled around his neck is a brown scarf. He makes an argumentative response to a voice from a person unseen. The camera doesn't pan, it holds on the two of them, unremittingly, carefully blindered, and by this time I've realized that we're watching the kangaroo trial.

You had a luxurious life, says the offscreen inquisitor. Meanwhile, he says, people starved. They received only a miserable ration, two hundred grams of sausage per day. You had the army to enforce your power. But now even the army has turned against you. Do you hear what we're saying? Stand up. Answer. You're a coward. With the army gone, you had your personal sharpshooters opening fire on the citizenry. Who were they, those snipers? Children died, old folks. Who gave the order? You're guilty of genocide, says the voice. Do you realize you're no longer president of Romania? What about the money in Swiss bank accounts?

Intermittently during this litany of accusations, Ceauşescu stares at the ceiling. He reaches across and sets his hand comfortingly on the hand of Elena, who shows no response. She has withdrawn into a gloom so cold it looks almost like boredom. Sporadically she rouses to haggle. The people love us, the intellectuals love us, she says; they won't stand for it when they hear you've arrested us. The people will fight against this treason! adds Ceauşescu, shaking his finger. As though to refute that notion, to preempt that remote possibility, the video lurches ahead (as the National Salvation Front did) to a point of irreversible finality: a freeze-frame image of Ceauşescu's fresh corpse, necktie still tied, a splash of blood beside his head. Silence. End of tape. Viciu turns off the machine.

Earlier I asked his opinion about what happened to Ceauşescu. Was it inordinately severe, or was it deserved? "It was murder," Viciu told me. Now, having seen this lame pantomime of a tribunal, added to things I've read and heard about the events of December 1989, I sense more vividly what he means. He means that the principal movers of the revolutionary junta had their own reasons—urgently expedient ones—for wanting Nicolae Ceauşescu not just deposed but exterminated. It wasn't the people's anger that killed him; it was a more coldly and narrowly taken decision. Alive, he would always be dangerous to those whose own political résumés weren't impeccably democratic. He would always be capable of stirring an uncomfortable recollection that, as Ion Micu said, the cult of the dictator requires a cadre of cultists and an institutionalized complicity, not just a dictator. More immediately, he might give hope and persistence to those loyalist snipers who were still killing peaceful demonstrators in the streets of Bucharest. Dead, he was instantly irrelevant—and certain people could hope that their own pasts, full of deft opportunism and other embarrassments, had been rendered irrelevant also.

Call it murder, or call it mortal politics of exactly the sort Ceauşescu himself had long practiced. If his execution was expedient, it was no more so than, say, war as an instrument of national purpose, or lethal injection as an instrument of criminal deterrence, or trophy hunting as an instrument of conservation.

THE TEETH
AND THE MEAT

52

The word "carnivore," in one of its usages, carries a capacious but very particular meaning. As a term of zoological classification, it refers to any member of the Carnivora, that order of mammals encompassing the cat family (Felidae), the dog family (Canidae), the raccoon family (Procyonidae), the mongoose family (Herpestidae), the family of weasels and otters and skunks (Mustelidae), the family of genets and civets (Viverridae), the hyenas (Hyaenidae), and the bears (Ursidae). Used this way, it doesn't imply that all bears are carnivorous all of the time. It simply indicates that they fall within a related group of species that, although latterly diversified into various shapes and habits, are descended from common, carnivorous ancestors.

The eight living species of Ursidae have diverged from that meat-eating heritage relatively far. Most of them are omnivores, dietary generalists who exploit a wide range of nutritional sources as opportunity or necessity dictates. The giant panda (*Ailuropoda melanoleuca*) is an exception, feeding almost exclusively on bamboo. The sloth bear (*Melursus ursinus*) specializes in licking up ants and termites (it's insectivorous, as distinct from carnivorous), occasionally supplementing that diet with honey and fruits. Even the polar bear (*Ursus maritimus*), more fully carnivorous than any other bear, with a strong predilection for ringed seal, sometimes succumbs to the temptation of bird eggs or berries. The brown bear is probably the least specialized ursid, willing to eat and able to digest a multifarious menu of items, ranging from grass to tubers to acorns to pine nuts to fruit to insects (larval or adult) to gophers to fish (especially spawning salmon and trout) to elk viscera (left by hunters) to moose calves (a delicacy, but risky in the face of a maternal moose) to carrion (such as the thawed, rotting carcass of a winter-kill bison) to peanut butter (from, say, a picnic box) to organic garbage (in Răcădău flavors, Yellowstone flavors, or any other) to human flesh.

People-eating by *Ursus arctos* is rare and anomalous, but it needs to be mentioned and set in perspective. On that subject, the records available in America are more vivid than the anecdotes heard in Romania. One famous case involved a young Swiss woman named Brigitta Fredenhagen, who hiked alone into the Yellowstone backcountry, in July 1984, and two days later was found to have been pulled from her tent, killed, and fed upon by a grizzly bear. No one knows exactly what happened, let alone why, beyond a few basic deductions from physical evidence at the scene. Fredenhagen's food had been hung between two trees, as caution in grizzly country dictates—though not hung high enough, and a bear had climbed up and gotten it. Her tent had been ripped open. Outside the tent was her sleeping bag and, nearby, according to one sober report, "a piece of lip and scalp with hair still attached." The rest of Fredenhagen's body, or what remained of it, lay eighty yards away. "Quite a bit of her soft tissue had been eaten," said the same report. The responsible bear was never captured or even identified, but inference tends toward the view that it was a subadult, probably male, and had become accustomed to human traffic on the trail near which Fredenhagen camped. She had made no discernible mistakes against recommended practice, aside from hanging her food too low and traveling alone in an area (near Pelican Valley) known to be heavily used by grizzlies. She took a chance, and she was unlucky.

William Tesinsky's luck failed him, also in Yellowstone Park, when he tried to photograph a grizzly at close range. This was two years after Brigitta Fredenhagen's death. Again there were no witnesses, and the event has been sketched inferentially from evidence. It happened in early October, just when bears eat voraciously to gain weight for hibernation. Tesinsky, an auto mechanic from northern Montana with an avocational passion for wildlife photography, had left his car hastily parked in a lot near Canyon Village and hiked up a draw toward a young female grizzly. Evidently he had spotted her from the road. Based on dig marks examined afterward, it seems the bear had been feeding peaceably on yampah roots and other plants. Tesinsky, pushing too near, must have provoked her. She abandoned the yampah and went for him. What occurred next is unknowable, but his photo equipment, as found later, suggested a moment or two of ugly mayhem. The camera was smeared with blood and hair, the tripod was bent, and Tesinsky's last exposed frame was an indecipherable blur. After three days a search party of rangers came looking for the owner of the abandoned car. The bear was still in the area, feeding on one of Tesinsky's legs. His torso was never found. The rest of his body, dismembered, had been variously scattered, consumed, and cached.

According to Scott McMillion, a Montana reporter and the author of a good, sensible book about bear attacks, *Mark of the Grizzly*, not enough remained of Tesinsky to determine the exact cause or manner of death. The autopsy report simply cited "injuries inflicted by a bear." It seems likely that the grizzly's initial attack reflected surprise and anger—her instinct for counterattack against perceived threats or competitors—more than hunger. "Then, once he was dead," speculated a ranger who took part in the search, "she realized he was meat." And meat was a precious resource. The bear, McMillion adds, treated Tesinsky "just like she would have treated an elk or a bison—as valuable calories to be buried, saved and guarded from other scavengers." John Petranyi died in Glacier National Park six years later, again during mid-autumn, the period of hasty gluttony before hibernation. Like Fredenhagen and Tesinsky, he was traveling alone. When found by another hiker, not far from a trail in the Granite Park area, he was in bad shape and maybe already dead. His body was covered with tooth and claw marks; his blood had flowed out in splashes and puddles; one arm and one buttock had been partially eaten. But he, or his corpse, was still warm. The other hiker backtracked to where he'd spotted Petranyi's dropped coat, thinking to cover him against shock or chill. But when the hiker returned five minutes later, Petranyi's body was gone. Whatever had triggered the initial attack, the bear now evidently felt a proprietary interest in the carcass as a bounty of meat. The hiker withdrew until he met other people on the trail, turned them back to get help, and meanwhile stayed in the area himself. Late in the afternoon, two rangers armed with twelve-gauge shotguns and big-slug ammunition arrived by helicopter. Guided by the hiker, they found Petranyi's remains about five hundred feet from where he'd first been seen. The body was now cold. Night fell, the rangers and the hiker retreated, and it was mid-afternoon of the next day before anyone got back to the site. Petranyi's corpse had been moved again, and the bear—or, as it turned out, *bears*, a sow and two cubs—had consumed more of it. The bears were now so jealous of their prey, in fact, that they had to be hazed off by helicopter before the body could be recovered. Even to the lead ranger, long experienced in bear management, this seemed odd.

"A grizzly bear had killed and eaten a man," writes McMillion. "It was an incredibly rare situation. In the ninety-year history of Glacier Park, where thousands upon thousands of people walk through grizzly country every year, Petranyi was only the ninth person to be killed by a grizzly bear. And in most of those cases, the bear left the body alone after the attack. During the same period, forty-eight people drowned in the park, twenty-three fell to their deaths from cliffs, and twenty-six died in car wrecks." But a traffic fatality is a banal misfortune that doesn't impinge on public consciousness so sharply as a homicidal, flesh-eating bear. It's one thing to be dead. It's another thing to be meat.

53

Meat-eating has its advantages. Most obviously, it provides riches of protein and fat. It also entails special demands and risks, such as the necessity of capturing prey, the task of killing what's captured, the high likelihood that a given hunt will end in failure, and the chance that a predator will itself be injured during the hectic business of predation. Claws and other appendages play roles in meat-getting, but the crucial tools for most carnivorous creatures are teeth. Teeth seize. Teeth hold. Teeth sever spinal cords and arteries. Teeth eviscerate. Teeth cut muscle and crush bone. Various sorts of teeth—differing from one animal to another, and from one zone of an animal's mouth to another—are adapted to these various functions.

Dentition is destiny, at least to some degree. You can deduce much about an animal's diet, ecology, and behavior from its dental morphology. Evolution has provided each species with a panoply of teeth that, besides reflecting what worked in the past, dictates what can work in the present and the future. The sloth bear has only forty teeth, two less than other bears; a pair of incisors have been dispensed with, leaving a gap convenient for termite-slurping by means of its long tongue and deft lips. The giant panda has massive molars and enlarged premolars, providing surface area for grinding the stems, leaves, and shoots of bamboo. The polar bear's canine teeth are elongated, enhancing the lethality of its bite; its molars are edgy enough for shearing soft tissue. The brown bear is dentally intermediate. Its molars and premolars are smaller than the giant panda's but large enough to grind corms and grasses; its canines are adequate as weapons but not so big and strong as to suggest purely carnivorous habits. The teeth of the brown bear equip it to be what it is: an omnivore, gobbling whatever comes.

This matter, dental morphology, isn't relevant only to mammals. Meat-eating reptiles and fish are no less dependent on their teeth, and the fit between dentition and feeding behavior shows well among alpha predators of all sorts. The great white shark (*Carcharodon carcharias*) bites with twenty-six upper teeth and twenty-four lower, each of them a triangular, serrated blade, well shaped for stabbing and slicing through the tough hide of a sea lion or a seal. The teeth are constructed from crystals of apatite (calcium phosphate) embedded in a matrix of gelatinous protein. The apatite crystals interconnect in a fibrous structure that provides strength, while the protein matrix offers some flex. Behind the front rank of teeth, upper and lower, are rank after rank of gleaming replacements, standing ready to shift forward whenever an active tooth becomes dulled (from chewing the skin and bone of marine mammals) or is lost. Roughly a third of the animal's tooth slots are in transition at any given time, a fact that reflects the high rate of violent turnover. So a great white shark can afford to be jarringly aggressive without concern for lost teeth. It owns an inexhaustible lifetime supply.

Its cartilaginous skull structure, in which the upper jaw as well as the lower one is mobile, allows the great white to open wide at the instant before impact, raising its snout into a sneer and protruding its chompers like buck teeth. Then it slams the upper teeth down and backward to intermesh with the lower teeth in a deep, scooping arc. While the lower teeth hold the victim pinned, the uppers slice away a sizable gobbet. An ichthyologist named John McCosker has given some thought to this big-bite tactic, and to the puzzling fact that, in most cases, there follows a passive delay that allows wounded prey to escape. McCosker has called it the "bite-and-spit paradox." The explanation he posits is that after such a punishing first strike, the great white generally backs off and waits for its victim to bleed to death. The mix of initial ferocity, caution, and patience saves it the risk of having one of its eyes (which are unprotected by nictitating membranes) clawed by a frantic sea lion.

The Komodo dragon (*Varanus komodoensis*), unlike most other lizards, doesn't swallow its prey whole. Instead it bites off sizable chunks of flesh from victims so large that, in their wholeness, they'd be more than a mouthful. It possesses a gobbet-snatching capacity similar to the great white shark's. Its teeth are hooked blades like linoleum knives, narrow from side to side, thick and strong from front to back, serrated along the inner edge of the curve. Each tooth has multiple replacements arrayed behind it. Some species of monitor lizard (smaller members of the genus *Varanus*) carry just one or two backups per tooth, but *Varanus komodoensis* has as many as five—another sharklike feature, hinting of rapid dental turnover in a life of heedless, violent mastication. The teeth are all similar in shape, but their size and arrangement incorporate two subtle modifications that help in cutting off big portions of flesh. First, the tooth profile of the upper jaw is convex, with shorter teeth at the rear, and slightly longer teeth at mid-jaw, all gradated in a gentle curve that automatically cuts deeper as the Komodo pulls away its head. Second, the teeth as seen in overhead view are each set at a slight angle—with the cutting edge outward, the trailing edge inward—so that they plow easily through tissue and spread the incision as they go. It's no wonder, given such armament, that a Komodo dragon can scoop pounds of flesh off a prey animal with a single bite.

The herpetologist Walter Auffenberg describes these adaptations in a book, *The Behavioral Ecology of the Komodo Monitor*. Auffenberg also notes that the big lizard (known as *ora* in the local Komodoese dialect) sometimes lays into a human. He mentions one case in which the victim was a fourteen-year-old boy, caught unwary while cutting wood in the forest. When the reptile charged, the boy ran—unfortunately, into a low-hanging vine. "The vine stopped the youngster for just a moment," writes Auffenberg, "and the ora bit him very severely in the buttocks, tearing away much flesh. Bleeding was profuse and the young man apparently bled to death in less than one-half hour." During my own visit to the island a decade ago, I spoke with a woman whose mother had survived a similar attack, wrenching herself away from a Komodo that had clamped onto her arm; as she wrenched, the Komodo's teeth stripped the limb of flesh. After a half-dozen years she had regained use of the arm. "Is not problem with bone," my translator relayed as we sat listening to the daughter's shy, squeamish account. "Just the meat."

Auffenberg adds that Komodo teeth "bear a greater resemblance to those of flesh-eating carcharid sharks and carnosaurian dinosaurs than to the teeth of living reptiles or mammals." It's an oblique reminder that Komodo dragons, despite the common reptile lineage and the superficial similarity, are very unlike crocodiles. Their different killing strategies are mandated by, and reflected in, very different forms of dental weaponry.

For instance, crocodiles don't have the advantage of cutting-edge teeth. Theirs are cone-shaped, suited to puncture and hold. Each tooth is set in an individual socket, and replacements arise (as with Komodo dragons as well as sharks) whenever old teeth are lost. The rear teeth are relatively short and blunt, useful for crushing prey under the immense pressures generated by powerful jaw muscles. The middle and front teeth are long, strong, more sharply pointed, and arranged so that uppers and lowers interlock. The fourth tooth on a crocodile's lower jaw is a tall dagger whose sheath is a notch on the upper jaw, in which it remains visibly pocketed when the mouth is closed. That conspicuous fourth tooth, in fact, is one of the diagnostic features distinguishing crocodiles from alligators. All these conical, overlapping teeth allow a crocodile to stab its prey and to grip it remorselessly, but not to dissect it.

If the victim is small, like a barramundi, the crocodile swallows it whole. Head raised, gullet open, the crocodile juggles it down the hatch with help from gravity. If the victim is large, like a deer or a human, the crocodile may pull it into deep water and drown it, using what some observers call the "death roll," a frenetic burst of twirling leveraged by the crocodile's tail. Once dead, the large victim must somehow be disarticulated, since the crocodile can't chew off a manageable bite. Shaking the prey violently, tearing and wrenching, are inelegant but effective means of doing that.

"Few of those who have experienced the crocodile's death roll have lived to describe it," says an Australian woman named Val Plumwood, who speaks on the subject with unenviable authority. Plumwood is an academic philosopher. Her published articles reflect an eclectic mix of interests, ranging from Plato to ecofeminism to crocodile attack. Being death-rolled by a crocodile is a trauma "beyond words," she reports. Nevertheless, she offers some: "total terror, total helplessness, total certainty, experienced with undivided mind and body, of a terrible death in the swirling depths." Plumwood's engagement with the subject happens to be personal as well as philosophic.

54

In early 1985, the start of the wet season for northern Australia, Val Plumwood was in Kakadu National Park, just adjacent to the west boundary of the Arnhem Land Reserve. Kakadu, a world-renowned protected area comparable to Serengeti or Yellowstone, encompasses roughly seven thousand square miles of marshy lowlands, warm rivers, rainforest, savanna, sandstone bluffs, stony gulches, and wildlife, as well as Aboriginal rock paintings reflecting forty millennia of continuous cultural tradition. During the rainy months it harbors vast aggregations of herons, spoonbills, ibises, jabiru storks, magpie geese, and other migratory water birds, as well cockatoos (five species), parrots, lorikeets, budgerigars, kingfishers, and hundreds of other bird species. There are also plenty of native mammals (including three species of wallaby, three wallaroos, three possums, and a bandicoot), numerous fish, snakes, lizards, frogs, and a robust population of salt-water crocodiles. As a national park it's open to public visitation, offering Australians and foreign tourists a glimpse of the ecological riches and cultural mysteries that lie, more closely held, across the boundary in Arnhem Land. Val Plumwood was staying in a trailer near a lagoon on the East Alligator River. On the morning of February 19, she set off alone in a borrowed canoe. She thought she'd seek out a certain rock-art site she had heard about, somewhere among the side channels of the river.

Although not a proficient canoeist, Plumwood was an outdoorsy woman with considerable bush experience, at least in the form of hiking and camping. A ranger had loaned her the boat and assured her that, if she stayed on the backwaters, avoiding the main river, she'd have no trouble from fast water or crocodiles. So she paddled carefully across the lagoon. After hours of exploring a maze of side channels, and still unable to find the art site, she paused for a quick, rain-drenched lunch. At that point she began perceiving a vague aura of menace from her surroundings. She felt she was being watched. She continued exploring, despite intermittent rain squalls, until an odd geologic formation—a single big rock balanced on a little pedestal—caught her attention. The very sight of the balanced rock increased her unease; it seemed some sort of mute reminder about the precariousness of life. In this dysphoric state, Plumwood turned back along a channel that, she hoped, would carry her home. Rounding a bend, she saw what looked like a floating stick. Weirdly, independent of the current, it converged with her canoe.

The stick was a crocodile. Either she hit it or else it hit her; anyway, it registered a surly thump against her boat's hull. Then a second thump. This was no accidental collision. "Again it came, again and again, now from behind, shuddering the flimsy craft," she recalled in an article published eleven years later. "I paddled furiously, but the blows continued." The canoe was under attack, she concluded. "For the first time, it came to me fully that I was prey."

The article, "Being Prey," is a valuable addition to the literature of carnivorous victimhood. It combines an account of the Kakadu experience with her later ruminations about the proper ecological as well as ethical relations between humans and nature. It was published first in an obscure journal, *Terra Nova*, and has since been reprinted several times. But before Plumwood could ruminate, before she could assimilate experience into larger ideas and literary form, she had to escape from her pickle.

Mistrusting the stability of the canoe, she acted on impulse—an impulse that proved almost fatally ill-advised. She stood, grabbed the lowest limb of a paperbark tree arching out from one bank, and tried to pull herself up. For a moment she dangled like a chunk of bait. The crocodile surged out of the water and seized her "between the legs," as she puts it, "in a red-hot pincer grip." She went down into the dark water.

Then came the death roll—actually, the first of several. Plumwood remembers it as "a centrifuge of whirling, boiling blackness, which seemed about to tear my limbs from my body, driving water into my bursting lungs. It lasted for an eternity, beyond endurance, but when I seemed all but finished, the rolling suddenly stopped." Crocodile metabolism allows only a short burst of furious exertion before the animal exhausts itself and must rest. The water was shallow. Plumwood got her feet down and her head above water, catching some breath; but, below, the crocodile hadn't turned her loose. After a moment it spun her again. Then again it paused, this time relaxing its jaws. She pulled free. She could move. Which way to go? The channel's bank looked too slick and muddy, so she reached for the paperbark limb. As she tried to climb up, the crocodile grabbed her again, now on the upper left thigh. It snatched her down into death roll number three—a canonical number, like something from the Old Testament. Jonah or Job, take your pick.

By her own later estimate, this animal was somewhere between eight and twelve feet long—more than big enough to overpower a strong person if it could just get the right leverage, hold her in deep water, drown the fight out of her. But Plumwood was more than strong; she was also tenacious and, as events would show, lucky. She had good lungs and a stiff will to survive. When the crocodile paused once more, she was still conscious. "I could see the crocodile taking a long time to kill me this way. It seemed to be intent on tearing me apart slowly, playing with me like a huge growling cat with a torn mouse." Probably it *was* intent on tearing her apart, not slowly and not for play, but because it had no other way of swallowing what it had caught. A shark or a Komodo dragon, by this time, would have bitten away a piece of her lower body, and she'd have been dying fast of blood loss and shock, like the Komodoese boy who got tangled in the vine. Instead, unable to disjoint her, unable to kill her with the energy and the water at its command, the crocodile again opened its jaws.

Now she tried for the bank, clambering up its muddy slope, slipping, sliding back, clawing desperately. Finally she reached the high ground. Gasp, wheeze, reconnoiter. She had escaped the crocodile, but she wasn't out of danger. Punctured and bloody, she hiked several hours toward the boat landing, collapsed on a bank, lay there in swampy darkness, and by good fortune and stamina lived long enough to be rescued. Thirteen hours later she was in a Darwin hospital.

Val Plumwood's account of all this, her analysis of it within the framework of her philosophic and political convictions, is by turns riveting, self-congratulatory, and trenchant. Since there were no witnesses to her ordeal, we can't know just how accurate all its precisely recounted, harrowing, heroic details may be. Yes, she was attacked and badly injured by a crocodile. Yes, she escaped, thanks largely to her own pluck. She therefore brings great personal force to her rebuttal of what she calls "the masculinist monster myth" and to her critique of the tradition within Western culture that views nature as separate from and inimical to humanity. Her "masculinist monster myth" is nothing other than the storybook demonization of alpha predators, cooked up and perpetuated for the sake of romanticizing their conquest by heroes (seldom heroines) such as Beowulf, Sigurd, Ashurbanipal, J. H. Patterson, and Nicolae Ceauşescu. Plumwood's own crocodile, to the contrary, did *not* seem "an implacable monster," she writes. It seemed merely an angry animal she had somehow offended, perhaps by intruding into its territory. Nature is not utterly profane, humanity not utterly sacred, she notes, and the boundary between them—perceived as stark—is illusory. It's a blurry and interpenetrating edge, subject to "boundary breakdowns" that challenge anthropocentric complacence. Making these sensible points with such experiential authority, Plumwood's "Being Prey" is an important document. Even when she's hampering her narrative with mystic intuitions and a touch of excusable braggadocio, a death-roll survivor commands attention.

"If ordinary death is a horror, death in the jaws of a crocodile is the ultimate horror," Plumwood testifies. Its deeper horribleness derives from those "forbidden boundary breakdowns," combining

decomposition of the victim's body with the overturning of the victory over nature and materiality that Christian death represents. Crocodile predation on humans threatens the dualistic vision of human mastery of the planet in which we are predators but can never ourselves be prey. We may daily consume other animals in their billions, but we ourselves cannot be food for worms and certainly not meat for crocodiles.

She's right. But she stretches too far, in that argumentative final sentence, bundling two ugly fates that deserve to be kept separate. The worms of the graveyard don't scare us the way the crocodiles of the East Alligator River do. We can contemplate being nightcrawler food without feeling primordial dread. Surrendering the cold carrion of our bodies to all manner of verminous decomposers, months or years after death, is an abstract indignity, not a living fear. It pales in comparison with the experience on which Plumwood is so qualified to speak: being prey.

55

The saltwater crocodile shares one other aspect of dental morphology with the Komodo dragon and the great white shark. In the jargon of science, their dentition is *homodont*, meaning "all teeth alike." For the crocodile, that's every tooth cone-shaped; for the Komodo and the shark, every tooth a flesh-cutting blade. In each case there's little variety within the animal's mouth, no elaborate modification into different tooth shapes for different functions. Homodonty: a simple arrangement, but it works.

Which is not to say that it couldn't be bettered. During the great course of time, as mammals evolved from their reptilian ancestors, dentition among other things got more complicated. Modern mammals are distinctly *heterodont*—that is, "different-toothed." Those differences between one sort of tooth and another show with particular clarity among the Carnivora. Within the jaws of a bear or a wolf or a hyena, teeth of dissimilar shapes and sizes are tactically arranged to perform dissimilar tasks.

Incisors, usually six of them across the front of each jaw, upper and lower, serve for nipping and cutting and tugging. Canines, one on each side behind the incisors, are enlarged and pointed for stabbing and ripping. Premolars and molars are variously adapted for cutting, crushing, or grinding. And the fourth upper premolar on each side, along with the first lower molar, have been modified at least somewhat, in all the Carnivora, so that they slide together as scissorlike blades. In that form, they're known as carnassial teeth. Highly modified carnassials are excellent tools for slicing flesh. Each kind of tooth fits a function, more or less closely, and the varied array within a given animal suggests its repertoire of hunting styles, eating habits, and foods. That repertoire may be generalized, like the brown bear's, or specialized.

Of all Carnivoran families, none is more specialized than the Felidae. Cats are predators, period. They don't browse the landscape as though it were a salad-and-cold-cuts buffet. They hunt, they kill, they eat meat. They are full-time carnivores, and it shows—in the shapes and arrangement of their teeth.

To begin with, cats have fewer. The tooth count for many species of Carnivora (including the canids, the ursids, and the viverrids) falls in a range from thirty-eight to forty-two. Felid jaws contain only thirty teeth or, in certain species (the lynxes, the golden cats, the caracal), twenty-eight. Dispensing with a dozen teeth allows the cats to have blunter faces and shorter jaws than a mongoose or a wolf. Their short jaws, rigged with stout muscles (notably the masseter and the temporalis) for closure, offer the mechanical advantage to deliver an extremely powerful bite. The missing teeth are premolars and molars, of which most cats have only fourteen (eight upper, six lower), compared with twenty-six in the typical canid. The absent premolars leave a large gap behind a cat's canine teeth, permitting those canines to be rammed deeply into a prey animal, like a dagger buried to its hilt. The carnassials (given room to expand, again thanks to the absence of other molars and premolars) are lengthened front to back and are high-ridged, providing a more efficient meat-cutting edge than in other carnivores. Near the midpoint of that edge is a V-shaped notch, which holds slippery tissue against the pressure of the slicing stroke. The canine teeth, two above and two below, are longer than doggy canines; they're also rounder in cross section, making them less suitable for tearing flesh but stronger against the risk of breakage. Tooth fracture is a serious problem for all the Carnivora. Broken bones mend, but when a mammal's tooth breaks, it's broken forever.

There's a peculiar inaptness to the very term "canine" as applied to teeth, since those have a bigger role, and a more extravagant evolutionary history, in the felids than in the canids. Compared with the canine teeth of dogs, wolves, coyotes, and their kin, the canines of cat-family species (both living and extinct) tend to be larger, more prominent, and more pointedly deployed in the killing of prey. The typical kill as performed by a small-bodied cat, upon a still-smaller prey animal such as a rabbit, involves a well-placed bite on the back of the neck. Driving one canine tooth like a wedge between two of the neck vertebrae, the cat forces those vertebrae apart until the spinal cord snaps, causing death. For the big cats who kill prey as large as or larger than themselves, though, a dorsal neck bite can be risky. So lions, tigers, cougars, and leopards usually rely on a suffocating bite, applied either to the underside of the throat or over the mouth and nostrils. Even the cheetah, which is more gracile than other big cats and has less formidable canine teeth, generally kills with a throat bite. The zoologist R. F. Ewer, in her classic book *The Carnivores*, mentions an interesting hypothesis about cheetah dentition: that small canine teeth with small roots allow room in the cheetah's face for its large nostrils, maximizing air intake during the animal's prodigious sprints. "Strange though it may sound," she notes, "it may therefore be true to say that the cheetah has small canines because it runs so fast."

Stalking and pouncing tactics have permitted other cats to get by with less wind and longer teeth. Among some extinct felids, canines were more conspicuous still. Most famously, they grew to immoderate sizes among the sabertooth cats.

56

Sabertooths are mystifying creatures that, for all their fame, aren't well understood—not by the general public, not even by the scientists who study them. Just how they used their oversize teeth is a question still subject to debate, and their phylogeny too is complicated. For starters, there's a distinction between sabertooth *cats* and sabertooth *mammals*, the former constituting a subset of the latter. Mammalian evolution has seen sabertooth marsupials and sabertooth creodonts as well as sabertooth felids. In fact, the phenomenon of sabertooth dentition arose four separate times, each time yielding a cluster of tusky species, and only one of those clusters lay within the cat family.

Since the four evolutionary episodes all converged toward a single style of tooth, this constitutes a textbook case of convergent evolution (or *iterative* evolution, to be more precise, since the different sabertooth groups occurred sequentially rather than simultaneously). About fifty million years ago, saber-shaped canine teeth appeared in an early mammalian order, the Creodonta, a group of widely distributed creatures that dominated meat-eater niches at that time. Much later, sabertooth forms turned up in another order, the Marsupialia, represented by a long-toothed, leopard-sized marsupial known as *Thylacosmilus atrox*, fossils of which have been found in Argentina. The upper canines of *Thylacosmilus atrox* were big—bigger in proportion to body size than in any sabertooth cat—and with the animal's mouth closed, they fit snugly down the outside of a deep flange on the lower jaw, like a sort of open sheath. About thirty-five million years ago came still another group of sabertooths, this time among the Nimravidae, a now-extinct family of carnivores that were superficially similar to cats. The nimravids, diverse and successful throughout Europe, Asia, and North America for almost twenty million years, included *Barbourofelis fricki*, a lion-sized beast with huge upper canines and another large flange on the lower jaw. *Barbourofelis fricki* survived until about six million years ago, at least in North America, by which time it was the last of the nimravids. Meanwhile the cat lineage had diverged from the other carnivores, with one branch generating its own set of sabertooth forms.

The great age of true sabertooth cats began about fifteen million years ago, in the midst of the Miocene epoch, with a genus of felids known as *Machairodus*. Five million years later, the species *Machairodus aphanistus* was common in Europe. It may also have spread across Asia into North America, where closely related species then arose. These included *M. giganteus*, distributed from Greece to China, and *M. coloradensis* in America, both of which showed a slightly more advanced stage of sabertooth modifications than *M. aphanistus*—their sabers were narrower, their carnassial teeth more bladelike. The genus *Machairodus* gives its name to the felid subfamily Machairodontinae, encompassing all the sabertooth cats. From the *Machairodus* line diverged another sabertooth genus, *Homotherium,* which appeared about 4.5 million years ago, yielding *H. latidens* in Europe and Asia, *H. ethiopicum* in Africa, and *H. serum* in North America. The several *Homotherium* species all shared a distinctive body plan, with especially long front legs as well as long teeth, as exemplified in one notable specimen, an intact skeleton from a cave in Texas. Other fossils in the same cave suggested that *Homotherium serum* was abundant thereabouts and that the cats had been feeding on baby mammoths.

Roughly contemporary with *Homotherium* was still another interesting genus, *Dinofelis,* containing a handful of species with teeth and skeletal features intermediate between those of modern cats and sabertooths. Sometimes labeled the "false sabertooths," they were just sabertoothy enough to be placed among the Machairodontinae. Their canines were knifelike rather than rounded, but not very long. One of those species, *Dinofelis piveteaui*, is known from a skull found at the Kromdraai cave site in South Africa that dates back 1.5 million years, at which time it shared the landscape with early hominids. Did that knife-toothed cat include big-brained primates among its prey? The ecological relationship between sabertooths and hominids is an obscure subject, well worth contemplating, but for the moment let's set it aside.

The most familiar of sabertooth cats is an American species, *Smilodon fatalis*, represented abundantly among thousands of bones extracted from the tar pits at Rancho La Brea, in Los Angeles. *S. fatalis* was just one species of a genus that spanned a range of body sizes and a broad sweep of American geography. *Smilodon gracilis*, its smaller cousin, appeared earlier in eastern North America. *Smilodon populator*, as big as a lion, with upper canines protruding almost seven inches from the jaw, evolved in South America, probably from an ancestral population of *S. gracilis* that had wandered down from the north. But *S. fatalis* looms more vividly in scientific memory than those others, based on the wealth of its recovered remains and the recentness of its disappearance. Until about ten thousand years ago, with humans moving through the region, *S. fatalis* survived at least in southern California. From the La Brea deposits alone, excavators have pulled 160,000 sabertooth bones and 1,775 teeth, representing at least twelve hundred individual cats. Such an abundance of *S. fatalis* presumably reflects the tendency of its preferred prey—giant camels, bison, horses, and other large herbivores—to get themselves stuck and find themselves vulnerable amid those tricky asphalt mires. The cats converged on the victims and, in turn, got fatally stuck.

The excessive enlargement of canine teeth among the sabertooth cats raises two simple questions, each divisible into further questions and correlative mysteries. First, why did evolution produce these structures? What adaptive value (if any) did they offer? What benefits outweighed the metabolic costs of growing such grossly elongated and backward-curved canine teeth, and the inconveniences of living with them? How did sabertooths kill and eat?

The second simple question is contrapuntal to the first: Why did sabertooth felids *cease* to exist? What caused their extinction after millions of years of success? Why did *S. fatalis* and its close relatives disappear during a time through which many other large cats—including *Panthera leo*, *Panthera pardus*, and *Puma concolor*, all of them still around today—managed to survive? Why didn't any sabertooths make it into the modern era?

The question about adaptive value and killing technique has no certain answer, despite 150 years of expert speculation. Beginning as early as 1853, when J. C. Warren published a paper in the *Proceedings of the Boston Society for Natural History*, the assumption was that sabertooths had used their big teeth for stabbing. At the turn of the twentieth century, another paleontologist added the suggestion that they had preyed mainly on thick-skinned herbivores such as mammoths, striking deeply with the initial bite, then ripping or gouging a large wound, through which the victim bled to death. These guesses are supported by the fact that sabertooth neck muscles (especially those that would have driven the teeth downward) were strong. A later variant of the deep-stabbing scenario had them plunging their canines into the back of the victim's neck or the base of its skull, in a sabertooth version of the dorsal-neck-bite technique of small cats.

There's a problem with this hypothesis. The canine teeth of most sabertooth cats were not only long and curved but also laterally flattened—that is, knifelike, as in the false sabertooths, though far bigger. Lateral flattening would have made the teeth more useful for tearing flesh but also more susceptible to breakage when they were rammed against bone or subjected to sudden sideways stress during the frenzy of a kill. This is the same problem for which the sheathlike lower jaw of that marsupial sabertooth, *Thylacosmilus atrox*, made some provision: the longer and thinner a tooth, the more fragile. But even a sheathing jaw flange, protecting the teeth when the mouth was closed, offered no help while the teeth were buried deeply into a struggling victim.

A statistical study of tooth breakage among carnivores, published by

Blaire Van Valkenburgh in *The American Naturalist*, showed that it's a common occurrence and possibly a serious problem, especially for bone-eating species such as hyenas, but also for lions, leopards, and cougars. Among those three species of cat, the teeth most often broken were canines. A related study, co-authored by Van Valkenburgh, found even greater incidences of tooth breakage among the large carnivores entombed at La Brea. But the data from that study contained one small surprise—that *Smilodon fatalis,* the best-known American sabertooth, had somehow minimized damage to its huge, vulnerable sabers. It had suffered broken incisors and premolars almost as often as broken canines. This suggests a killing technique different from the wild, risky, neck-bite methods of modern felids.

An early alternative to the deep-stabbing hypothesis was the carrion-slicing hypothesis, which circumvented the tooth-breakage problem with a notion that sabertooths weren't so much predators as scavengers. Maybe they lived at the fringes of the action, using their big, edgy canines for ripping into half-rotten carcasses. The carrion-slicing idea had its adherents for while, but it lost standing in the 1940s and 1950s as further analysis of skulls and teeth seemed to confirm that the sabertooth cats had been predatory. Nowadays the consensus favors sabertooth predation over sabertooth scrounging, but researchers still offer hypotheses and counterarguments about how it might have been done. They generally agree that the big upper canines must have been used to inflict long, shallow, arcing bites rather than deep ones.

But how would such bites serve the purpose of killing prey? A scientist named William Akersten argued in 1985 that a "shear-bite" to the abdomen of a victim, held immobile by the sabertooth's front paws, would have caused enough blood loss—along with, possibly, bringing on shock—to put the victim out of commission. In Akersten's illustrative scenario, a pride of sabertooths stalk a herd of mammoths; one sabertooth, noticing an unprotected juvenile, dashes in and wrestles the little mammoth off its feet; the cat delivers a shear-bite, opening a bad wound in the juvenile's abdomen, then withdraws before the maternal mammoth can stomp it silly. "The pride regroups at a distance," Akersten posited, "and waits for the critically injured juvenile to die and for the rest of the herd to leave." If you want a modern analog to this predatory strategy, he added, take a look at the Komodo dragon. He might also have mentioned McCosker's bite-and-spit hypothesis about the great white shark.

A recent variant on Akersten's idea, from Alan Turner and Mauricio Antón, suggests a shear-bite not to the abdomen but to the throat. Such a throat bite would have inflicted "massive and, more importantly, rapid damage to the windpipe and major blood vessels." This hypothesis, even more than Akersten's, assumes that the sabertooth had some means of stopping a prey animal, flipping it topsy-turvy, and restraining it sufficiently to expose the underside of its neck. Skeletal evidence reveals that *S. fatalis* and other sabertooths did have the means: exceptionally strong forelegs and retractile claws.

Retractile claws—these are the other great weapon of felid predation. Among all the Carnivora, only cats and a few species of viverrids possess them. Retraction allows the claws to grow long and remain sharp, drawn up amid the soft protection of the paw pads when they're not in use. A pair of retractor ligaments between the second and third phalanges (toe bones) pull the claw back into its rest position, and against this constant tension a protractor muscle (attached to the underside of the third toe bone, on which the claw is mounted) delivers a countervailing tug, swinging the claw out whenever the cat contracts that muscle. Because the muscle contraction is voluntary and the ligament pull is automatic, as R. F. Ewer explains, it would be more accurate to say that cat claws are *pro*tractile, not retractile, but the net effect is what matters: front paws that convert from running pads to grappling hooks instantly, like the snap of a switchblade.

Dogs, bears, and hyenas possess no such ingenious mechanism, which is why their claws, constantly scraping against the ground, remain relatively dull. Among sabertooth mammals, the nimravids had retractile claws; so did *S. fatalis* and the other sabertooth cats. This association between retractile claws and extreme dentition, recurring from one group of species to another, hints further of a link between sabertooth advantages and sabertooth hazards. A predator armed with those long, slicing canines, capable of deep penetration but susceptible to breakage, may have had greater need for grasping, restraining, and steadying each squirmy prey animal. Saberlike claws could have provided the crucial complement to saberlike teeth.

Strong forelegs, deft paws, retractile claws, scimitar canines, muscular necks. With such an arsenal of assets, why did they die out?

The popular notion that sabertooth cats were ultimately doomed by the very hypertrophy of their teeth—finding themselves overspecialized, unable to bite properly, unable to eat—is wrong. For millions of generations, their outlandish dentition had served them just fine. The urge to find a single answer to the whole extinction riddle, a single cause for their species-by-species disappearance, is probably also misguided. The sabertooth cats (unlike the dinosaurs?) didn't get conked on the head by a killer asteroid. Over a span of fifteen million years and across five continents, they came and went, variously prospering when the regional climate was good, when the landscape was hospitable, when the prey animals were abundant and accessible; variously growing rare, rarer, and finally extinct when those favorable circumstances changed for the worse. They were eliminated, as so many species in the history of life have been, not by a sudden hammer-stroke of doom but by the slow, ineluctable abrasion of time and change.

At the end of the Miocene epoch, about five million years ago, climatic changes brought important transformations in the vegetation structure of European landscapes. About the same time, more than a hundred genera of European land mammals vanished, including three large sabertooths from the genus *Machairodus*. We can guess that those cats were suffering secondary effects, such as starvation, from the loss of their prey species. But we can't know for sure.

In eastern Africa, beginning around three million years ago, another wave of climatic and vegetation changes seems to have shifted the mammalian herbivore fauna toward greater representation of fast-running antelopes. Four groups of species within the cat family coexisted in the region at that time: true sabertooths such as *Megantereon eurynodon*, some remnant members of the *Machairodus* lineage, intermediate forms within the *Dinofelis* genus, and smaller-toothed felids such as we know in Africa today, namely the lion, the leopard, and the cheetah. Then the African populations of sabertooth went extinct. So did the *Dinofelis* group. The smaller-toothed species, which were quicker on their feet, survived to become stars of the postmodern photo safari.

Competition with those speedier cats, as well as the changes in vegetation structure and prey availability, may have contributed to the disappearance of the African sabertooths. "Another factor," suggested C. K. Brain, a South African paleoanthropologist, "was almost certainly the rise of human intelligence and technology." Brain's notion derived from his studies of hominid fossil sites, such as Kromdraai. "Sabertooths, like all large carnivores, must surely have posed a threat to early man," he wrote. "One may be sure that human hunters would have taken steps to minimize this threat." But *how* sure may one really be? His implication is that human hunting bands, partly in self-defense, may have gone some way toward exterminating the sabertooths. It's just logical speculation, not a hypothesis supported by data, but it comes embedded within Brain's painstakingly empirical 1981 book, *The Hunters or the Hunted?*

In Europe too, lions and leopards shared landscape with the last of the sabertooth cats, at least until *Homotherium latidens* dropped away. And in America, the overlap of sabertooth and smaller-toothed cats lasted longer than anywhere else—to the end of the Pleistocene epoch, about eleven thousand years ago. By that time, unlucky individuals not just of *Smilodon fatalis* but also of the cougar (*Puma concolor*) and of what paleontologists call the American lion (*Panthera atrox*, very similar if not identical to *Panthera leo*) were getting themselves bogged in La Brean tar. It seems odd now (but maybe not inappropriate, given some current bigger-than-life features thereabouts) that the world's last surviving sabertooth cat was a resident of the Los Angeles area.

57

Eleven thousand years ago, those lingering American sabertooths may have had some contact with the first American humans, but the pits of La Brea haven't yielded evidence of any such event. In Africa, where the sympatry of sabertooths and hominids occurred 1.5 million years earlier, as reflected in fossils from the Kromdraai cave, the nature of their interactions is also unlimned by solid evidence. But some researchers have been bold enough to speculate. An anthropologist named Curtis W. Marean has posited that the presence of sabertooth cats actually benefited our Pleistocene relatives, such as *Homo habilis*, by providing a supply of incompletely consumed herbivore carcasses from which enterprising but timid hominids could scavenge the occasional free meal.

The supposition behind this idea is that sabertooths killed big-bodied prey but didn't consume them completely—not so completely as a modern lion does—because their teeth were unsuitable for disarticulating skeletons or crushing bone. Scraps of flesh and rich cores of marrow must have remained, offering precious nutritional resources to whatever hyenas or vultures or jackals or omnivorous primates could exploit them. When the habitat changed and the sabertooths went extinct, according to Marean, those hominids may have been forced toward a riskier and more aggressive form of scavenging, such as stealing carcasses from lions and leopards who wouldn't willingly

328 MONSTER OF GOD

relinquish them. That shift, in turn, may have entailed new survival demands that helped push hominid evolution toward larger body size (from *Homo habilis* to *Homo erectus*) and more effective social cooperation. Prehumans became more nearly human, by this hypothesis, when circumstances compelled them to live like hyenas.

C. K. Brain suggested a different possibility. His study of the Kromdraai cave site, and of several other South African caves, led him toward a vision of the hominid *Australopithecus robustus* (which he called *Paranthropus robustus*) as a frequent victim of predation by several species of cat. The caves contained large numbers of fragmentary primate fossils, both from *Australopithecus robustus* and from baboons. Those fragments seemed to represent the chewed-up remains of carnivore prey, whole animals that had been dragged back to a safe haven for eating. Among Brain's crucial bits of evidence is one australopithecine cranium, known by its specimen code as SK 54. This cranium, the top section from the skull of a child, is perforated by two mysterious punctures. In spacing and size, the punctures match almost perfectly with the lower canine teeth of a leopard. Brain speculated (in a journal paper, then again in his book *The Hunters or the Hunted?*) that leopards commonly preyed on australopithecine hominids, hauling their kills into trees (as leopards still do) in order to eat them without being harried by hyenas. Because some leopard-perch trees were located just above cave openings, Brain argued, Kromdraai and those other sites captured much of what fell, becoming bone middens for hominid-eating leopards.

He added that "more than one species of cat was involved in the predation process. Besides leopards, it is highly likely that *Dinofelis* cats also participated in this process." *Dinofelis* cats, you'll remember, were the "false sabertooths" with bladelike canines of intermediate length. They lived in the area at the time. They were capable of killing terrestrial primates. Maybe, Brain hazarded, they had even become specialized for that task—they, and perhaps also others. Besides the false sabertooth *Dinofelis piveteaui*, the true sabertooth *Megantereon euryn-odon* left behind fossil evidence of its presence at Kromdraai.

So it's possible that sabertooth cats loomed as monsters of special menace within the dawning consciousness of early hominids. But the evidence, to be frank, is skimpy and circumstantial. C. K. Brain's chain of argument is attenuated. And our relatives at that distance, more than a million years ago, haven't passed down any eloquent hints. We'll never know whether their fear of big-toothed felines was acute, reverentially muted, or dulled into routine amid a welter of other dangers. Survival on the landscape was difficult, epic poetry hadn't yet been invented, cave art was a thing of the future, and death by predation must have seemed ordinary. No one had escaped the awareness of being meat.

PERESTROIKA

58

I've come to the Bikin River valley, in the Sikhote-Alin mountain range of the Russian Far East, because I want to talk with the last of the people who have an intimate relationship with the greatest of the last great predators. The people are Udege, members of an indigenous tribe belonging to the Tungus-Manchu racial group. With their broad, strong cheeks below crescent-shaped eyes, they look vaguely Manchurian or Mongolian. Their ancestors have hunted, trapped, and fished along the Bikin for more than a thousand years. Russia, to these folk, is a colonial abstraction imposed from afar, and the Union of Soviet Socialist Republics is only a memory involving state-run fur-buying outposts. The predator is *Panthera tigris altaica*, casually known as the Siberian tiger.

Less casually, this cat is the Amur tiger. It's a more accurate label, given that Siberia (as strictly construed by Russian geographers) is a tigerless inland region that doesn't actually include the Russian Far East (though by a looser and more common construal, Siberia does extend to the Pacific), whereas the Amur river drainage encompasses almost all of the historical range of the *altaica* subspecies. Among traditional Udege the tiger is called Amba, a respectful name for an almost deified character about whom their attitudes are complicated, ambivalent, and in flux.

The culture of the Udege has endured, and the tiger has endured alongside it, for one reason: because the Bikin valley is a very remote place. The river itself is barely larger than a modest trout stream. It flows west and southwest out of the Sikhote-Alin range, which forms the arched backbone of Primorskiy Krai (Maritime Province) at Russia's extreme southeastern edge. Meandering some 350 miles, emerging from the mountains onto an open plain, the Bikin empties into the Ussuri River, which drains to the Amur, which circumvents the mountains in flowing northeastward to its mouth on a strait near the Sea of Okhotsk. The Ussuri basin is much developed with agriculture, industry, and towns such as Dal'nerechensk, but the Bikin still holds the largest intact forest—roughly 1.2 million hectares of fir, spruce, white birch, Korean pine, and other species—on the western slope of the Sikhote-Alin. The nearest big city is Khabarovsk, a notorious provincial capital downstream at the Ussuri-Amur confluence. If none of these names sounds familiar, it's because few of us ever glance at a map of Asia and run our eyes toward the hulking, inscrutable mainland just northwest of Hokkaido.

To get there you fly the Pacific, bounce briefly at the Seoul airport in South Korea, and land again in Vladivostok, the old Soviet naval port on the Sea of Japan. In a small plaza along Vladivostok's Okeanskiy Prospekt, overlooking the bay, you can see a huge bronze tiger, cast at the local ship-repair plant as an icon of post-Communist civic pride, but the real tigers that once prowled into the city's neighborhoods—eating dogs, reminding Vladivostokians that they lived on a wilderness frontier, and giving rise to municipal labels such as Tiger Street and Tiger Hill—have been killed off or driven out. Their habitat no longer enwraps the metropolis. Although one tiger was shot at a Vladivostok trolley stop as recently as 1986, having wandered there under God knows what ill-starred impulse, the main population of *Panthera tigris altaica* is now confined to the Sikhote-Alin slopes and valleys, of which the Bikin is among the most northerly, isolated, and pristine.

Escaping the city, you drive north on an icy highway along the east side of the Ussuri, on the west side of which lie the Manchurian provinces of China. After ten or twelve hours of harrowing two-lane traffic, relieved occasionally by a pause for sausage rolls and Coke at a truck stop, you turn east onto a gravel road that ascends the Bikin. Within a few miles the road becomes a snow-packed track, which receives just enough traffic to keep it polished slick as the pate on an alabaster bust of Lenin. You follow this ribbon of white into darkness, winding upward toward a village called Krasniy Yar, the main settlement of Udege along the middle Bikin. If your tires are good and you left Vladivostok early, you might reach Krasniy Yar in time for a mid-evening dinner of vodka and fish soup and potatoes. If your tires are as smooth as curling stones, like the ones on an old Toyota belonging to a Russian biologist named Dmitri Pikunov, your passengers will walk the hills, throwing their shoulders against the back fenders whenever your wheels begin spinning futilely. As moments of traction allow, you'll surge forward in second gear, leaving your pushers to straggle behind, where they can savor the exhilaration of a frosty nighttime hike through tiger habitat. On the flats you'll wait for them, and if luck holds you'll top the last rise without having slid the Toyota into a ditch. In that case you may still reach the village, as Pikunov does, and I with him, in time for a late but triumphant dinner of vodka and fish soup and vodka.

Dima Pikunov is a burly, impetuous man of vehement moods and quiet charms, now in his early sixties but still fit enough and stubborn enough to track tigers through snowy mountains. His eyes are pale blue, his hair is sandy and thinning, his chest swells high above an ample belly. Excitable, obdurate, and brusque, he's also generous and impassioned, a rowdy one-of-a-kind fellow who will alternately amuse you and bully you, like a cross between Mel Brooks and Nikita Khrushchev. Born in the Urals, he took his early training in wildlife management at an institute in Irkutsk, a Soviet city just north of

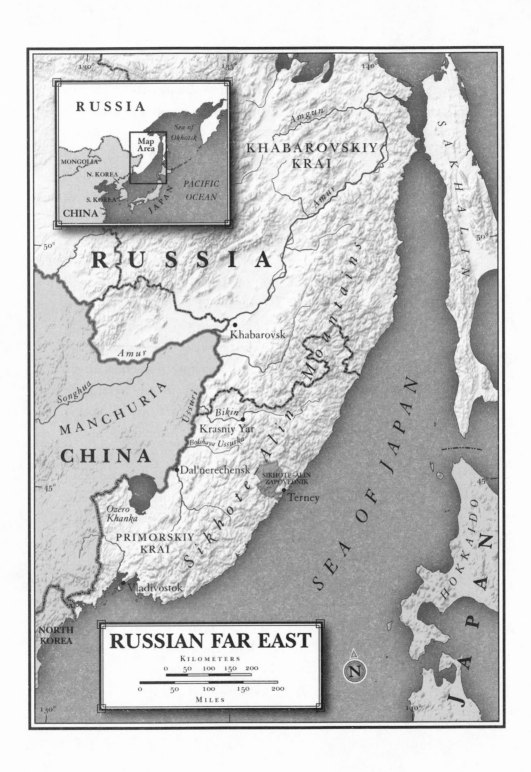

RUSSIA

KHABAROVSKIY
KRAI

SAKHALIN

Sea of
Okhotsk

MONGOLIA

N. KOREA

S. KOREA

CHINA

JAPAN

PACIFIC
OCEAN

Map
Area

Amgun

Amur

RUSSIA

Khabarovsk

Amur

Songhua

MANCHURIA

Ussuri

Bikin

Krasniy Yar

Bolshaya Ussurka

Dal'nerechensk

SIKHOTE-ALIN
ZAPOVEDNIK

Terney

CHINA

Ozero
Khanka

PRIMORSKIY
KRAI

Sikhote-Alin Mountains

SEA OF JAPAN

HOKKAIDO

Vladivostok

NORTH
KOREA

JAPAN

RUSSIAN FAR EAST

KILOMETERS

0   50 100 150 200

0   50   100   150   200

MILES

N

Mongolia, then continued his eastward migration to the Far East. He arrived in 1961. Part of his early work focused on the game species that provided sport-hunting opportunities for military personnel. He first visited the Bikin valley in 1969, assigned under a branch of the Soviet Academy of Sciences to assess its populations of large mammals, some of which were economically important for their meat and their furs. He wrote a dissertation on leopards, then in 1977 found support for what became his life's major work, a long-term field study of the Amur tiger. One dimension of that study has been pragmatic: What impact does tiger predation have on standing stocks of commerciable wildlife? Another dimension is more personal: Dima Pikunov is fascinated, as a scientist and an animal lover, with the magnificent cat itself. For the past twenty-five years he has spent part of nearly every winter in the Bikin, learning the landscape and absorbing a sense of how tigers live in it. He works the old-fashioned way—by following sign, reading clues from the prints and the scrape marks and the fed-upon kills. Throughout those decades of fieldwork he has rarely been blessed with a glimpse of a living tiger.

For the first dozen years, he didn't dart them, trap them, or radio-collar them. He relied on non-intrusive observation and inference, not because he was a romantic or a Luddite but because to a wildlife biologist in the Soviet Union no other methods or tools were available. Among his published scientific works is a study of the tiger's eating habits, based on the examined remains of 720 prey animals. He once followed a single tiger for more than forty-five days, traveling on skis with a heavy pack, resupplied intermittently by helicopter, feeding himself from the leftovers of tiger kills when his food stock got low. He carried a small tent, a stove, a radio as big as two loaves of bread, and, for lack of space, no sleeping bag. Always he moved patiently, hanging back, avoiding visual contact, calibrating his own pace against the freshness of the tracks, so as not to disturb a tiger from its natural rhythm of travel. He saw the population rise back from a historical low. He saw tigers reoccupy habitat that had been vacant. Continuing his work in the new Russia, trying to gather a uniform, long-term set of data, he still didn't dart, trap, or radio-collar any cats. He remained a tracker. Nowadays he enjoys the convenience of a Buran snowmobile, but much of the real business is still legwork, done on skis and on foot. I've come to the Bikin River with him because few other Russian researchers can match his on-the-ground knowledge of the Amur tiger, and because no other tiger biologist is informed by such a long-standing, attentive association with the Udege people. I've come because he was enough of a temerarious, trusting soul to invite me.

The old Toyota was crowded during our long slippery ascent to Krasniy Yar. In addition to Dima and me, and our bags of groceries and field gear, and a broken-handled shovel, and Dima's feckless black dog, it carried one other traveler, an expat American named Misha Jones. Long-haired and bearded, hefty as a linebacker, Misha is a liberal-spirited Oregonian who has lived in Russia for two decades—sufficient time to become a fluent interpreter and a valued liaison man for a variety of short- and long-term clients, of which I'm the latest, and to accede to the permanent transmogrification of his first name, which was originally Michael. Besides serving me as translator, Misha is making his first trip back to Krasniy Yar since the years he spent here with a Udege girlfriend, attempting to broker some community-development support from international organizations. On the middle Bikin, as elsewhere, plenty of new economic pressures and seductive moneymaking opportunities materialized after the fall of the USSR, but what Misha envisioned was different: savvy, light-handed assistance of a kind to leave the culture and the ecosystem intact. He tried to deliver. It was an effort that, for reasons beyond his control, ended sourly. His misgivings about returning now, several years later, are mitigated by the opportunity of sailing in under my flag rather than his own. Helping an ignorant American writer ask hunters about tigers is far less delicate than hemming the edges of a civic misunderstanding or a torn romance.

The total surviving population of Udege, here and elsewhere, is almost as small as the population of Amur tigers. The tribe numbers about eight hundred souls, maybe fewer. Fewer still are those Udege who live by the old ways, shooting deer and boar for meat; trapping sable, squirrel, otter, and Siberian polecat for pelts; spending much of each winter in solitary cabins strung out along the Bikin and its upstream tributaries. The village of Krasniy Yar, with its wooden bungalows and shacks fronting rectilinear lanes, is an artificial concentration of what once was a forest-dwelling people. It reflects forces exerted by Soviet planning (though similar effects might have come from a frontier market), including the mandated extraction of furs, paid for and redistributed through a system of hunting collectives known as Gospromkhoz. "We have turned the Udege into professional, industrial hunters," Dima says about that arrangement. The very commercialization of their woodland skills has tended to draw them away from the woods. Too many Udege, he adds, now spend too much of their time cooped up in Krasniy Yar, drinking vodka. But among the winter residents of the village are a few old men, retired and semi-retired hunters, enfeebled by time and not alcohol, who can speak about Amba, the tiger, with a sense of connectedness rooted deeply in the past.

Dima warned me, before we left Vladivostok, against idealized preconceptions of the Udege as a sylvan tribe. We were at his apartment, having a get-acquainted talk through the intermediation of Misha Jones. Dima, wearing a denim shirt with a Dallas Cowboys logo over the left breast, had welcomed me with his strong handshake and his crease-edged, skeptical eyes. We sat in his office. He was cordial but, clearly, a man with no time or patience for bullshit. What sort of writer was I, and what did I want from him? As Misha translated my explanations, I took a few seconds to glance across the tiger books, the bear books, the Russian scientific journals filling his office shelves, among which I noticed a paperback copy of *Grizzly Years*, by Doug Peacock.

That's a good sign, I thought. *Grizzly Years* is an extraordinary memoir of Vietnam and afterward, by a former Green Beret sergeant who found healing from his war experiences among the grizzly bears of the northern Rockies. Peacock is not a biologist and doesn't pretend to be; he's just a fierce, bullheaded, large-hearted man who loves dangerous beasts as much as he loathes the decadent pieties of late-modern civilization, and who understands those beasts in ways irreducible to ethological metrics. Seeing his book on Dima Pikunov's shelves, like a wink from an old friend, inspired me to articulate my purposes in a way that might suit the Russian's own prickly sense of mission.

What did I want from him? I mentioned my interest in traditional relationships between native peoples and the big predators with whom they share habitat. *Traditional* relationships? Humph, let me throw a little cold water on your thinking, Dima said bluntly.

Even among the Udege, he said, things are changing. Tradition is eroding. Circumstances are shifting. For instance, take the trapping of sable. It's an extremely important cash-earning activity, formerly practiced only by men on homemade skis, each of whom worked his trapline in an allotted area, basing himself for the winter at a simple cabin. Trapping areas were passed down from father to son. There was room for everybody, because "everybody" constituted very few people. But recently, Dima said, the Udege have discovered that snowmobiles make trapping easier. Inflation has struck—the inflation of prices and of desires. Just a few years ago, a snowmobile could be had for twelve sable skins. Now the cost is ten times that. Imagine, paying one hundred twenty sable for what was once accomplished with a pair of skis! Given such temptations, some Udege have started to think: *If I killed a tiger . . .*

Dima didn't need to tell me that dead tigers are precious for their pelts, their organs, and their bones, all of which can be sold to stealthy middlemen who find ways of smuggling them out to China, South Korea, and Japan. Some sense of the poaching problem, as it faces the

Amur tiger in particular, has leaked into the conservation literature. China supports a lively trade in supposedly medicinal tinctures and pills made from tiger bone, and a packet of bones goes at a high rate per pound. A skin can bring four thousand dollars. Nowadays the tiger is protected by Russian law, and the law is enforced at least partially by anti-poaching teams. But the anti-poaching teams are few and they can't cover every forest road, every village, every lonely spot throughout Primorskiy Krai where a tiger carcass might change hands. Under such circumstances, and given the fact that Udege people are as human as any others, it's no wonder that ancient attitudes toward Amba are becoming vitiated by new attitudes toward ready cash, technological conveniences, and distilled spirits bottled in distant cities. Dima didn't mention all these factors in our first chat, but he hit the main points. One hundred twenty sable!

On the other hand, an eroding culture isn't a lost culture, and the old ways of trapping, hunting, coexisting with Amba still abide, not just in memory but in practice, at least among a small subset of Udege on the Bikin River. They're mostly old men. Dima knows them well, he explained, from his own decades of fieldwork up there. Yes, their voices are worth hearing. If I want to meet them, Dima said, then okay, he will take me. Evidently my pitch had been satisfactory, or my preconceptions weren't too off-putting. Maybe my friendship with Peacock brought me luck. Anyway, for some reason he was willing to invest a little effort in my education.

But there were practicalities to consider, Dima said. Winter conditions on the Bikin are severe. Difficult even for him. He customarily travels upriver, above Krasniy Yar, by snowmobile and skis. No easy transport exists for visitors. He'll need to make some arrangements. And of course it will be cold. Seriously cold.

Not a problem, I thought blithely. Skis? Sure, let's go. Snowmobile? If necessary. Cold? Hey, I'm from Montana. Little did I imagine.

59

There's an old ecological principle known as Bergmann's Rule, which is not really a rule, just a description of what some scientists take to be an empirical pattern. The pattern, they say, is that body size among animals tends to be larger in cold regions than in warm ones.

More precisely articulated, Bergmann's Rule applies not to differences between species but among the individuals of a given species. It's traceable to an obscure German publication from 1847, of which the key point is: "Races of warm-blooded vertebrates from cooler climates tend to be larger than races of the same species from warmer climates." House sparrows living in Canada, for example, are generally bigger than house sparrows in Central America. The inferred reason for this pattern is that largeness, and therefore a relatively low ratio of surface to volume, makes it easier for creatures to conserve heat. Innards generate heat; skin dissipates heat; the more innards per skin area, the greater the heat retention. A few dissenting biologists have dismissed Bergmann's Rule; they've challenged, in fact, both the inferred mechanism of adaptation and the very notion that any such pattern exists. But whether it exists or not, and whether the imperative of retaining heat in cold climates accounts for it, Bergmann's Rule is sometimes cited to explain why the Amur tiger grows extraordinarily big. It's logical to suppose—though this is unproven—that tigers in the snowy mountains of southeastern Russia might need larger bodies than tropical tigers, thereby exposing less surface per volume, to keep themselves warm.

The species *Panthera tigris* seems to have originated in eastern Asia, possibly as long as two million years ago. From there it dispersed across the Asian landmass. Eventually, regional concentrations of tigers became isolated into separate, far-flung populations. To what extent such isolation dates back into the Pleistocene (reflecting the emergence of natural barriers, such as the Taklimakan Desert) and to what extent it has occurred within recent centuries (owing to the fragmentation of tiger habitat by human activities) is hard to know. Scientists have recognized eight subspecies, each localized in a habitat zone, the zones scattered from one end of Asia to the other—that is, from the southern coast of the Caspian Sea, down through India and Malaysia, onto the nearer Indonesian islands (which were connected to Malaysia by land bridges during the low-water conditions of recent ice ages, allowing tigers to colonize what are now Sumatra, Java, and Bali), up through Indochina and China, as far north as the Amur River basin and the coastline of southeastern Russia. Some researchers have lately questioned the validity of those subspecies, both on genetic and on morphological grounds. Genetic work suggests that the eight-way isolation is recent, too recent for subspecies to differentiate. A morphological study, looking at body size, skull characters, coloration, and striping patterns, shows that variations in such parameters are more gradual than discrete. The author of that study, a cat expert named Andrew Kitchener, finds no more than a "very poor scientific basis for the eight currently recognized subspecies of tiger." He adds that *Panthera tigris* is critically endangered no matter how the species is parsed, and that illusory taxonomic distinctions might have the bad effect of leading conservation planners, with their finite resources, astray.

But the recent history of the species is still written in terms of those eight subspecies, or races, or whatevers. Three went extinct during the twentieth century: the Caspian tiger (*P. tigris virgata*), the Bali tiger (*P. tigris balica*), and the Javan tiger (*P. tigris sondaica*), the last of which disappeared from its final refuge within Java's Meru Betiri National Park in the early 1980s. Another subspecies, the South China tiger (*P. tigris amoyensis*), is now nearly if not totally extinct in the wild. With three subspecies gone and one very doubtful, that leaves the fate of the tiger dependent on the status of four other subspecies: the Indian tiger

(*P. tigris tigris*, sometimes called the Bengal); the Indochinese tiger (*P. tigris corbetti*, named for Jim Corbett, though the man-eaters he hunted were in India, not Indochina); the Sumatran tiger (*P. tigris sumatrae*, last of the island populations); and the Amur tiger. Together these four aggregations, plus the few if any individuals in South China, total somewhere between five thousand and seven thousand animals, the entire sum of *Panthera tigris* surviving in the wild.

From the evidence of specimens left behind, it's possible to say that the Bali tiger was much smaller than most other subspecies, and that the Javan subspecies was smallish too. This reflects the well-known phenomenon of insular dwarfism. Largish mammals, when isolated on islands, tend to evolve into miniature forms. Pygmy hippos on Madagascar, midget mammoths on the Channel Islands, a dwarf elephant on Sicily during the Pleistocene, undersize tigers on Java and Bali—don't get me started. Near the opposite extreme is the Indian subspecies, which grows almost as big as the Amur tiger, though presumably not for the same reason.

Indian tigers have little need to conserve heat. In lieu of Bergmann's Rule, they may be subject to Geist's Rule, a more recent notion that links enlarged body size with seasonal peaks in abundance (as opposed to unvarying year-round availability) of food. Andrew Kitchener suggested that link in his article on variations in tiger morphology. Then again, Geist's Rule derived originally from observed patterns among herbivores, not among predators, for whom seasonal fluctuations of food supply tend to be far less extreme. Plants burgeon and then die with the seasons; plant-eating creatures may well be adapted (by large body sizes, which allow them to binge, then go hungry, and survive) to such cycles. But because prey populations don't rise and fall so abruptly, applying Geist's Rule to predators is more of a stretch.

The most reliable size comparisons among tiger subspecies are probably the ones derived from measuring skulls and teeth. Kitchener made such comparisons and found that body size among tigers "appears to vary clinally with increasing latitude, reaching peaks in the north of the Indian subcontinent and the Russian Far East." In plain talk, he means that average size changes gradually, not abruptly, from one area to another. Kitchener's published paper includes some spiffy coordinate graphs, showing tooth and skull measurements from dozens of specimens plotted against the latitudes where the specimens were taken.

Of course, you and I don't care greatly to hear that the crown length of the fourth upper premolar was 34 millimeters in this dainty Balinese animal versus a whopping 38 millimeters in that Siberian behemoth. We prefer a more vivid, holistic impression of size. How large were the tigers themselves? Well, reports vary—which isn't surprising, given that a freshly killed tiger or a momentarily tranquilized one doesn't lend itself to measurement quite so well as a skull. According to an authoritative source, males of the Indian subspecies top out at about 560 pounds, whereas a male Amur tiger may exceed 670. But even "authoritative" publications sometimes incorporate hearsay and exaggeration. Another source, a Russian researcher named Igor Nikolaev, with excellent field credentials and despite the incentive of national pride to entice exaggeration, is more conservative. Nikolaev testifies that he has never seen or heard of a wild Amur tiger above 650 pounds. Still other researchers, after years of trapping and weighing tigers, report that every male they have handled was well *under* 500 pounds.

Nikolaev's testimony comes via Peter Matthiessen, in his graceful book on *P. t. altaica* and its native landscape, *Tigers in the Snow*. Matthiessen himself adds some useful perspective on the matter of size: "While it is true that in its winter pelage, *altaica*—once called *longipilis* due to its long hair—appears more massive than the Indian race, it is only two to four inches taller at the shoulder than those mighty Bengal tigers, *P. t. tigris*, from the northern subcontinent under the Himalayas." Still, four inches' difference at the shoulder is no insignificant bit. On felids in the quarter-ton range, it could represent a sizable difference in weight. But never mind. Be conservative. Put the maximum for *P. t. altaica* at, say, just over 600 pounds. Put the average size of a grown male between 400 and 500. Beyond question, that's still a very large pussycat.

Fur coloring also varies among tiger populations, and these variations too correlate with geography. The background color ranges from darkish red to pale yellow, with darker shades more characteristic of tigers in Southeast Asia and lighter orange or yellow more typical of Amur tigers, especially in winter. Do the differences in fur color reflect differences in habitat and the imperatives of camouflage? Probably. Do those differences track any single measurable parameter—say, humidity, as it varies from the wet Indochinese rainforests to the dry temperate forests of Russia? Possibly. There's even another fancy bit of labeling for that notion: Gloger's Rule. Stripe color varies too. But again, with stripe color and background color as with body size, there aren't discrete differences from one subspecies to another. Kitchener, having looked as carefully at color as he did at skull size, saw more variation from one animal to another within a regional population than between populations in different regions. He also studied striping patterns—thick stripes or thin? how many? do they end in spots?— among which he found a wealth of variations that not even Bergmann, Geist, and Gloger, all ruling together, could explain.

Kitchener's analysis of size, color, and other morphological variables tends to suggest that, not many millennia ago, tigers were widespread across the Asian continent. They seem to have interbred fluidly and adapted to local conditions incrementally, without acquiring the more conspicuous evolutionary peculiarities that come from prolonged local isolation.

One further point about color. A careless assumption, embraced by people who don't know better and some who should, is that Siberian tigers are white. The correlative assumption is that white tigers are Siberian. They aren't, and they aren't. This imaginary whiteness of *P. t. altaica*, presumably based on the supposed camouflage value against snowy landscapes, only further confuses the subjects of tiger evolution and tiger conservation. Polar bears are white, yes, and the tundra wolf (*Canis lupus albus*) is white, and the arctic fox (*Alopex lagopus*) turns white in winter. All of them are carnivores, whose stalking tactics seem to be aided by snowy camouflage. But Siberian tigers, under whatever name, are not white—except conceivably in the rarest (no instance is recorded, so far as I know) of circumstances.

White tigers are mutants, not members of a white subspecies or race. They are aberrant individuals manifesting a rare genetic disorder. Their problem is that they are homozygous for a certain recessive allele, which deprives them of pigmentation and often entails the incidental affliction of strabismus. Stated more plainly, they're blanched and, in some cases, cross-eyed. As partial albinos, they have blue irises and pink noses as well as creamy white fur marked with smoky stripes. Where do they come from, if not Siberia? In 1951, a white male tiger was captured at a place called Rewa, in central India. Caged within a palace, given the name Mohan, bred to a captive female and then bred again to his own daughter, he fathered a litter of four white cubs. Mohan thereby became the progenitor of most of the white tigers ever seen in captivity. His line, with the signature whiteness, was perpetuated by further incestuous crossings of animals bearing the mutant allele. In later years many of them were produced, as a sort of box-office gimmick, by a breeding program at the Cincinnati Zoo. When you see a white tiger, check the animal for crossed eyes. Then remember: Its extraordinary fur color, and its sadly befuddled gaze, may tell you something about bygone eras of palace life in India or zoo management in Cincinnati, but they say nothing about adaptive conditions in the mountains of eastern Russia.

The winter coat of *P. t. altaica* tends to be paler than its summer fur, though not nearly so pale as the fur of a polar bear or an ermine. It blends well amid brown tree trunks and leafless brush. That seems to be camouflage enough; the forest, after all, is only snow-white from knee height down. The tiger's fur grows out shaggy, as Matthiessen notes, providing a cold-weather adaptation that would be worse than irrelevant in the tropics. It makes life possible for a big cat—but not easy, never easy—in the Sikhote-Alin mountains.

60

On our first full day in the village of Krasniy Yar, Dima and Misha and I tramp from one house to another, through the crystalline cold of a bright February day, to call on four elderly men, each of whom in his prime was a masterly hunter and trapper. From them I get a sample of memories, beliefs, and attitudes that reflect the complexity of human-tiger relations, which are richly ambivalent, even in such a remote bit of outback as the Bikin. I also hear, voiced and echoed in various plaintive forms, the same theme I'd expect from four elderly men of any indigenous culture—no, make that any culture, period—on the planet: the corrosion of their familiar world by forms of mystifying, hurtful, ever-accelerating change. We start with a visit to Su-San Tyfuivich Geonka, who lives in a house of drab planking and giddy yellow shutters, sharing the place with his pretty granddaughter, a black spaniel puppy, and a calico cat. How's your wife? asks Dima, making small talk to lubricate our intrusion. She died in November, says Su-San Tyfuivich.

The old Udege wears a green sweater and a sad, pinch-lipped expression faintly resembling a smile. Settling himself at a little kitchen table to receive us, he plugs into his left ear a hearing aid from which a small antenna wire dangles down. A rug hangs over the door frame for extra insulation. A ceramic-block stove, standard style throughout the Russian Far East, pushes some warmth into the room. The granddaughter scrubs dishes as we talk.

Su-San Tyfuivich began work as a professional hunter in 1934, he tells me. Collectivization had come to the Far East on a cold wind blowing out of Stalin's Moscow, and the Collective of the Seventeenth Party Congress, named for an especially chilly 1934 gathering, held sway on the Bikin. Sable furs were the premium commodity hereabouts, but of course a man could also take red deer, musk deer, wild boar, squirrels, chipmunks, and other species. One chipmunk pelt was worth twenty squirrel pelts, he says—twenty kopeks for a chipmunk versus one kopek for a squirrel. (Why chipmunk pelage should have been so much more valuable than squirrel, he neglects to explain. The aesthetic appeal of striped markings?) Later, it went to forty kopeks per chipmunk. For them he used a small rifle. For the big animals, a shotgun. He delivered furs and meat to the collective in exchange for commissary credits that would buy other items. In summer he punted upriver in a flat-bottom boat. There were no roads. He kept bees, cows, horses. He planted a little buckwheat and oats. In winter he lived in a tent, nicely furnished with a wood stove and a floor of pelts. "It was very convenient," Su-San recalls. The furs and meat he brought out by sled.

Chinese traders sometimes came upriver in those years, seeking furs and ginseng, musk gland from the musk deer, penis from the red deer, deer tail, deer velvet, all manner of forest products. And tiger bones, tiger skins? No, not tiger. Not that Su-San Tyfuivich was aware of, or that he cares to remember.

His own relations with the big cats were distant and respectful. "I saw them. But I never shot them. It was not permitted to us to shoot tigers." Apart from the Soviet regulations, that was proscribed by Udege beliefs. "If you shoot a tiger, then fate will come back and do something to you," he says. Fate's instrument would be the tiger's own spirit, harrowing you, exacting retribution. "Nothing good will come from the killing of a tiger." Like other Udege, Su-San Tyfui-vich occasionally hunted with dogs, but unlike others, he never lost one of his canines to a tiger. He kept away from tigers as much as possible, even making it a scruple not to swipe their kills. "If you take a tiger's prey, that tiger will surround you and will not give you any peace." Respect the tiger and it will respect you, that's his dictum. Otherwise, you suffer ugly consequences. He can cite cases, name names. "Those who shot tigers don't have a long life," he says. Dima, the rational scientist, doesn't buy this notion—though even Dima has noticed, he admits, that tigers seem to have an uncanny gift for settling scores. "A tiger will find its way to get revenge," Su-San Tyfuivich insists. "Because a tiger is an enchanter. It will enchant you."

Nikolai Alexsandrovich Semonchuk tells a different story. White-haired and squinting, he's nearly blind from cataracts, which spares him from having to view the midden of empty vodka bottles and other trash that his sons and daughters-in-law have allowed to pile up in the foyer of his little house. Dima and I sit on a sagging bed, Nikolai Alexsandrovich on a stool, as he offers near-random recollections of one year and another—the year when he shot fifteen hundred squir-rels, the year when wolves killed his dogs, the year when a dog went blind from fright after being chased by a tiger. During all his decades in the forest, he says, he had little direct contact with tigers. He never even heard one roar, except a single time, distantly. Still, he begrudges tigers the toll they take on wild boar and red deer, especially now that the boar population has been so badly reduced. (The current scarcity of boar is a concern throughout Krasniy Yar. Has it resulted from tiger predation, overhunting, or subtle ecological factors triggering a natu-ral fluctuation? That's a complicated issue of interest to Dima but not to Nikolai Alexsandrovich.) His career as a professional hunter, hav-ing begun in 1957 under the Gospromkhoz collective system, ended in 1993, just as the post-Soviet vacuum was being filled by new forms of authority and opportunism, including a sudden increase in tiger poaching. Whether incipient cataracts prevented Nikolai Alexsandrovich from seeing those changes is a question I don't ask.

"The tigers sort of squeezed me out," he says. They've gotten too abundant, he thinks. They eat up the boar, they make life difficult for the hunter. "I was out there scaring off the tigers all the time. The tigers came back to scare me, so I left." This defeatist grumble seems inconsistent with his testimony of having seen and heard them so seldom. Then again, Nikolai Alexsandrovich is not obliged to be consistent. He's a sightless old man, a reservoir of memories and attitudes, who hasn't asked for my attention, and he can say whatever he damn pleases.

Ivan Gambovich Kulindziga is even less willing to sacralize *Panthera tigris altaica* as some mystic spirit of the forest. A handsome seventy-three-year-old, slightly dapper in a black felt vest and rose-and-blue paisley shirt, he smokes his cigarettes in a pipe-shaped little holder. He's a full-blood Udege, Ivan Gambovich declares proudly. As a boy he went to school downriver in a village filled with transplanted Ukrainians; the Ukrainian kids called him "slant-eyes," so he was forced to beat them up. In 1962 he started work as a hunter for the collective, then got drawn into administration. "I have thirty-six years of Party membership," he says, another point of pride. In the late 1980s, suffering heart troubles, he abandoned the apparatchik role for a more tranquil occupation—hunting again, among tigers, in the forest. Now he appears fit and strong, a Udege elder statesman, distinguished-looking enough to pass for governor of Manchuria.

His friendship with Dima is warm but edgy, and he seems to enjoy honing that edge with dismissive comments about Dima's favorite animal. "What does the tiger give us? What good is it?" Well, it's a native animal, says Dima. So is the wild boar! declares Ivan Gambovich. This exchange leads back to the question of whether tigers bear responsibility for depleting consumable and commerciable wildlife. "You can't eat 'em. They don't produce anything. They steal our game." Ivan

Gambovich, armed with verbal intelligence as well as forest skills, is a formidable, shifty arguer. One moment he advocates open season on tigers. A moment later, no no, he was just kidding. Still another moment passes, and he's pressing again. Your tigers are all right, he tells Dima condescendingly, patting his friend's knee—but do we need so many? Something should be done.

I ask Ivan Gambovich whether he has ever shot a tiger. Oh, he declines to talk about that in front of Dima, but . . . changing his quicksilver mind, he does talk. "The damn thing took four of my dogs. Four of my best sable dogs." In another case, on some similar provocation, he killed a tiger and buried it. Buried it, even the meat? Especially the meat, says Ivan Gambovich. Have you ever skinned a tiger? he asks me. It stinks. Takes a week to get your hands clean. Rising to a brazen flourish, he admits to four instances. "The first time I killed a tiger was because it got in the way when I was going to shoot a red deer. The second case was when a tiger was following my son around on his trapline. The third time was this one where I shot it and buried it." And the fourth, that was the dog-killer.

Dima isn't shocked. Most likely he has already heard these reports, if not directly from Ivan Gambovich, then on the breezes of village gossip. Anyway, his policy—a sensible one—is to remain engaged with his Udege friends, whether or not they share his fondness for *P. t. altaica* and comply with Russian-imposed strictures against shooting it. When this indigenous culture dies away and the native hunters die with it, he knows, whatever replaces them—timber entrepreneurs, dredge miners, road-building crews, recreational snowmobilers on vacation from Seoul or Minneapolis—is likely to be worse for the forest, the river, and the inconvenient creature formerly known as Amba.

As for me, I'm not shocked either. The testimony of Ivan Gambovich is discouraging, yes. It's a splash of cold water—just what Dima promised me, back in Vladivostok, as corrective for my ingenuous (as he saw it) curiosity about relations between the Udege and the tiger.

But that's not to say I find it surprising. I find it notable. I find it real. I find it complicated, in a familiar way. The world is full of cold water, and ever since I began traveling to investigate the status of alpha-predator populations that share landscapes with humans, I've been feeling chilly and wet.

The fourth of our Krasniy Yar codgers is Vladimir Alekseevich Kanchuga, whose very nickname is Amba, suggesting some special affinity between him and the tiger. I've heard about this guy, famed as one of the supreme Udege hunter-trappers within living memory, capable of taking more than a hundred sable in a year. His house, just down the lane from Su-San Tyfuivich's place, is a brown frame with blue trim and a hayloft attic. The great Amba turns out to be a round-faced, narrow-shouldered man of about eighty, husky but tiny, with gray hair that stands long and upright, like the coiffure of Einstein or Don King. He's mending a fishnet when we arrive.

In old times among his people, Amba tells me, the tiger was considered something of a god. But understand, he says: No single big *God* occupies a central position in the Udege belief system. Instead they recognize many of what might better be called . . . spirits. Important spirits. The tiger is among those. It has a symbolic connection with the hunt. A hunter might pause at certain spots in the forest and offer prayers, reverently acknowledging its presence. And there were other forms of observance, such as the one his mother taught him, involving a strip of cloth. You carried this in contingency and, upon encountering a tiger in the forest, you tied the cloth around a tree, bowed, and withdrew. That showed respect; it also brought luck. Amba himself practiced the cloth-strip obeisance, he says, which partly accounts for his exceptional success in trapping sable. As for killing a tiger, only rare circumstances would justify that—if the animal was old and decrepit, say, or otherwise disabled—and even then the person would pray, asking forgiveness.

About his nickname? Well, as an eighteen-year-old, Vladimir Alekseevich got very sick. He doesn't name the ailment, but it was serious enough to afflict him for a year. When he finally came through it, he says, he'd lost fear of everything. And so he was given that nick-name, in reference to the fearlessness of the tiger.

During the prime years of Amba's hunting career, tigers were rela-tively numerous. Every winter, he saw tracks. Almost every winter, he encountered a living tiger. His own hunting territory was once occu-pied by seven tigers with overlapping ranges—a large male, a female with cubs, several younger individuals. The male was a smart animal who would sometimes trail along as Amba hunted, following his snowmobile track, alert to the chance of an easy meal. But it caused no mischief. Amba himself, though occasionally spooked by a tiger, was never attacked. His mother's cloth-strip practice must have helped. But his dogs, who paid no such respects, were less lucky. "I probably had six or seven dogs eaten by tigers. I went out one time with six and came back with only two." And it wasn't just dog losses that he begrudged. It was also the wild game, for which tiger and hunter com-pete. Yes, Amba agrees with those who blame declining ungulate pop-ulations on the abundance of tigers. There are too many of them nowadays, he thinks.

He remembers the era in which tigers were scarce. Back in the 1930s and earlier, no law prohibited shooting them. Chinese traders visited the Bikin, eager to buy every tiger bone they could find. Prices went high. "If you killed a tiger and sold the bones," Amba says, "you were set for life."

Then the Chinese were expelled—in the wave of paranoia, xeno-phobia, imperial consolidation, recriminations, denouncements, betrayals, and purges that one historian has wryly called "ideological perestroika." Of course it all originated from the febrile brain of Josef Stalin, carried outward and to extremes by his henchmen and reaching the Far East in late 1937. Amba doesn't go into political details, but history records that Stalin personally dispatched Genrikh Lyushkov, commissar of state security, on a mission eastward to extirpate foreign elements and disloyal inclinations. Roughly nineteen thousand Chinese were deported or otherwise relocated (some may have been packed off toward the Siberian interior, in boxcars, along with equally unwelcome Koreans) by May 1938. This was just three years after a huge *zapovednik* (a strict nature reserve) had been designated in the heart of the Sikhote-Alin range. Mandating protection for 1.8 million hectares (almost seven thousand square miles) of wildlife habitat, the Sikhote-Alin Zapovednik wasn't primarily intended for tigers, but it was destined to benefit them incidentally.

As in the case of Nicolae Ceauşescu and his greedy arrogation of Romania's bears, a weird ambivalence adheres to the various effects of Comrade Stalin's ugly regime in the Far East. "If the Soviet authorities hadn't become the power in the region," Amba tells me, "it might have been the end of the tiger."

61

Amba's memory of tiger scarcity in the 1930s jibes with reality as reported by other sources. That was the low point for *Panthera tigris altaica*, which had once thrived over an area spanning the Korean peninsula, most of Manchuria, the Amur basin on both sides of the main river (including its Ussuri and Bikin tributaries), and the eastern slope of the Sikhote-Alin range, which drains directly to the sea. A rough estimate by one Soviet-era biologist suggests that in the period just before economic development took hold, somewhere between six hundred and eight hundred tigers inhabited the Russian Far East. By 1940, according to another estimate, the tiger population throughout eastern Russia had fallen to no more than thirty.

For decades these estimates, and the studies supporting them, were unavailable to anyone who didn't read Russian, but lately they have leaked into English, at least in the form of abstracts and typewritten translations. Among the earliest was a paper by Nikolai Apollonovich Baikov, a former tsarist officer who had come east to serve as a guard on a railroad line transecting Manchuria (a shortcut from Irkutsk toVladivostok) and, having transformed himself into a naturalist, explorer, and ethnographer, made wide-ranging journeys through the eastern Manchurian wilds. Baikov, with a strong antipathy for the Soviet regime, settled in the Manchurian city of Harbin and from there, in 1925, published a compilation of natural history and cultural lore featuring what he labeled "the Manchurian tiger," *Felis tigris mandchurica*. Notwithstanding the nomenclature, which has been rejected by modern taxonomists, he was writing about *Panthera tigris altaica*.

Baikov made no guess at the population size in 1925, but he estimated that fifty to sixty tigers were killed annually in Manchuria and another twenty-five or so in Russia, all of which went to the Chinese market. Professional hunters used trip-wire guns set on ridge trails along which tigers, during winter, left their tracks in the snow. Hunters had also lately begun to poison tigers with strychnine and to blow out their brains (leaving the body and skin intact) with blasting caps placed in bait carcasses. A tiger skin might bring hundreds of dollars if the animal had been killed in its winter fur. Other body parts were valued variously. "The whiskers, heart, blood, bones, eyes and liver as well as the sexual organs of an adult male tiger bring an especially high price," Baikov wrote. Besides making powders and pills from the organs and bones, Chinese customers ate the meat. "It is rather tender, tasty, and odorless," he reported, from either hearsay or personal experience; "fried, it has the color of lamb and tastes like beef or pork."

In contrast to such gourmandism, he sketched the more reverential attitudes prevailing among what he called "semi-barbarian inhabitants" of the taiga (the northern coniferous forest), which included Tungus peoples such as the Udege, as well as Manchurian native tribes. "Consciously recognizing their weakness in the face of this mighty predator, and realizing the uselessness of doing battle with him," Baikov wrote, "these naive children of nature in ancient times created a special cult of the tiger." The cult was perpetuated across campfires and in remote yurts and cabins within the forest. Only interlopers from China and elsewhere, immune or oblivious to this dimension of respect, were impertinent enough to target tigers. Native trappers and hunters, Baikov explained, "under the spell of the threatening 'tsar of the taiga,' not only avoid hunting it, but take every possible measure to protect it from outsiders (i.e., the Chinese and the Russians), who view the tiger as nothing but a profitable hunting trophy." The "tsar of the taiga" was more honored in these "semi-barbarian" circles than the tsar of the Romanovs had been. Pious old Manchurians refrained from calling it by its name, so as not to seem disrespectful and stir up its wrath.

"If a tiger eats a human," Baikov wrote, then the trappers declare that "in an earlier life this person was probably a pig and the predator recognized this by his smell; if he hadn't been a pig, then in any case, a dog, for otherwise the tiger would not have bothered him." It sounds like a variant of that hard-hearted fatalism among the Turkana of Lake Rudolf: Getting eaten by a predator signified not so much pitiable misfortune as divine justice. These traditional Manchurian forest peoples venerated the tiger almost as a deity, Baikov added, offering it sacrifices—sometimes even human sacrifices, such as a dia- pered baby tied to a tree—in gratitude for good hunting, or to placate a particularly bold animal that had already killed humans. If Baikov had his facts straight and they really *did* offer human flesh to rogue tigers, by the way, it probably only made their man-eater troubles worse.

Even the veneration of tigers by native peoples, as Baikov described it, was tempered with fetishism. Tiger claws were carried as amulets against evil eye; the bones of a tiger's front paw were believed to bring protection in risky situations; dried tiger-eye (taken internally or applied topically? he didn't specify) conferred second sight; and a tiger skin calmed the nerves of a person lying on it. But where did all these parts come from, if the Tungus and other indigenes observed a strict taboo against killing tigers? The unspoken implication is that maybe it wasn't only profane outsiders—Chinese and Russian hunters, with their trip wires and their strychnine—who accounted for fifty to sixty dead tigers per year.

Given such hunting pressure, as well as human population growth and landscape conversion, the tiger population declined steadily in Manchuria. By the mid-1990s, more than fifty-five million people inhabited the two northernmost Manchurian provinces, Heilongjiang and Jilin, amid whom survived, somehow, no more than about a dozen tigers.

On the Russian side of the border, things seemed to be going almost as badly for *P. t. altaica*. At the end of the nineteenth century, one estimate put the annual kill at 120 to 150 tigers. During the 1920s, as the Bolshevik regime consolidated its power, immigration eastward brought more people into the region. Further tendrils of infrastructure went out, and foreign concessionaires were welcomed with long-term contracts, under the blessing of Lenin himself, to open commercial interests such as lead mines and gold fields. Military presence became a crucial factor beginning in 1929, with the creation of the Special Far Eastern Army (OKDVA), led by a forceful commander named Vasily Konstantinovich Blücher. In the 1930s, as Moscow grew nervous at Japan's occupation of Manchuria, Blücher received more troops and greater discretionary power to promote settlement and economic development. Collective farms were established, in some cases by hauling in whole assemblages of peasants dragooned from the central provinces. "A million people (not including forced laborers) came to the Far East during 1931–1939," John J. Stephan writes in his book *The Russian Far East*, a comprehensive political and social history published in 1994. That eastern influx amounted to "one out of every three who crossed the Urals." Demobilized soldiers were encouraged to remain in the region; army communes were given stock animals, timber rights, and freedom from taxation. Blücher believed that the Far East "should develop an independent economic base," Stephan says, and that "farmer-soldiers" could help the province achieve self-sufficiency in producing food. "To institutionalize an OKDVA link with the land, he proposed to the Politburo a plan to create a special collective farm corps, offering soldiers a chance to bring their families to the Far East." Between 1926 and 1939, the regional population doubled. Tigers were pushed from their old territories by agricultural conversion of lowlands, timber extraction on higher slopes, the growth of cities, and the expanding transport network.

Furthermore, tiger hunting was still legal. Peter Matthiessen mentions that the cats were "hammered hard by Communist Party nimrods," perhaps in the same spirit of privileged excess as Nicolae Ceauşescu when he slaughtered Romanian bears. Another disruptive factor was the live capture of cubs and juveniles for sale to the world's zoos, among whom the very rarity of *P. t. altaica* made it a desirable acquisition. The live-capture trade was brisk in several villages along the Bolshaya Ussurka River, a Ussuri tributary draining west from the Sikhote-Alin mountains not far south of the Bikin. Within those villages, expert tiger trappers emerged from certain families of hunters— the Trofimovs, the Kalugins, the Ivaschenkos, the Cherepanovs. They used dogs to track a tiger cub and corner it, forked poles to pin it down, and tough fabric to shackle its paws, and by one account they managed a good record of not breaking cubs' teeth. But kidnapping young tigers was a frenetic process, whatever the technique, and sometimes it entailed killing a protective mother. Baikov had warned, back in 1925, that "the tigress is madly courageous when protecting her young as she is cautious, cunning and prudent in raising them. Despair and mother love will make her attack hunters in mad fury and lose her usual caution." Extracting cubs from the wild could scarcely be done without serious harm to the adult population.

By 1940, according to L. G. Kaplanov, who was the first researcher to do systematic survey work, only a few isolated pockets of occupied tiger habitat remained. The largest of those was in the upper reaches of the Bolshaya Ussurka drainage, partly overlapping the area that had then recently been set in reserve as the Sikhote-Alin Zapovednik. Kaplanov studied tigers by snow-tracking there, covering more than 750 miles (presumably on skis) during the 1939–40 winter. He found evidence of just twelve to fourteen tigers. In 1940, three of those were shot. It was Kaplanov who cautioned that the total population in the Russian Far East was no more than thirty, and possibly as few as twenty. Almost no tigers had been seen on the eastern slopes of the central Sikhote-Alin range in two decades. The Bikin valley also seemed to be empty. Kaplanov's study didn't appear until 1948, partly because in the meantime he himself had been killed by a poacher. He was thirty-two years old. Before dying, he'd put on paper his proposal (posthumously published) for a five-year ban against shooting adult tigers and trapping the young.

Tiger hunting was prohibited in 1947. The live capture of cubs continued. This skimming away of young tigers wasn't just inimical to population renewal and fatal to the occasional tigress; it was also wasteful, as reflected in the 1955–56 season, when fifteen cubs were caught and thirteen of those died soon in captivity. Then came a ban on trapping, but that was only temporary. Conservation policy vacillated; protective measures were successively enacted and withdrawn. For instance, the Sikhote-Alin Zapovednik was statutorily shrunk in 1951 to barely one-twentieth of its original area, and later enlarged by a factor of four, to its present size of 400,000 hectares (about 1,500 square miles). The tiger population fluctuated in response. Mainly the trend has been upward since the low point registered by Kaplanov in 1940. A report published in 1961 spoke of ninety to a hundred tigers throughout the Far East, of which thirty-five were in Khabarovskiy Krai, a more northerly province embracing the upper end of the Sikhote-Alin range, and the majority in Primorskiy Krai, which contains the mountain heartland and the richest of the strict reserves. That report, by K. G. Abramov, also argued the necessity of a comprehensive, methodologically consistent census. In the winter of 1969–70 such a census was done, at least within Primorskiy Krai, and found evidence of about 130 tigers. Another census, from the mid-1980s, yielded an estimate of 240 to 250 tigers throughout the Soviet Far East. The principal investigator on the latter effort was Dima Pikunov.

The numbers were up, but Pikunov wasn't sanguine. He knew that the Amur tiger was far from secure. Although it had passed through a severe population bottleneck (as had the Asiatic lion), rebounding from just a few dozen individuals to about ten times that many (again, like the lion), it still wasn't numerous enough, genetically diverse enough, or broadly distributed enough to be safe against the various threats that could drive it toward extinction. Some of those threats are essentially random—such as epidemic disease, punishing weather, forest fire, fluctuations in food supply resulting from natural ecological cycles, inbreeding depression, and accidental imbalances in the gender ratio. A second category encompasses the intentional (and therefore predictable, not to say controllable) activities of people. Those threats include habitat destruction, habitat fragmentation, depletion of tiger prey by human meat-hunters, commercial poaching of tigers, and incidental tiger-human conflicts, which often end with the death of the tiger. Conservation biologists know the two categories as stochastic and deterministic factors. The rest of us can just as well understand them as bad luck and bad practice. Each individual factor is liable to be exacerbated by one of the others—for instance, tiger-human conflicts become more likely as habitat shrinks. Together they represent a maelstrom of negative influences, amid which 250 individuals of various ages are probably too few to constitute what the experts consider a viable population—that is, too few to give prospect of long-term survival.

In this context, the ongoing tension between humans and tigers, along with the continuing loss of habitat by logging, seemed especially worrisome to Dima Pikunov, who reported his census figures in a brief paper. He warned that thirty to forty tigers were being shot annually, and that the population was once again in decline.

Something had to be done. Pikunov offered recommendations. Enlarge the Sikhote-Alin Zapovednik, he argued. Do the same for the Lazovskiy Zapovednik, another strict reserve. Complement those reserves with two large new zones of protected landscape, one in the northern Sikhote-Alin and one in the south. Set the boundaries of the new protected zones to embrace another 2.4 million hectares (9,000 square miles) of tiger habitat, within which timber extraction, meat hunting, industrial development, and tiger poaching would be forever prohibited. Limit the flow of human settlers into those areas. Deal with the tiger's population growth, its tendency to re-colonize beyond protected-area boundaries, by trapping spillover cubs and sending them to zoos. And finally, in order to monitor population trends, conduct a comprehensive census of tigers throughout the USSR no less than once every five years.

Dima Pikunov's professional home was the Pacific Institute of Geography, within the Far East Branch of the Soviet Academy of Sciences, based in Vladivostok. In 1988 he traveled to Cincinnati, of all places—not to admire mutant white tigers but to attend the Fifth World Conference on Breeding Endangered Species in Captivity. He himself was no captive breeder. He left that subject to others while giving the conferees his views on the present status and future prospects of *P. t. altaica* in the wild. Choosing his moment, and presumably aware of the audience at home as well as the one sitting before him in southern Ohio, Pikunov made the recommendations mentioned above.

"Preservation of the Amur tiger, the largest and integral center of which is located in the Soviet Far East," he concluded, "would be impossible without international cooperation." Pikunov was fishing in dark water. For the time being, he and the other Soviet tiger biologists had no international partners.

62

At about the same time as Pikunov's talk in Cincinnati, a colleague and sometime collaborator at the Pacific Institute of Geography, Anatoly P. Bragin, came at the subject from a different angle. In a paper that appeared in 1989, Bragin and a co-author, Victor V. Gaponov, declared that the "old-fashioned attitude to and protection of the tiger is inconsistent with the demands of *perestroika*, the implementation of which is creating a novel ecologo-economic situation in the region." Tiger protection inconsistent with *perestroika*? An old-fashioned attitude versus "a novel ecologo-economic situation" in the Far East? Was this scientific writing, or was it trendy political jive? Well, a bit of both, and the gummy opacity of that sentence should not be blamed entirely on translation. What did it mean?

Complicated forces were in motion that would soon cause drastic changes, not just to tiger conservation but to every aspect of resource management, governance, and daily life in the USSR. Those forces, temporarily obscured beneath the upbeat semantic bunting in which Mikhail Gorbachev had draped them—that is, the terms *glasnost* and *perestroika*—were traceable to deeper and less tractable realities such as industrial stagnation, political bankruptcy, military hypertrophy, and the effect of increased global connectivity upon social expectations stultified by eight decades of brutal sequestration. "Similar to

all other fields of national economy, taiga management needs a radical *perestroika*," wrote Bragin and Gaponov, "the aim of which is to
take the taiga away from the departments and give it back to people."
Give it back to *people*? What did *they* know about taiga, except how
to live in it?

Perestroika means, of course, "restructuring." And the restructuring
of conservation strategy proposed in this paper was as radical as the
authors claimed, notwithstanding their cautious opportunism in wrap-
ping it in Gorbachev's fashionable byword. It was based on a single,
stark realization: The Amur tiger, to survive for the long term, needed
more habitat than the system of protected areas in Russia allowed.

The distribution of tigers in the Soviet Far East, according to Bragin
and Gaponov, spanned about 22 million hectares of forested landscape.
That total included both the zapovedniks and other protected areas
(amounting to just a few million hectares) as well as larger expanses
of forest carrying no protected status at all. As time passed, as human
presence throughout the region increased and commercial exploitation continued, much of the unprotected forest would disappear, or at
least be severely degraded. Tiger habitat would shrink. If it shrank by
as much as three-quarters of the total but no farther, Bragin and
Gaponov asserted, and if the remaining quarter were placed under "a
special nature management regime," *P. t. altaica* could survive at its
current population level, about 250 adults. That level, they believed
(though some conservation biologists would disagree), should be large
enough to sustain long-term viability against the various sorts of stochastic threats. The critical habitat threshold, the minimum amount of
hospitable landscape necessary for supporting 250 adult tigers, they
placed at roughly 5.7 million hectares. "Without this minimum area,"
the two authors warned, "the maintenance of a viable population of
the Amur tiger will be impossible." But just what did they mean by "a
special nature management regime"? That was the crux.

What they did *not* mean was more zapovedniks. Political and economic realities made it hopeless to expect the establishment of strictly protected reserves encompassing so much terrain. Therefore tigers would need to claim habitat not just within such areas but also outside them, where humans make their own claims on the landscape. (This was a profoundly important and far-reaching point, which remains relevant not just to *P. t. altaica* but to all alpha predators: They will not survive into the distant future unless they are allowed to survive *amid*, not just apart from, areas of landscape that are occupied and exploited by humans. The world's national parks and wildlife sanctuaries generally aren't big enough to achieve long-term survival of such species within their boundaries.) Tiger activities and human activities, Bragin and Gaponov understood, had to be made to overlap. But how?

They offered several suggestions. The first concerned administrative mechanisms. Bragin and Gaponov proposed that "zones of a special economic regime" should be established, much larger than the strictly protected areas but complementary, on which limited forms of commercial extraction (including meat hunting for deer and boar) would be permitted and on which resident tigers would be tolerated, even encouraged. How to arrange that? What manner of land-management perestroika could reconcile all the demands of tiger conservation—such as monitoring, prevention of poaching, habitat protection, maintenance of prey populations, and mitigation of tiger-human conflicts—with the economic imperatives of human society? One means might be to grant long-term leases on such zones to enterprising individuals, or to membership-based hunting associations, who would then handle most of the policing, monitoring, and resource extraction for their own entrepreneurial and recreational gain. Leave the forest standing, but privatize the role of protecting it.

That much was new and risky. Another idea floated by Bragin and Gaponov was riskier still: Privatize the lives of the tigers themselves. Since a total ban on killing *P. t. altaica* might seem onerous and unnecessary to leaseholders, "it would be most reasonable to give the tiger on these areas a status of commercial game-animal." That is, let the lessees sell the right to shoot tigers within their leased areas, just as the Romanian Forest Department sells the right to shoot bears.

Who would buy such a right? Bragin and Gaponov were realistic. "For the purpose of obtaining the maximum level of income from the realization of bagged tigers it is necessary to go out to the foreign market," they conceded, "since our home market owing to the force of certain circumstances has a limited purchasing capacity." In other words: Let's invite wealthy trophy hunters from the United States, Germany, Switzerland, and elsewhere to pay big money for the chance of killing a big cat and taking home the pelt.

Privatize the Siberian tiger? Sell its life on the international market? The very suggestion was heretical—no less so to Soviet dogmas of resource management than it may seem to cat conservationists now. But in the climate of 1989, with structures changing and powers shifting all across Eastern Europe and into the USSR, Bragin and Gaponov were emboldened. They even tossed in a comment on the merits of (nondialectical) materialism: "If in the present conditions the leaseholder of hunting land will have a material incentive in tigers living on his ground, he will not only protect them but, as one game-manager has expressed himself, 'will feed them out of his hand.' It may be then that tigers will be more numerous." Of course the lease-holder wouldn't actually feed them "out of his hand"—unless he was prepared to trade that hand for a prosthesis. But we get the idea: An affluent tiger rancher is a happy tiger rancher. Grahame Webb, amid his crocodiles down in Australia, would agree.

Still, provocative brainstorming was one thing, which Bragin and Gaponov had delivered; implementation was something else. To create a new system of nature-protection zones, with a new code of private incentives and devolved enforcement, would have been difficult under any political conditions, and conditions just then were especially tricky. Within two years, Mikhail Gorbachev's incremental perestroika went into the dustbin of history. The Soviet Far East became a much different place for tigers as well as for people when, in 1991, the word "Soviet" disappeared from Far Eastern maps.

63

The isolation of Soviet tiger biologists ended, almost by coincidence, at about the same time the USSR did. International support, as called for by Dima Pikunov during the 1988 conference in Cincinnati, began to arrive. That was the good news for *P. t. altaica* in the early 1990s. The bad news was that conservation measures—specifically, effective suppression of the black market in tiger parts, and firm restraint against unsustainable (sometimes downright illegal) timber harvesting—collapsed along with the authority, the rigor, and the operating budgets of Soviet institutions.

The abrupt decline of subsidies from Moscow, and the slump in domestic trade, according to the historian John J. Stephan, "sent the Far East into an economic free fall starting in the autumn of 1991. Acute fuel shortages crippled utilities, curtailed services, and cut off supplies to outlying areas." The army set up emergency field kitchens to help city dwellers left desperate by power outages. Unemployment and malnutrition rose. By 1993, Stephan reports, "the region was an economic basket case." Then privatization of services and resources turned the basket into a cornucopia for greedy, unscrupulous grabbers. "After lurking on the fringe for decades, hustlers, speculators, and blackmarketeers occupied center stage, their ranks swelled by 'honest' entrepreneurs who had discovered, often at no small expense, that business could not be conducted without recourse to bribes." Primorskiy Krai was one of the provinces hit hardest. "Yet most Far Easterners showed

their proverbial resilience when it came to food," Stephan adds, "growing and pickling produce from their dachas. Poaching, long an unofficial sport, helped fill the protein gap." He means poaching of ungulates
for meat. Poaching of tigers for hard currency, meanwhile, helped fill
the gap between rubles and purchasing power. These destructive market forces came into play almost simultaneously with several promising
international collaborations, and a race was on.

The market forces got a small head start. The Hyundai Corporation
swept up from South Korea, in 1990, to sign a thirty-year logging
agreement with the Primorskiy Krai government. Hyundai's workers
were soon cutting in the Sikhote-Alin range, near a village called
Svetlaya, which had once been a gulag camp. Boris Yeltsin, having
assumed the presidency of a transmogrified, post-Soviet Russia, ceded
authority over resource extraction to regional authorities. Given such
loose regional oversight, the Hyundai operation continued for the next
few years to take seven million cubic feet of timber per year, despite
negative environmental-impact reports at the federal level. It was
putatively an international joint venture, but the manner of enterprise—exporting raw logs, mainly to Japan—didn't leave behind
much of the harvested wealth, and Hyundai's Russian partners meanwhile looted what in-country assets there were. In 1992 the same joint-
venture group tried for similar cutting privileges in the upper Bikin
basin, but protests from an alliance of Udege people and conservationists blocked that gambit, at least temporarily.

Such selective cutting of oak, ash, and Korean pine in the Sikhote-
Alin mountains indirectly affected the tiger population. Poaching
affected it directly, and faster. By 1990, an encouraging new estimate
had suggested that there might be three hundred to four hundred
tigers in the Soviet Far East, but then came the drastic political and
economic changes, and the population trend tilted back down. Budget
deflations for big centralized agencies, including the Ministry of
Environmental Protection and Natural Resources, in combination

with the collapse of the ruble, resulted in layoffs and huge reductions in real income for workers in the provinces, such as forest guards at zapovedniks in the Far East. With border controls also relaxed and the regional economy depressed, the geographical location of Primorskiy Krai—adjacent to northern China, smooshed against the northern end of North Korea, a short hop from Japan—became acutely relevant. Here, not far from the world's hungriest markets for tiger parts, lay the world's last sizable population of Amur tigers. Demand sucked at supply, and by 1993 about sixty tigers per year were being illegally killed. At that catastrophic rate, the entire population would be gone within a decade. Even if the kill rate declined as tigers became vanishingly rare, such losses could quickly depress *P. t. altaica* to a level from which recovery might be unlikely. That the economic benefits were minuscule, on a regional scale, didn't lessen the pressure. Yes, tiger poaching was small stuff compared with the vast timber resources coveted by domestic entrepreneurs and foreign corporations, but a single dead tiger still represented a big payoff at the village level, and it only took one man to pull a trigger.

Although most of the tigers now lived within Primorskiy Krai, much of the smuggling seems to have gone through the city of Khabarovsk, farther north. According to a report by two researchers from the Global Survival Network, an international anti-poaching group, Chinese traders in Khabarovsk were paying five thousand dollars for a tiger carcass. Bones alone brought more than $130 per pound. One channel of the trade involved crooked officials in Khabarovsk who sold dead tigers, skinned and frozen, to Russian-Korean residents in that city. In another case, the central figure was a truck driver for a logging company. "He had full access to the taiga by the roads his company created and the vehicle he drove," says the GSN report. "He would carry a rifle with him during his work and poach deer and wild pig, and tigers on the side." This truck driver sold his dead tigers through two contacts, a corrupt customs officer in Vladivostok and a shipping-company employee in a small port town on the Sikhote-Alin coast. "The tiger parts were easily concealed on board the ships amidst commercial containers and tons of Russian logs," the report explains. The forest and its preeminent predator, slabbed out and packed for export, sailed away on outgoing market tides.

But with help from GSN and other international groups, beginning in late 1993, the Ministry of Environmental Protection (under the latest of its evolving names) launched a serious effort, called Inspection Tiger, to curtail poaching. Mobilized in new vehicles, dressed in imposing uniforms, rangers made spot checks along taiga roads. Agents went undercover to investigate the trade. At least one dealer, from the Bikin valley, was jailed. In the summer of 1995, Prime Minister Victor Chernomyrdin issued National Decree number 795, "On Saving the Amur Tiger and Other Endangered Fauna and Flora of the Russian Far East," which signaled political support for enforcement and prosecution. Seven people were indicted within the next year and a half, and poaching declined. The present rate, estimated at twenty to thirty tigers per year, is still a bad drain from a population of four hundred, but not disastrous.

While that effort dealt with poaching, another international collaboration advanced the study of living tigers. This one began back in 1989 at a campfire in the mountains of central Idaho, around which Howard Quigley, a post-doctoral biologist at the University of Idaho, was hosting a delegation of visitors from the Soviet Academy of Sciences. Quigley's post-doc fellowship came through Maurice Hornocker, a legendarily steely field man who had done pioneering work on Idaho's cougars in the 1960s. Uncomfortable within the strictures of a university department, Hornocker had established his own outfit, the Hornocker Wildlife Research Institute, through which he supported and led conservation-oriented ecological studies of big mammals in various parts of North America. Howard Quigley, before joining Hornocker's institute, had studied cougars in

Idaho, jaguars in Brazil, and giant pandas in the mountains of central China. When the campfire talk turned to Siberian tigers, the leader of the Soviet group tossed out an invitation: Come visit and we'll discuss the possibility of a joint effort. By January of 1990, Hornocker and Quigley were on an Aeroflot flight into Vladivostok. They were met by a delegation of stern-looking Russians, among whom the first to step forward was Dima Pikunov. After days of conversation, then a field visit to the Sikhote-Alin Zapovednik, and then many months of delicate trans-Pacific wangling, the Siberian Tiger Project was launched. Its goal was to collect the best possible data on tiger ecology and to apply those data to tiger conservation. Its methodological approach was to combine the radiotelemetry and tranquilizing techniques of the Hornocker Institute with the tracking skills and the local ecological knowledge of the Russians.

In February 1992, the field crew trapped and collared their first tiger, a young female to whom they gave the name Olga. By that time another American biologist, a burly and easygoing young man named Dale Miquelle, had arrived to serve as Hornocker's field coordinator on the project. Miquelle grew up in the Boston suburbs, graduated from Yale, and did graduate work on moose populations in Michigan and Alaska, as well as assisting for a time with tiger research in Nepal. His doctorate came through the University of Idaho, where he knew both Quigley and Hornocker. Although the university happened to be in a small town called Moscow, in the panhandle north of Boise, Miquelle had never set foot in the other Moscow, or visited eastern Russia, or imagined that he might ever want to, until Quigley approached him about working on the Amur tiger.

At first Miquelle was reluctant, he recalls, "but it was a job I couldn't say no to." A whole list of discouraging factors made him initially wary: the low density of tigers, their inherent elusiveness, the political and cultural difficulties of functioning in post-Communist Russia. Balanced against them, he saw a tantalizing challenge, "an opportunity to either fail miserably or possibly do something worthwhile." He came, began learning the Russian language, and settled in a small fishing village, Terney, just outside the Sikhote-Alin Zapovednik. He would give himself three years, Miquelle figured, to see what might be doable. His main Russian counterpart was Evgeny N. Smirnov, chief tiger biologist for the zapovednik. Smirnov was an elfin man, with large eyeglasses and a warm smile, who had come to this remote place as a mouse researcher in the 1960s and gotten interested in tigers without ever losing interest in mice. Miquelle's field assistants were men such as Kolya Reebin, sad-eyed and laconic, indefatigable and steady. Reebin had driven a taxi in Ukraine before reinventing his life in Primorskiy Krai, first as a forest guard and then as a tracker of tigers. Like any frontier zone in a majestically severe landscape—the Yukon of Robert Service, the Empty Quarter of Wilfred Thesiger, the Ogooué River in West Africa as seen by Mary Kingsley in 1893—the Sikhote-Alin mountains of the Russian Far East offer scope for, and invitation to, personal transformation. A decade later, Dale Miquelle is still in Terney, living in a trim little house with his Russian wife and stepson.

He's in Terney, that is, when not drawn away to some other part of the region. The institutional aegis within which he works made its own transformation in the year 2000, when the Hornocker Institute merged with a more powerful partner, the Wildlife Conservation Society, of New York, and since then Miquelle's own scope of activity has expanded. In a given week he might be handling bureaucratic chores in Vladivostock, or assisting a tiger-and-leopard study in eastern Manchuria, or inspecting a hunting lease near the border, or meeting the other far-flung responsibilities that now fall to him as director of the whole Russian Far East program of WCS. On a day in late January, returning from just such a trip, he has driven nine hours with only one passenger for company by the time his four-by-four Toyota crosses a boundary into the zapovednik. The mountains roundabout look polished and ancient, their lines softened beneath two feet of snow, their whiteness shaded to creamy brown by a fuzz of bare birches and oaks. Westward, beyond the near slopes, are higher peaks and ridges dark with spruce and fir. Eastward, invisible but close, are ocean cliffs overlooking the Sea of Japan.

"This is some of the best tiger habitat in Russia," Miquelle tells me as we top a rise on the gravel road leading into Terney.

64

Terney is a humble, ramshackle town of five thousand people, its out-lying neighborhoods draped across foothills, its center on the flats, where a small stream called the Serebryanka (Silver) River exits sud-denly from the mountains to the sea. Its chief enterprises are fishing, fish canning, timber, and (since Terney is the seat of the *raion*, or county, as well as headquarters for the Sikhote-Alin Zapovednik) the business of government. The fishing trade seems to be stable; timber threatens to boom; government isn't what it once was. The Siberian Tiger Project, funded largely by nongovernmental organizations over-seas, has brought some additional income but no glaring changes of civic demeanor. The local school resembles a whitewashed factory, and if you search carefully, or get directions, or can recognize the word *magazin* in Cyrillic letters without any hint from neon, you'll find sev-eral small grocery shops, fronted by somber steel doors. Inside these shops you can get cheap vodka, decent beer, bland cheese, mysterious sausage, chocolate, dried apricots, and some excellent pickled salmon. Four dollars in rubles buys as much as you'll be able to carry.

The town's unpaved lanes, at least where the slope is steep, where the mud in springtime makes them unclimbable, are lined with wooden walkways—careless arrangements, like scrap boards tossed down across a mucky construction site. Brave little bungalows, sided

with weathered planking and roofed with corrugated asbestos, sit clustered into irregular blocks. Amid the general grayness and brownness,
a few houses are brightly painted—rather, they *were* brightly painted
at one time. With passing years, their brightness has turned dim but
not subtle, leaving drab neighborhoods stippled in a weird palette of
colors suggesting vagaries in the supply of Soviet paint. What once
may have been chartreuse now looks more like wasabi. Some sort of
vivid cerulean has faded to cheery but flat shades of blue, the blue of
wilting party balloons and elderly parakeets.

Indoor plumbing is a rarity in Terney. Within sixty steps of each
house is a well, from which water gets carried into the kitchen and out
to the *banya* (a sauna shed, for weekly bathing) by the bucketful. In
another corner of each yard, at about the same distance, stands an out-
house. Sixty steps is a long way to go in Siberian winter—I speak now
from experience as well as empathy—for a bucket of dishwater or a
moment of bowel release. When a blizzard hits, every home feels like
a fort under siege. Yet the people of Terney are generous, hearty, open,
and gracious, at least when you come properly introduced in a stove-
heated kitchen; they're only stolid and wary if, as a stranger, you
encounter them on the street. The electricity goes off at ten in the
evening. It's a quiet and undramatic place, a small town like many others in Russia and elsewhere, except that occasionally a tiger strolls
through, pausing at this household or that one, like an angel of death,
to kill and eat a dog.

65

My ten days in Terney and its environs are devoted to three sorts of
activity: tracking tigers across snowy slopes and icy stream bottoms at
the heels of Dale Miquelle, Kolya Reebin, and others; talking with

local authorities such as Evgeny Smirnov, the tiger-and-mouse man, about matters of science and conservation; and standing witness to the ongoing, one-sided conflict between tigers and dogs.

The dog business intrigues me particularly. The jeopardy of these canines, at the uneasy interface between tigers and humans, represents a symptom of larger tensions. Yet it doesn't seem an entirely bad thing. A person might legitimately wonder: Is dog-eating by tigers a form of friction—or is it a form of lubricant?

On the evening of our arrival from Vladivostok, for instance, Miquelle receives word that a radio-collared tiger has dipped into town and killed a pooch belonging to Kolya Reebin himself. It happened just last night. The dog, known as Dick, started barking around four A.M. and then fell silent. In the morning Kolya found him, dead and half-eaten. What remained of Dick (his head, his neck, a bit of shoulders) was still attached to his chain. Now, for a tiger to kill the watchdog of the tiger project's ace tracker—that seems like more than coincidence, doesn't it? With only a small leap of unscientific imagining, it can be viewed as an act of retaliation, a rude feline counterstrike reflecting the fact that this particular tiger (as identified by his radio frequency) is a young male named Fedya, whom Kolya himself darted for capture last November.

Fedya originally drew the project's attention by killing a cow and a dog just north of the river. Following standard capture-and-release tactics, a field team set a foot snare beside the cow's partially eaten carcass, on the chance that the same tiger, still hungry, might return. When it did, Kolya shot the animal with a tranquilizer-filled dart. Carefully examined, collared, given his name, Fedya was then released on the site, notwithstanding its proximity to Terney. In the weeks after, he crossed the river, passed through town without incident, and made himself comfortable on a forested hillside behind Kolya's back door. He remained a good cautious tiger, living off wild prey, avoiding the further enticement of cows and dogs, until last night.

Having heard the news, Miquelle and I report promptly to Kolya's house, joining a stakeout in progress there. The group includes Kolya's younger brother, Sasha, who is also a tracker for the project; Anatoly Khobatnov, a stern-looking but gentle man with a dark beard, wearing a camo vest, carrying an AK-47, who serves on the government's Tiger Response Team; and John Goodrich, a lanky American biologist who joined Miquelle on the project six years ago. Anatoly has booby-trapped the uneaten remnants of Dick with two fireworks rockets and a flare. These little surprises, all set on a trip-wire trigger, are meant to give the tiger a scary dose of aversive conditioning if he ventures back for another meal. Will he succumb to that temptation? In the front room of Kolya's tiny house, a radio receiver chirps steadily on Fedya's frequency, like a cuckoo clock without hands, promising to squawk but not saying when.

The receiver's signal has an unearthly timbre, familiar to anyone who has radio-tracked animals or listened to sonar in a submarine: *teek . . . teek . . . teek . . . teek*. Although its rhythm is slow and its volume loud, indicating that Fedya has bedded nearby, the only way to learn whether he has had his fill of dogmeat is by waiting.

So we wait. But, like Fedya, we don't wait with empty stomachs. This is Russia, and people have gathered; therefore, food. At the insistence of Kolya's wife, Lubya, we all belly up to a supper of borscht, carrot salad, fried fish, sausage, and fish-sperm dumplings. Afterward, more firmly ballasted to the furniture, we again sit and listen. Another hour. Two hours. *Teek . . . teek . . . teek . . . teek*. For diversion there's a video of *The Pelican Brief*, starring Julia Roberts and Denzel Washington—but dubbed in Russian, which makes its programmatic Hollywood melodrama slightly more interesting to me, since I don't understand a word of the dialogue except *da* and *nyet*. We watch the movie in Kolya and Lubya's bedroom, all of us sprawled companionably across the bed and the floor. Meanwhile the receiver tolls. Occasionally the signal changes tempo—quickening to *teek teek teek teek*—which indicates that the animal has stirred. This spurs Kolya to rouse himself and carry the gizmo outside for a careful scan. Making slow sweeps with a directional antenna, he can home onto the signal and fine-point its bearing. But after each trip outdoors, he returns with an inconclusive report—signal again slow, Fedya still resting. And I ask myself: Let me see now, are we disappointed or glad? Disappointed, I suppose, given that Fedya is unlikely to cure himself of the dog-eating habit without a rude lesson. Besides, we all crave some action.

But tonight's not the night. Late in the evening, after Denzel and Julia have eluded bullets and careening cars and sinister white men in pinstripe suits, and Fedya has shown his indifference to booby-trapped dog leftovers, we adjourn. This tiger is reckless but not gullible.

Next day I call on Anatoly Astafiev, director of the Sikhote-Alin Zapovednik. Having heard of our previous night's vigil, he mentions that tigers were present on that same forested hillside—behind where Kolya Reebin now lives, yes—during the late 1970s and early 80s. Not infrequently, they came into town. "Of course they ate a lot of dogs," says Astafiev. More than two dozen, by his recollection. But there were no real problems with people, he adds, just a steady drain on the supply of domestic canines, which seems to have been easily enough countenanced and recouped.

Several days later I go tracking with Sasha Reebin. He's a cheerful twenty-five-year-old, born in Ukraine like his brother, trained at a technical institute to build cars, then briefly employed at a factory that made heavy mining equipment for export, until he came out here in search of a more sylvan life. With his sublime, youthful tolerance for exhaustion and cold, Sasha is well suited to tiger work in these mountains. He speaks only a few words of English, and I speak fewer of Russian, so we can scarcely communicate except with gestures and nods, but that's okay, since most of this day's work will be visual, aural, and physical. Along a roadside within the zapovednik, we spot fresh prints in the snow. From the receiver we get a strong signal.

"Lidya," he says.

"Lidya," I echo, understanding that Sasha has recognized another of the five cats currently collared. John Goodrich has briefed me on this roster, which includes Fedya the Naughty and four females: Olga (a sentimental favorite, since she was that first tiger collared by the project, in 1992), Nellie, Ludmilla, and Lidya. Three other males—Alec, Boris, and Misha—and several additional females have come and gone, some of them killed by poachers. Boris was shot dead near a town to the south, after he snooped his way into a shed and then got caught on the way out. Olga has avoided trouble, knock wood, and is now wearing her fifth battery-refreshed collar. Fedya, if he doesn't reform, stands risk of ending like Boris. And, as for Lidya, she's the current tenant of a territory that—possibly because the coast road transects it, allowing access for poaching—has been fatally unlucky for a couple of earlier collared tigresses.

The strong *teek teek teek teek* says that Lidya is nearby, somewhere amid the forested slopes and brushy draws rising westward. A day earlier she was east of the road, along the sea cliffs, where John and I followed her for the length of a frigid, wind-scoured afternoon. Overnight, she seems to have crossed through here. Tracing her movements by radio, triangulating to her positions, plotting them onto a map, following by foot to read sign on the ground—her beds, the remnants of kills, the subtle clues that tell of her other activities— are the tasks at which Sasha and his brother spend their days. Such evidence, collected with help from field assistants, allows Goodrich, Miquelle, and Smirnov to construct an impression of how *P. t. altaica* lives within its habitat. The process is incremental, not epiphanic. Stamina, patience, acuteness of observation, and accuracy are the requisite traits. Data are mainly indirect. Glimpsing an actual tiger happens rarely.

The wind is sharp again today as Sasha and I begin walking. Beneath a blue sky and a dazzling, tepid sun, it strips heat off our bodies like a fire hose. I squint against the blast, savoring one couplet from a poem that lurks indelibly in my memory and always brings me perverse cheer in mortally chilly situations:

> If our eyes we'd close, then the lashes froze till sometimes
>   we couldn't see;
> It wasn't much fun, but the only one to whimper was
>   Sam McGee.

We walk for six hours, forward and backward along Lidya's tracks, tracing her like a specter in the snow.

At first, as we follow her forward, Sasha pauses regularly to listen with the receiver. Where the drifts are thigh-deep, we posthole along in our boots and wool pants, lifting knees high as we step. In shallower snow across open hillsides, we move quicker but stumble occasionally, losing traction on oak leaves layered dry and slippery beneath the bottommost crust. In some places the night wind has blown over Lidya's tracks, leaving them softened and blurred but still deep, big, unmissable. Since she places her hind paws exactly into her front-paw prints, four legs make scarcely more impression than two, and by the spacing her stride looks almost human. But on the slick, yellow-blue surface of an ice-blocked stream, each pug mark shows clearly, big as a grapefruit, embossed there by the animal's weight and heat: a palm pad and four ovoid toes.

We top several ridges and wallow through draws, climbing onward, adjusting our line as the *teek teek teek teek* grows louder. Finally, as we pause on the spine of one ridge, Sasha gets a very strong signal indeed. He disconnects the antenna and *still* gets a strong signal. "Three hundred meters," he says, pointing ahead. "Inactive. She is inactive. She is . . ."—here he struggles for an English word—". . . *sleep?*" Whether in fact she's asleep, or just basking in a sunbeam—or maybe crouched behind a boulder, to spy on the lunkheads invading her territory—will remain unknown. This is close enough, Sasha decides. We don't want to make her nervous, do we? No, let's not do that, I agree.

We withdraw. From the road again, we back-follow her earlier tracks, tracing them into last night and yesterday. Sasha ignores the receiver now, relying on old-fashioned skills. He shows me a spot at the base of a tree where Lidya has left a scent marking. Even with his guidance, I can barely smell her astringent signature. He points out a smallish burrow into the snowy bank of a creek, and in Russian names the animal that made it. A mink? Maybe a sable? I can only guess. We see sign of sika deer—their tracks, their parsimonious little piles of scat. Traversing slopes and over ridges, lifting snow with each step, Sasha trudges inexorably. I work hard to keep up, panting and sweating inside my bulbous down coat. The wind has died, but the afternoon hasn't warmed. Despite my core heat and sweat, my fingers go numb whenever I pause to write notes. We don't stop for lunch. I gobble some jerky as I walk. Sasha, in his light wool jacket, eats nothing and shows no sign of fatigue.

Eventually we descend into a creek bottom clotted with ice, brush, golden tussocks of grass, and drifted snow. Sasha begins making diagonal zigs and zags to re-cut Lidya's trail as he loses it, finds it, and loses it again. Then she's gone for good, her tracks erased by twenty-four hours of wind-driven powder. We walk for another hour, heading downstream along the creek, before intersecting a path that should lead back toward the road. By now I'm staggering. Exhausted and famished. My thighs can barely lift. My toes have gone numb and reawakened more times than I can remember. Somewhere in my pack is a bit of chocolate. Collapsing at last into Sasha's truck, I'm amazed to find that my canteen isn't frozen solid. It has been a typical day in tiger habitat.

Lidya is a fortunate and well-adapted animal. Her territory, spanning dozens of square miles southwest of Terney, constitutes excellent habitat—rich in prey, rich in water, vitiated only by the transecting intrusion of the road. Other territories are more problematic. Other tigers are less aloof. The next morning, as John Goodrich and I share coffee beside his kitchen stove, we hear that Fedya has killed another dog.

66

"Cat Bites Dog" isn't headline news in the world of the Amur tiger. Source after source, in the Russian scientific literature, alludes to the chronic conflict between *P. t. altaica* and hunting dogs, farm dogs, watchdogs, even village dogs with no high-hazard duties, and most of those anecdotes end with one or more canines converted to cat food. It's understandable, given that the weight ratio of a medium-sized adult tiger to a large domestic dog is roughly similar to that of a fox to a rabbit.

The physical outmatching is made worse by the fact that dogs, unlike rabbits, carry at least some genetic predisposition to stand and fight. If they're working animals, conditioned to a sense of mission as well as bred for loyalty to a master, their resistance may be stalwart and suicidal; if they're feckless, undisciplined pets, not necessarily. "Dogs not trained for tigers are useless in the hunt," Nikolai Baikov wrote, back in 1925, "for at the first sight of the predator's fresh tracks, or at the appearance of the animal itself, they are overtaken by panic and fear, and are afflicted with a nervous stupor. This is when dogs most often become the tiger's prey." Such a loss of nerve can be dangerous not just for the dog. K. G. Abramov reported in 1965 that, among the tiger attacks upon humans that he'd studied, almost all came when a tiger had been wounded, or otherwise provoked by the person, or "when some cowardly dog near a hunter, in trying to save itself, climbed between the legs of its master, setting the scene for a confrontation. A tiger never backs out of a chance to take on a dog, even one whose master is well-armed." Abramov mentioned the case of a hunter named Belonosov, who got himself lost in the wintry woods along a drainage called Shanduiskiy Creek and, apparently, was stalked by a tiger. Of his three dogs, the tiger took one, then left Belonosov unharmed—or at least left him to suffer another fate. He froze to death. When his corpse was found, the two other dogs were sitting beside it—faithful forever, and smart enough to have chosen discretion over valor.

A recent monograph from Igor Nikolaev and (posthumously) his former research partner, A. G. Yudakov, lists the tiger's favored prey species along one stretch of the Bolshaya Ussurka River, as documented over four winter seasons: wild boar, red deer, roe deer, musk deer, Asiatic black bears, and dogs. No domestic animals other than canines were killed by the tigers Nikolaev and Yudakov studied. (The omission of cows, goats, and sheep is probably more accountable to a dearth of livestock amid the mountains of the upper Bolshaya Ussurka, where hunters and their dogs predominate over herders, than to any tiger distaste for beef, goat, or mutton.) During the early 1970s, according to E. N. Matyushkin and two co-authors, "a young tiger regularly hunted dogs in the outskirts of the village of Kievka—even moving between the different houses." One of Matyushkin's co-authors was Evgeny Smirnov, the tiger man at the Sikhote-Alin Zapovednik. Smirnov also published a paper in which he noted—in fact, quantified—the importance of dogs in the tiger's diet. According to data he gathered during his early years at the zapovednik, when the tiger population was resurging from historical lows, dogs constituted twenty-two percent of tiger prey.

Another recurrent theme in the Russian scientific literature, which seems oddly counterposed to the cat-eats-dog phenomenon, is that Amur tigers seldom attack humans. Things may have been different during the late nineteenth century (when, according to one source, tiger-on-human attacks were common in the Russian Far East), but throughout the modern era there has been a notable scarcity of tiger-caused deaths and injuries. Episodes of direct conflict involving healthy tigers (as distinct from maimed or infirm ones) and of predation without provocation are fewer still. "Cases of uninjured tigers attacking a human are generally rare," Baikov wrote in 1925, "and occur only when it is pursued for a long time and is not given a chance to cat or rest. In such cases the irritated and enraged predator often hides himself and unexpectedly springs on the hunter who is following along the animal's tracks." From the time of Baikov into the 1950s, another source testified, no man-eating tigers appeared in the Russian Far East. That record continued until February 18, 1976, when a tractor driver was killed by a tiger along a road near the village of Lazo. The tractor driver hadn't committed any notable provocation, and the tiger lingered to feed on the corpse.

Evgeny Smirnov cited another case, of a man killed in November 1980: "The attack, by a tigress accompanied by cubs, was not provoked." Then in February 1992 a tiger pounced on and killed a hunter who had been checking his trapline in the Samarga River drainage, at the north end of the Sikhote-Alin range. Two years later, in the mountains west of Terney, another woodsman was attacked while circuiting his traps. This tiger had been resting in a day bed, beneath the roots of a fallen tree, when the man walked by and then paused, possibly to take aim at a squirrel. The tiger burst out—on impulse?—and came at the man in a trot that stretched into running leaps. The man's mittens, along with his squirrel gun, still cocked, were found in the snow. Igor Nikolaev and another co-worker, Victor Yudin, have deduced a rough scenario of the event from tracks, blood, and other evidence. "The man was killed in the tiger's fourth bound, and had no time even to jump behind a tree," they report. From the kill site, the tiger moved its prey to some spruce trees a hundred feet away. "One-quarter of this distance it carried the man in its mouth, holding him up from the ground, as determined from the snow trail and stains of blood found 77 cm high on a tree trunk. The man's corpse was eaten fully by the second time the tiger came to its prey." Nikolaev and Yudin describe that case, among others, in a recent survey of tiger-human conflicts throughout Primorskiy Krai between 1970 and 1996. They mention also a fatal incident in January 1996, near the city of Partizansk, after which the victim's body "would have been eaten entirely if the tiger was not shot." Amid this modest list of recent attacks, the tractor driver's case from 1976 raises a question: Was it an anomaly, or did it represent the first hint of a changing trend?

Nikolaev and Yudin take pains to set tiger-human conflicts into a larger context, one that's far different from the Russian Far East of fifty years ago. "Tiger numbers have increased substantially, and the animal's behavior has changed as well. Increased contact with man and man's activities has led to reduced fear on the part of the tiger, resulting in a growing number of conflicts," they posit. The rate of poaching rose markedly during the early 1990s and, along with poaching, the incidence of wounded, crippled, and otherwise aggravated tigers. "Not only are such animals likely to become involved in livestock depredations, they are also likely to hunt people." Given such circumstances, the two scientists consider it remarkable that they found so *few* cases of attack on humans, and fewer still that were unprovoked. Dogs may fall within the tiger's mental rubric of prey animals, but humans ordinarily do not. On the contrary, Nikolaev and Yudin write,

> any deviations in a tiger's normal behavior when encountering people or domestic animals are usually caused by previously incurred wounds, or by people's inappropriate actions in a given situation. It is worth noting that it is usually armed people who are attacked. Weapons enable a person to defend himself, and to attack, but inaccurate shooting in such quickly developing situations usually results in a wounded animal that will attack.

Since the 1980s, with more humans than ever and more tigers than formerly sharing the forest, hundreds of encounters between the two species have occurred, most of which ended peaceably—and all of which *would* have ended peaceably, Nikolaev and Yudin argue, if the humans had been unarmed. "With this knowledge it can be concluded that there are no man-eaters in the Amur tiger population." But of course what they mean by "no man-eaters" is qualified: no tigers that prey routinely on human victims, or that feed on human flesh except after some unfortunate convergence of mistakes.

Dale Miquelle adds his own caveat: that no human has been killed by an Amur tiger since 1997. That may signal a decrease, he suspects, in tiger-poaching attempts by unscrupulous hunters and the fatal episodes of comeuppance to which they sometimes lead.

Other experts, informed by knowledge and experience of *Panthera tigris* in other regions and periods of history, share the conviction that tigers don't naturally regard humans as prey. Among those experts is Charles McDougal, trained originally as an anthropologist, and at one time a tiger hunter himself, who has spent much of the past forty years studying the species and advocating its protection in Nepal and India. "The normal tiger exhibits a deep-rooted aversion to man, with whom he avoids contact," McDougal wrote, in a 1987 paper, "The Man-Eating Tiger in Geographical and Historical Perspective." The basis of that aversion is hard to identify, he acknowledged, but it probably dates well back into the evolutionary past, long before the invention of firearms. "At some stage during the tiger's prehistorical interaction with humans, avoidance of bipedal man became an adaptive behavioral strategy." Then again, there were always exceptions, when human activities disrupted the relationship between tigers and their natural prey. The most interesting thing about McDougal's look at the subject is that he found a high degree of difference in where those exceptions occur.

Man-eaters seem to have been virtually unknown among the

Caspian subspecies of the tiger, he reported, at least along the southern edge of the Caspian Sea, in what is now Iran. Burma, Thailand, and the Malaysian peninsula contained many tigers during the late nineteenth century and into the twentieth, but few accounts of man-eater trouble emerged from those countries. Tribal people in the interior of Vietnam had little fear of tigers, which they were accustomed to meet uneventfully in the forest. And in Sumatra also, a century ago, man-eating was rare though tigers were abundant.

At the other extreme, according to McDougal's research, were three places where man-eating had been acutely problematic, at least until the resident tigers were all but exterminated: southeastern China, the island of Singapore, and Manchuria. The problem in China had drawn comment as far back as Marco Polo's time, though it may have grown worse in the early twentieth century, as the last tigers were constricted within the last remnants of forest, surrounded by humans and farms and livestock. "Sixty persons were killed in one village during the course of a few weeks in 1922," McDougal wrote. On Singapore, a tiny place where the conquest of landscape happened a century earlier, up to three hundred people were killed by tigers each year. In Manchuria, during the late nineteenth century, as railroads were built across wilderness, "man-eaters entered huts to carry off Chinese and Russian settlers." Just beyond the Manchurian border, villagers in Korea also suffered from man-eating tigers. In Russia, man-eaters may or may not have been a serious concern.

The Indian subcontinent is so rich in tiger history and tiger travails that McDougal gave it a category all its own. Here the scope of the man-eater situation (which is only one aspect of the long, bloody chronicle of tiger-human conflict in India, with tigers generally on the losing side) can be inferred from a few statistics. In 1877, almost eight hundred people were killed by tigers throughout British India. One individual animal, known as the Champawat tigress (for a village near the Nepalese border where she appeared), was thought to have claimed two hundred victims in Nepal before crossing into India, in 1905, and then another 236 before Jim Corbett shot her. Between 1902 and 1910, the average number of tiger-caused human fatalities for all of India was 851 annually, and in 1922 that figure rose to 1,603. It declined again in the late twentieth century, as tigers became relatively scarce, but it didn't go to zero. As late as 1982, tigers were still killing about forty-five people per year in the Sundarbans swampland at the mouth of the Ganges, and about twenty per year in the Kheri District along the north-central border, up near where Corbett had his adventures. Just across the border in Nepal, where Charles McDougal's own firsthand experience was mainly rooted, tiger-attack fatalities increased sharply in Royal Chitwan National Park after 1980, coincident with the park's new policy of allowing local people to enter and cut grass for thatch. Thirteen people were killed and eaten, in and around Chitwan, within subsequent years.

The proximal causes of man-eating behavior in these places and times were various, McDougal noted, but almost all cases shared one common circumstance: a shortage of natural prey for the tigers, "either due to the disturbance of tiger habitat by humans or the dispersal by tigers into areas of peripheral habitat." In other words, the underlying causes were habitat constriction, absence of wild prey, and dire necessity. Man-eating tigers were often females—one expert claimed so, anyway—because maternal tigresses became especially desperate when a shortage of prey threatened their cubs with starvation. Tigers impaired by wounds, old age, or other physical infirmities (such as the Champawat tigress, with two canine teeth lost, evidently to an earlier gunshot) also faced special disadvantages, driving them toward reckless behavior when their natural prey was scarce.

Is it surprising, in a world where *Homo sapiens* teems ever more abundantly and squeezes ever more tightly around the few remnants of tiger habitat, that man-eating continues to occur? Charles McDougal didn't think so. In another paper, co-authored with an eminent tiger biologist named John Seidensticker, he suggested the opposite. "Tigers do not kill human beings in numbers proportionate to their availability and their potential vulnerability," Seidensticker and McDougal wrote. "One of the puzzling aspects of tiger predatory behaviour is why tigers do not kill far more people than they do." This paper was presented to a symposium on mammals as predators, in London, and published among the proceedings in 1993. It anticipated the view expressed by Igor Nikolaev and Victor Yudin, in Russian, just a few years later—that the most noteworthy thing about tiger attacks on humans is how *seldom* they occur, and what a small fraction of those attacks are unprovoked.

Seidensticker and McDougal went further than Nikolaev and Yudin, proposing an explanation. Their paper covered the whole subject of predatory behavior in tigers—including the way they select a victim, the way they approach it, the way they kill it—and the hypothesis about attacks on humans came as an aside. Seidensticker and McDougal found that tigers generally strike their prey from behind, after a short dash from good cover while the victim is moving away, and that the kill is made either with a crushing bite to the nape of the neck, which severs the spinal cord, or with a strangling bite to the throat. The crucial variables in determining just when and how a tiger will strike are the availability of adequate cover for the tiger and the movement and posture of the prey. The part about posture was especially suggestive. "Walking in a normal upright posture, a person does not represent the 'right' form for a prey animal," they noted. "A standing person's head and neck are in the wrong place and most adult human beings are taller than many large prey species." Those humans who do suffer attack have in many cases made themselves more accessible, more enticing, by some sort of movement bringing the neck closer to the ground, such as "rubber tappers who go out in the early morning dark and bend down to make their cuts, and grass cutters who are bending over, and people who go out at night to relieve themselves." An upright adult person, the two authors argued, represents a very different sort of visual signal to a tiger—a signal that doesn't beckon appetizingly like the sight of a roe deer, a boar, a buffalo calf, or a barking dog.

On the other hand, if a tiger has been wounded by gunshot or otherwise provoked to do battle (which is a different mode of behavior from stalking prey), the bipedal posture of a human constitutes no sufficient deterrent. That point was made clear in the older literature. "When a tiger openly attacks a person," Nikolai Baikov wrote, "it executes one or several jumps, beating the human along the side of the head or shoulders with blows of its paws." A single blow is often enough to shatter the skull or break a collarbone. With the human toppled, "the tiger digs its claws as deeply as possible into the head or body, trying to rip off the clothing. It can open up the spine or the chest with a single whack." Without a firearm, Baikov added, the human doesn't have a chance.

But the various Russians disagreed about how dangerous to people, or not, a normal tiger might be. K. G. Abramov concurred with Seidensticker and McDougal that, "as a rule, a tiger avoids attacking humans except when compelled by circumstances." Abramov himself had served as the first director of the Sikhote-Alin Zapovednik and, during his thirty-five years of tiger studies, had encountered not a single verified case of tiger-on-human attack in the Russian Far East. He acknowledged, skeptically, the man-eater lore purveyed by his predecessor Baikov. "But then how do you explain that the Udege, with whom I have had ample opportunities to talk about tigers, know nothing about man-eating tigers? All the cases that they give have to do with abducting dogs."

Abramov made that comment in 1965. Another thirty-five years have passed before I get my own opportunity to speak with the Udege, by which time their attitude toward Amba is just one cultural tradition among many being challenged by an onslaught of new external forces. Russia itself is a very different place, and although global perestroika hasn't yet radically redefined relations among tigers, dogs, and native hunters along the Bikin River, change is coming, as change always does. The only real questions on that account—the ones that have lured me to the Bikin—are whether the next phase of change here will be gradual or cataclysmic, and whether it will leave any chance for the continuing existence of *Panthera tigris altaica* in the wild.

67

Have I mentioned that Siberia is cold? Yes, all right, "Siberia" is a broad and imprecise geographical label, and I've been visiting (at most, by the looser construal) only its southeastern corner. Have I mentioned that the Russian Far East, in February, can be chillier than martinis in a meat freezer? Have I adequately conveyed that the Sikhote-Alin mountains enjoy a winter climate beside which Montana seems tropical?

After speaking with those Udege elders in retirement, Dima and Misha and I prepare for an excursion upriver from the village of Krasniy Yar, into a stretch of the Bikin valley where younger Udege still hunt and trap. It's a part of the drainage that Dima knows well from his own regular visits (twice each winter) to monitor tiger presence by checking for sign along a grid of trails. We plan to be up there for just two days and a night. We hope to bivouac in a hunter's cabin. Dima has contacts. We'll travel by snowmobile, a guilty convenience that will allow us to move fast and see a handful of hunters at their far-flung camps; it will also give me a new appreciation for the concept of windchill.

On the morning of our departure, we dress as though we're headed to the frozen backside of Mars. For me that means two pairs of socks, long johns, thick wool pants, ski bibs atop the wool, padded trousers (borrowed from Dima) atop the ski bibs, a long-underwear shirt, a wool shirt, a pile sweater, a pile jacket atop the sweater, a Gore-Tex parka atop the pile, a neck gaiter, an expedition-quality down coat, a balaclava, a hat with dorky earflaps, the hood from the Gore-Tex parka pulled over the hat, a pair of snowmobile boots (borrowed from Dale Miquelle), and a pair of pile mittens with windproof shells, inside which my hands are roughly as dexterous as feet. In other words, I've donned every piece of warm clothing I brought to Russia, plus more. I can barely move. To piss is an arduous operation that could result in a sprained wrist.

It has taken us half the morning to suit up, thaw the engine of Dima's snowmobile, and then load jerry cans of extra gas, a bit of food, a bit of vodka, and our sleeping bags onto a toboggan-like cargo sled hitched behind Dima's machine. The snowmobile is a 32-horse Buran, powerful enough to move all three of us plus our supplies at a discomfitingly brisk clip. We roar out of the village at ten A.M., with the sun bright and the temperature well below zero. Dima, on the saddle, wrapped in a sheepskin coat and vinyl wind leggings and a fur hat, half-blinded by a wool scarf tied across his face, drives like a snow-crazed Cossack while Misha and I hunker clumsily on the sled. Misha has offered me the front-facing position and himself taken the caboose lookout, which seems generous until we're five minutes under way and I feel the cold air ripping over my cheeks like varnish remover. Neither of us has much to hold on to. We clutch at a rope here, a flange there, bracing ourselves against the jerry cans, the sleeping bags, the sack of potatoes. The sled, just a curl-nosed slab of steel, features no shock absorbers and no windshield, let alone a steering device, a handrail, a brake, or a way of signaling Dima to slow the hell down. From where I sit, it feels as though we're a cross between the Joad family, all loaded for California, and a third-string Bulgarian luge team gone off course. Some miles outside the village, Dima swings right, plunging us down an embankment onto the frozen Bikin River. Then he cranks the throttle, and we scream along like a jet-powered dogsled across the snow-covered ice.

After half an hour we stop for a shakedown check and discover that one of the gas cans, with a faulty cap, has leaked a gallon or three into our load. Dima's sleeping bag is soaked; mine, by good luck, is only lightly marinated. Dima, throwing a funk, stalks around in his growly mode, as though there might be someone to blame but himself. He rips off the canvas tarp, opens the load, discards a soggy piece of cardboard, repacks, and then, with his characteristic quick recovery of ebullience, announces merrily: Never mind, at least the vodka is okay! Onward.

We pass a bridge. It's a sleek, modern structure of concrete pylons, steel beams, and slabs, recently completed as part of an on-again, off-again strategic scheme to cut an alternate route (away from the troubled Chinese border) between the city of Khabarovsk and a port near Valdivostok. In one direction, the road from here does lead out to Khabarovsk; in the other direction, across the bridge, it stabs eight miles into the forest and then stops. Someday, maybe soon, when the balance of politics and opportunism and market incentives shifts again, that far-side road will stab deeper. To the traditional Udege, and to their friends such as Dima and Misha, the bridge itself represents a menacing breach of the valley's blessed remoteness, across which ruination can be expected to arrive. Some people will grow rich from extracting the Korean pine, white fir, and other timber resources of the upper Bikin, but those people won't be the Udege, who will earn wages at best and, at worst, see ancestral hunting grounds converted to clear-cuts, skidder trails, gold-mining camps, and further access roads allowing all manner of outsiders to join in a scramble for what remains of the game meat, the sable, the river's fish. It doesn't take an ecologist to comprehend that if those transformations occur, the Bikin will be no place for tigers.

At midday we stop again on the ice to coax warmth and sensation back into our fingers, toes, faces. By now we've crossed into the area of Dima's long-term tiger work. It was back in November of 1969 that he made his first visit here, as a young wildlife manager on a reconnaissance survey of meat and fur resources, with only a dawning interest in *P. t. altaica* and no official mandate to study it. As for the other wildlife—he had never before, anywhere, seen such an abundance of game. Plenty of deer, plenty of sable, and enough wild boar to plow up entire hillsides in their rooting for food. On his second day in the field, he shot a boar that topped 250 pounds, a fine animal with a three-inch thickness of fat; it kept him in meat, boot grease, and cooking oil for a month. Today, on the other hand, we've seen only a couple of sets of deer tracks crossing the ice. Whether or not the populations are down, Dima posits, the increased human traffic by snowmobile is probably scaring them away from the river.

Besides the packed snowmobile trail we've been following, and the scarcity of game tracks, we note another sign of human visitation. Along the bank on our right, against a rocky cliff rising thirty feet above the river, leans a wooden ladder. Above the ladder's reach is a small animist shrine, like a dollhouse, set onto a narrow ledge. Near the base of the cliff is a sapling festooned with bits of bright cloth and paper. It's a reverential site of the sort mentioned by Vladimir Alekseevich Kanchuga, the old man nicknamed Amba, at which hunters and other travelers pay their respects to the tiger and solicit good luck by way of a prayer and a physical offering. A statue of the Virgin at roadside in Mexico captures roughly the same spirit of devotional trust—except that the tiger on the Bikin is indigenous in a way that Madre Maria in Oaxaca is not. Tucked onto one branch of the sapling is a Prima cigarette pack, gaudily red and white. Elsewhere an empty cigarette pack would look like disdainful litter, but not here; here it's a token of the woodsman's piety. With encouragement from Dima, I climb the ladder and place ten rubles into a crack in the rocks. Then again we go howling onward, like three demented polar explorers in a drawing by Ralph Steadman.

Around mid-afternoon, Dima crawls the snowmobile out of the riverbed and into a camp hidden among trees on the south bank, where two middle-aged Udege brothers welcome us into their tiny cabin. They're nephews of Su-San Tyfuivich Geonka, the old man with the antenna-wire hearing aid who told me that "a tiger is an enchanter." This parcel of hunting territory has been held by their family for most of a century, but lately, says the older brother, things aren't like they once were. It has been a bad year for sable, for instance. Deer are less abundant. "There's lots of strangers in the forest," he complains. "This place is like a revolving door. That's why there's no fish anymore." He smokes Primas, I notice, from a red-and-white pack.

After a short visit—just long enough to exchange news and discuss tigers and trends, while we three thaw ourselves at the brothers' stove—Dima herds everyone outside for a group photo. It's his practice, during these regular calls on his Udege friends and informants upriver, to bring small gifts in the form of food, vodka, and snapshots from earlier occasions. The brothers pose proudly with their home-made skis, which are wrapped with deer hide for traction, and their new puppy. The older brother wears a watch cap trumpeting the New York Knicks.

We visit several other camps before dark, including one operated by Ruslan Nikolaevich Kanchuga, a young relative of Amba himself. Ruslan, a small man with a sparse mustache and the flattened, askew nose of a prizefighter, consents stolidly to Dima's request for a night's lodging. He was alone yesterday, he'll be alone tomorrow, but yes, why not, we may share his cabin tonight. The log walls are decorated with pinups clipped from Russian tabloids. The furnishings consist of a woodstove, two plank beds, a plank table cantilevered out from one wall, and a gravity-valve water canister mounted above a slop bucket. The doorway is five feet tall, high enough for Ruslan to pass through without crouching, low enough to hold in heat. Snug and well-chinked, the cabin is warm—probably warmer than most of the houses and public buildings in Krasniy Yar, Terney, and Vladivostok.

Misha, a helpful guest, splits wood to feed the stove until morning. I air out my gas-splashed sleeping bag (Dima's is beyond airing, but in this cabin he'll do fine without it) and study the pinups and other items tacked to the walls. I see a newspaper photo of a man standing beside a stretched tiger pelt, with a caption that translates: "Victor Gaponov, in charge of a hunting competition, with an Amur tiger skin, given to the winner." It can only be the same Victor Gaponov who co-authored that 1989 paper with Anatoly Bragin—the one advocating "a radical *perestroika*" of forest management that might include demoting the tiger to game-animal status and selling high-ticket licenses to foreign hunters. As laws presently stand, Dima assures me, it would be flagrantly illegal to offer a tiger skin as a hunting-competition prize. Maybe the clipping represents satire, he suggests. Or wishful thinking, like the pinups, is my hypothesis.

Night wraps us in cozily as Ruslan rustles up a dinner of rice and boar hocks. Dima produces a bottle of Armu vodka, his preferred brand, along with the little set of brass shot glasses that travel everywhere in his field kit. Ruslan is on the wagon, as it turns out; in fact, he has had himself hypnotized in an effort to cure a booze problem. So he drinks tea while we knock down the vodka, toast by toast, in the Russian way. The boar hocks and rice disappear, and then I proffer a big bar of Alpen Gold chocolate, a luxury against which Ruslan displays no hypnotic aversion. The evening unrolls slowly, like an air mattress inflating with talk.

Ruslan is a *serious* hunter, Dima has told me meaningly. As his father before him was a *serious* hunter. Dima intends it as a compliment: Here is a man who knows the forest as no dilettante. And yet, for a Udege of the postwar generation, Ruslan has had a complicated, eclectic life. In 1971, having finished primary school, he began coming upriver into hunting camp with his father. He was sixteen. Four years later he became a staff hunter for Gospromkhoz, paid by the govern-

ment for his woodsman skills. Then he spent two years in the army—
trained in electronic communications at an institute in Vladivostok,
assigned afterward to signal-corps duty up in Khabarovskiy Krai. It
was during his army hitch that his father retired from hunting, and so
the family's hunting area, absent Ruslan to claim it, went to someone
else. He resumed the hunting life in the 1980s and now spends most of
his time out here in the forest, but he's forced to share a modest area
with two other claimants. Three hunters along a Bikin tributary of no
more than twelve miles—it's crowded.

He shoots boar, red deer, and bear. Some hunters will find a denning bear and then "sell" it to one of the new Russians who come for
sport shooting, he tells me, but that sort of cheesy opportunism doesn't
tempt Ruslan. He'd rather kill the bear himself, take the meat, sell the
gall bladder. A gall bladder brings three dollars per gram, unless it's
big, in which case the gram price is higher still. Bear paws are saleable
too. He traps sable, otter, lynx, mink, badger, and a few other species.
Want to see a sable? he asks, and hands me a dead one, whole and
limp, guts in, its fur silken brown. He trapped it yesterday. It's still
thawing. He'll skin it tomorrow. At his belt, I notice, Ruslan wears a
discreet little skinning knife. No, this hasn't been a good year for sable.
Nor for trapping generally. If he had to rely on his trapline, he'd sit
around sucking his finger, unable to make ends meet. Things have
been so bad, he admits, that in December he poached a doe. He has
pressures, responsibilities, a slew of relatives down in Krasniy Yar who
expect him to deliver meat.

At this point I ask Ruslan about tigers, and the tenor of the conversation shifts. He says something that seems amazing, yet quite believ-
able in its simple candor. He says, "I've never seen one. I've never seen
one eye to eye."

After all his winters of hunting and trapping, all his travels
throughout this forest, he has never set his eyes on a living tiger. (Dima
stands as voucher to the plausibility of this statement, having earlier

admitted similarly, "Nobody ever asks me how often I've seen a tiger. I've actually seen a tiger very few times.") The ghostly elusiveness of the animal, though a sweet frustration to Dima, suits Ruslan just fine. "I need a tiger like I need a hole in the head," he says. Actually, that's a loose rendering of his comment, as interpreted politely by Misha, who then explains that Ruslan has used a crude idiomatic expression. A more literal translation would be "I need a tiger like I need one on my dick."

He certainly *doesn't* need a tiger there. Nor does he need one crossing his path in the forest, or eating his dogs, or competing with him for wild boar, and he doesn't hanker for one lying dead at his feet either. Tiger poaching, no thanks. Yes, true enough, he would like to own a Yamaha snowmobile, like the fat dumb new Russians who come sporting around. But a dead tiger doesn't buy him that, Ruslan says. A dead tiger buys only hassle. Some people claim that a tiger carcass can be worth 60,000 rubles. But if you shoot one, then you've got to sell it, quickly and quietly. To whom? Ruslan doesn't have those connections, and besides, a guy could get caught.

So he's content at his lack of close encounters with *P. t. altaica*. "My grandpa used to say, You go your route, and he'll go his route. You don't bother him, and he won't bother you." The danger of walking through tiger-haunted bottomland here on the Bikin, Ruslan figures, is no greater than the danger of walking through traffic-jammed, crime-ridden Vladivostok.

Ruslan Nikolaevich Kanchuga is not a sentimentalist. He's a hard-headed Udege man from an old Udege family, positioned in socioeconomic space with a stance that's wide but steady—one foot planted firmly in tradition, the other foot thrust forward into a much different terrain of possibilities, desires, and costs. Late in the evening, he mentions passingly that, notwithstanding his middle-aged bachelorhood and his isolation up here on the Bikin, "I still wanna get married. I'm gonna look for a wife on the Internet."

What kind of wife? I ask. Any kind. The sole stipulations, he says, are that she must love to eat fish and that she is able to stomach raw meat. I begin wondering just where and under what circumstances Ruslan will be able to advertise his availability online. Laptops and modems haven't yet invaded this stretch of the Bikin River.

And the tiger? I ask. Does it possess a special sort of power or aura? Is it sacred in some sense? Is it the spirit of the forest? Is it godly?

"It's an animal," Ruslan says.

SCIENCE FICTION
ENDING

68

Once there were lions across southern France. There were lions in Germany, along the bottomlands of the Rhine. There were lions in Britain, having gotten there dry-footed during an interlude of lowered sea level and established themselves as far west as Devon. There were lions in Poland. Dispersing northward out of Africa, lions seem to have reached Europe about 900,000 years ago, and to have become widespread and relatively common throughout the continent. They endured the ice ages and they thrived during the interglacials, preying on a rich fauna of native ungulates—reindeer, Irish elk, aurochs, bison, wild horse, ibex, and other species. In at least some European locations they survived to the end of the Pleistocene, just 11,000 years ago, long after modern humans had arrived.

Bony remains from this big European cat—giving evidence of its distribution, its geological age, and the size and shape of individual animals—have been found in glacial gravels and caves. Partly for that reason, it's loosely known as the cave lion. Whether it had a strong, species-wide propensity toward denning in caves, or only did so opportunistically in special circumstances, is uncertain. According to one informed reading of the fossil data, "In the mountains it used caves as refuges, but on the plains it survived without them." Still, the association between lion bones and caves may be unrepresentative and misleading, insofar as the caves served as natural mausoleums, preserving samples of bony material that would have been destroyed by weather and scavengers if left outside. So the term "cave lion" is imprecise, but it parallels the term "cave bear" as applied (more appropriately) to the Pleistocene species *Ursus spelaeus*, that huge vegetarian bruin with a habit of hibernating in caves. Most experts use "cave lion" informally without fretting about ecological precision.

In life, the cave lion resembled the African lion as known today, although on average it was larger. The late Finnish paleontologist Björn Kurtén, an authority on Pleistocene mammals, called it "gigantic," even possibly "the largest felid that ever existed." Scientists have argued for more than a century about its taxonomic relationship to other extinct and living cats, and those disagreements are reflected in vagaries of nomenclature; it has been variously labeled *Panthera spelaea*, *Felis spelaea*, *Panthera leo spelaea*, and *Panthera atrox*. Alan Turner and Mauricio Antón, in their recent study of the big cats, dismiss the size differences as inconclusive and place the cave lion alongside the modern lion within *Panthera leo*. But the argument isn't settled. To other experts it's still *Panthera spelaea*, an extinct Pleistocene species, larger than the African lion, larger even than the Amur tiger, and conspicuously maneless.

Of course, a lion's mane is never preserved in the fossil record. The maneless condition of males as well as females among the European cave lion is inferred from another medium of evidence—the same one that testifies to the overlap of lions and humans on the late-Pleistocene European landscape. That medium is art, as created tens of thousands of years ago on the walls of limestone caverns. Maneless lions are a dominant motif in one of the most spectacular and surprising assemblages of Paleolithic cave paintings ever found. This ancient gallery, sealed by a rock fall for dozens of millennia and discovered only in 1994, is known as la Grotte Chauvet.

69

On the afternoon of December 18, 1994, three amateur spelunkers exploring a cliff face along the Ardèche River, in southeastern France, opened a new sight line into the history of art. They did it by probing just a little farther, a little more persistently, in an area with which they and other cavers were already familiar.

The cliff face was part of the Cirque d'Estre, a gorge wall looming over the Ardèche. The three friends followed an old mule path that traversed upward to a ledge. Below they could see vineyards and a road. Pushing through some vegetation, they located a narrow slot, barely bigger than a porthole, in the white rock of the cliff. They climbed through the slot into a small, sloping chamber within the rock, at the far end of which was a pile of stone rubble. The chamber had been previously explored, during which someone had noticed a draft of air through the rubble, like breath from a huge underground lung. On that day the venturesome trio decided to investigate the source of the draft.

Digging out the rubble by hand, they found a natural duct, a tubular shaft that fell away vertically like a drain, barely large enough to allow a person to wiggle down it. One by one, headfirst, they went, following the tube as it descended and then bent back upward and around. After twenty feet of squirming, they emerged onto a ledge above a larger chamber. By the light of their helmet lamps they could see the chamber floor, thirty feet below. "To measure the resonance of the echo, we shouted," they wrote later. "The noise carried far, and seemed to get lost in the immensity of the cave." They had no idea what they'd found, except that it seemed to be a large, pristine cavern. They climbed back out the tube, fetched a rope ladder from their van, and returned to explore further.

One of the three spelunkers was Jean-Marie Chauvet, a local resident and a cave enthusiast since his boyhood, who now worked for the Ministry of Culture as a guard at painted caves in the Ardèche valley. The Ardèche caves, of which twenty-seven had come to light, weren't so stunning or famous as Lascaux, Altamira, Les Trois-Frères, or some others; in the world of Paleolithic art, the Ardèche was just a modest sideshow to those greater attractions of western France, the Pyrenees, and the Cantabrian coast of northern Spain. Jean-Marie Chauvet and his friends were destined to change that. They unrolled their ladder, climbed down, and quickly made three discoveries.

The first discovery was that this cave, opening out from one chamber to another beyond reach of their helmet lamps, was much larger than other caves of the region. The second was that it had served as a refuge for the cave bear, *Ursus spelaeus*. Bear bones and teeth littered the floor, and the three cavers took care to avoid crushing them underfoot. The third discovery came when their lamps swept across a smeary red pattern on a spur of rock hanging down from the ceiling. Looking closer, they recognized the shape of a mammoth, sketched in ochre pigment. "We were overwhelmed," they wrote. "Henceforth our view of the cave would never be the same. Prehistoric people had been here before us." The mammoth was only a tiny sample of what they were about to see.

During the rest of that evening and in a second visit several days later, Chauvet and his companions explored the length of the cave, which stretched for almost two thousand feet through a succession of five major chambers. The place was full of art. They found a leopard done in red ochre—the first leopard figure ever encountered in any Paleolithic gallery. They found horses, bison, aurochs, more mammoths, reindeer, ibex, giant deer, and many rhinoceroses, including two rhinos depicted head to head in a fight. Rhinos are rare in European cave art, and none had ever been seen in the Ardèche. They found bears, strong-browed and burly, outlined in ochre or charcoal. Most dramatically, they found lions.

At first they saw only a few scattered lion images, one of which "was very strange and seemed a bit of a failure" because of its misshapen mouth; another was notable for its shapely head and a strong eyebrow. Each of those lions was maneless but unmistakably feline. So were three large figures portrayed, side by side, in full-body profile. Then, in the deepest chamber, approaching a dead end, Chauvet and his partners turned toward the left wall. "Suddenly our lamps lit up a monumental black frieze," they recalled. "It took our breath away as, silently, we played our torches over its panels. There was a burst of shouts of joy and tears. We felt gripped by madness and dizziness." They were gazing at a huge, dynamic mural that showed a dozen lions, again all maneless. The cats were gathered in a cluster, like a pride on the hunt, staring intently toward a mixed group of other animals—bison, rhinos, a horse, and a gangly young mammoth. Some of the lions seemed to be craning forward, shoulders low, noses out, in postures that suggested stalking. Their heads were well-shaped and deftly shaded, evoking bone structure and depth. They were done with the assured touch of a gifted artist who must have observed living models closely. They were realistic. They were poised. They were beautiful.

Among students and connoisseurs of Chauvet Cave, that innermost mural is now known as the Lion Panel. Wherever the cave is described or discussed, in articles or books, you can expect to see the Lion Panel reproduced. Nowhere else in the galleries of Paleolithic art is there anything like it.

In the years since 1994, Chauvet Cave has been protected by the French government and carefully studied by a team of scholars and technicians under the direction of Jean Clottes, a renowned expert on prehistoric European art. Clottes has said that Chauvet ranks alongside Lascaux as one of the most spectacular caves ever opened. More than four hundred animal images—paintings, drawings, etchings into the soft surface of rock—have been found and catalogued, of which many portray rhinos, mammoths, bears, and lions. This is unusual among aggregations of cave art, which are normally focused on the big herbivore species that early humans hunted for meat. The bison of Altamira, the deer and horses of Lascaux, are more typical. In the Trois-Frères cave is a stiff-bodied lion, and Le Tuc-d'Audoubert contains a lion attacking a horse, but such images are rare and seem to be marginal to the dominant visual themes. In Chauvet, on the other hand, Clottes and his team have found seventy-three lion images, more than the total from all other decorated caves in Europe. Uniquely among Paleolithic sites, Chauvet is the grotto of dangerous beasts.

And yet it isn't a record of fear and loathing. The rhinos are graceful. The bears are hulky but not menacing. The lions are handsome, stern, and majestic. Whoever painted these images—the best of them, anyway—did so with a skilled hand, a calm heart, and an attentive, reverential eye.

The manelessness of the Chauvet lions raises some questions. Are they all females? If so, why did the Chauvet artists decline to portray males? Or was the species (*Panthera spelaea*) or the subspecies (*Panthera leo spelaea*) a maneless form in both sexes? Those questions seem to be answered by a pair of overlapping charcoal images on a wall of the penultimate chamber. The whole composition, known as the Panel of the Horses, is layered with perspective. In the frontmost layer are three black-maned horses. Behind them is a lion and, behind that one, another, both seen in profile, facing opposite directions. The background lion seems to be snarling; the foreground lion seems to be sniffing at the other lion's rear end. A plausible interpretation, offered by Craig Packer, a field biologist who studies African lions, after visiting the cave himself, is that these paired drawings depict a horny male lion (the sniffer) being rejected, at least temporarily, by an unwilling or unready (anyway, snarling) female. Jean Clottes has noted that another lion image—a full body figure in the deepest chamber, portrayed without a mane and *with* testicles—provides clear evidence that European cave lions, male and female, were maneless.

"Exact dates for the paintings are still a matter of conjecture," Clottes wrote in 1995, just after his own first inspection of Chauvet. Many of the other celebrated art caves—Altamira, Les Trois-Frères, Niaux, and more—date from a period of Paleolithic culture known as the Magdalenian, which ended about 10,000 years ago. But in the case of Chauvet, Clottes added, "some details point to a pre-Magdalenian period—the Solutrean, about 18,000 to 21,000 years ago." Lascaux, another base of comparison, is an oldish cave characterized mainly by an archaic Magdalenian style, traceable back about 17,000 years, almost contemporaneous with the late Solutrean. The caves known as Cougnac and Pech-Merle, also in western France, are still older, dating from 24,000 years ago, and a cave called Gargas, in the Pyrenees, is as old as 26,000 years. Those earlier caves display primitive stylistic conventions, suggesting a gradual development of sophistication toward the sensibilities and techniques that eventually appeared in such great Magdalenian showcases as Altamira. Chauvet contains some elements of style, subject, and coloring that seem to connect it with Solutrean caves of the Ardèche valley, and other details hinting at an affinity with the archaic Magdalenian of Lascaux. All in all, based on stylistic analysis, Clottes tentatively placed Chauvet in a range between 17,000 and 21,000 years old, or possibly older. And he hedged his estimate with a prudent reminder that the new cave, in its "powerful originality," might yield more surprises.

Radiocarbon tests were then done on carbon samples from Chauvet. The results came as a delicious shock. At least some of the images seem to be 35,000 years old—that is, older than any other rock art known on Earth.

And yet, ancient as it is, Chauvet isn't primitive. The sophisticated techniques—spatial perspective, inner shading, scraping around the outside of an image to add relief—are already there. "That changed our whole conception of the evolution of art," Jean Clottes recently said in a public lecture. The gradualist paradigm was no longer tenable. The new understanding, forced by the evidence of Chauvet, was that about 40,000 years ago modern humans (as distinct from the Neanderthal race, which they supplanted) must have arrived in Europe with their artistic sense quite fully developed. Their skills and their vision were already sophisticated, and would scarcely advance further over the next 20,000 years. So the evolution of art had in fact been a spiky phenomenon, Clottes concluded, not a gradual one. It was no small part of Chauvet's significance, he said, that this single, wondrous cave delivered up such a perplexing secret into the late twentieth century.

Beyond that issue, another mystery emerges from Chauvet—one that even the radiocarbon tests, the stylistic analyses, and the archeological contextualizing haven't yet solved. Why were those ancient artists so dreamily fascinated by lions? Where was the fear? Where was the loathing? Where was the angry, competitive resentment of one meat-eater toward another? How could the Paleolithic people of the Ardèche valley afford such aesthetic admiration, and such serene spiritual appreciation, for a species of large predator that, at least occasionally, must have included *Homo sapiens* within its diet?

70

On the geological scale, 35,000 years is a very short time. On the scale of human evolution, human culture, human psychology and memory, it's long. Chauvet Cave proves that, for at least thirty-five millennia, we humans have embraced lions as an important part of our metaphysical universe as well as our ecological one. And it does more than that. No one can view a reproduction of the Lion Panel without sensing that the artist or artists who drew it not only knew lions, and respected them as formidable predators, but in some sense also treasured them. The lions of Chauvet are too knowingly, lovingly rendered to be explained away as bogeys, negative effigies of an enemy or competitor, expressions of abhorrence or dismissal.

Then what *do* those images express? Nobody knows, not even the experts. Jean Clottes has refrained from offering easy, quick interpretations for the peculiar menagerie of Chauvet. He suspects that Paleolithic cave art in general reflects shamanistic beliefs, and that the artists themselves were the shamans, socially selected or self-selected to enter those deep caves, put themselves into trances, create or revisit the magic pictures on the walls, and thereby achieve various ends that included "healing the sick, foretelling the future, meeting spirit-animals, changing the weather, and controlling real animals by supernatural means." How does that general notion apply specifically to the lions of the Ardèche? "Chauvet Cave reveals that big cats played an unsuspected and important role in the local bestiary," Clottes has written. But *what* role? "These animals doubtless symbolized danger, strength and power." That much is inarguable, but still vague. Is it possible that the artists tried to "capture the essence" of that power in their images, Clottes wonders, and thereby gain control over "the danger they represented and the domination they exerted over their environment?" Maybe. And maybe not. "All speculations about this problem," he admits candidly, "can only be guesses."

If the answer isn't obvious to Jean Clottes, then the answer isn't obvious, and the rest of us are left with our puzzlement, our subjective responses to the art, and our own guesses. As for me, I don't have an easy, quick theory about Chauvet either. I haven't seen the place firsthand (a tempting prospect, but considering the fragility of that underground environment, and the necessary restriction of access, I decided against trying for an invitation), and I'm no authority on the semiotic vocabulary of Paleolithic art. All I know from contemplating photos of the Lion Panel, the triple profile, and the other Chauvet cats—as I've been doing now obsessively for several weeks—is that the people who created these images recognized more than danger, strength, and power. They also saw grace, grandeur, lordly confidence, quietude, mercilessness, keen attention, and some sort of all-driving primacy in those ferocious beasts; and they took pains to register, to preserve, even somehow to adopt what they saw through the medium of charcoal on rock. Call it shamanism, call it totemism, call it idolatry. Call it, simply, art. Anyway, they succeeded. *Panthera leo spelaea* is gone, but Chauvet remains.

71

Thirty-five thousand years, then, is the minimum verified duration of the important psychic relationship between *Homo sapiens* and alpha predators. That stretch of time is a vector from past to present. What about the vector from present to future? Can we expect another thirty-five thousand years of ambivalent, consequential coexistence? Not hardly. Can we expect a thousand? Again, no. Can we expect . . . oh, say, another three hundred years during which humans will continue to share landscape with large, dangerous beasts? I doubt it. My guess, a regretfully gloomy one, is that the last wild, viable, free-ranging populations of big flesh-eaters will disappear sometime around the middle of the next century. I see the year 2150 as a probable end point to the special relationship between us and them, which was already ancient when the first sketch went up at Chauvet. That's not far off—less than eight human generations. It's just time enough to encompass a welter of uncertainties, along with one weighty inevitability: the continuing growth of human population and consumption.

I don't mean to be too fatalistic. The future status of each of those predator species, subspecies, and populations, in their various ecological and political contexts, will depend on many interacting factors—some global, some local, some tangible and measurable and controllable, some not. For instance, the wholesale prices paid for premium-grade crocodile skins, by luxury leather-goods manufacturers in Italy and France, will have a direct impact on the viability of commercial crocodile farming in northern Australia. But farmed crocodiles are one thing, wild crocodiles are another, and the flow of dollars and euros toward pricey crocodile-leather handbags may or may not directly affect the incentive structure for private landowners along the Mary River, or for Aboriginal clans along the Liverpool, to conserve streambank habitat for the nesting of *Crocodylus porosus* in the wild. In western India, the benefits of modern medicine, electrification, groundwater pumping, and education may raise the life expectancy and infant survival rate among Maldharis, including those still resident within the Gir forest, which may in turn increase the total pressure by them and their buffalo on the same grazing and browsing resources needed by the wild ungulates that serve as natural prey for the lions. Then again, the attractions of town and the possibilities revealed by education may entice young Maldharis away from traditional life on the nesses, gradually emptying the forest of buffalo herders and leaving it (at least for the short term) to the chital deer, the lions, and the ecotourists. In Romania, as the country tries to construct a functional economy out of post-Communist rubble, the demand for bearskin trophies may be eclipsed by the demand for beech timber, fir, spruce, and oak, and the privatization of forest lands may turn the Carpathians into a raggedy patchwork of clear-cuts, paved roads, cow pastures, and weekend homes. Anything could happen in the Russian Far East, where the modest archipelago of zapovedniks and other protected areas is necessary but not sufficient to support an enduring population of tigers, and the crucial lands in between may soon suffer commercial liquidation. The privatization of hunting rights on at least some portion of those lands offers a modestly promising alternative to the more destructive forms of forest-product extraction. But should hunting rights include the right of auctioning off a small number of tigers to be killed for trophies, as proposed by Bragin and Gaponov (and, from a distance, by Grahame Webb)? Although other tiger biologists in Russia are still wary of that idea, it remains a discussable option. Meanwhile, the black-market value of a tiger carcass in Seoul or Beijing can be expected to rise whenever the supply in the Sikhote-Alin mountains declines.

No one can predict how these regional situations, with all their political, economic, and geographic variables, will play out, and to bundle them into a global prognosis for alpha predators in general would be, if not impossible, at least hasty and foolhardy. I won't be so presumptuous as to try. Instead I refer you to another body of prediction that—so it seems to me—makes all the short-term calculus of poaching rates, conservation incentives, habitat protection, harvest quotas, and "sustainable" exploitation almost irrelevant. I'm talking about what comes from the Population Division of the United Nations.

On my desk at this moment are two white-paper documents, each no bigger than a magazine but crammed with statistics. The more recent is titled "World Population Prospects: The 2000 Revision." It's a concise extract, the "highlights" version of a multivolume report prepared by the Population Division, which is housed within the UN's Department of Economic and Social Affairs. It draws on demographic data and trends going back to 1950, and it projects forward to the year 2050. Such a revision, with updated inputs and estimates, is produced every two years. The other report, published in 1998 as part of a less frequent series, takes a longer view. This one is called "World Population Projections to 2150," adding a century to the scope of outlook. One hundred fifty years is a big leap for demographic forecasters, but those at the UN seem to be judicious and erudite as well as bold.

Both reports consider such factors as changing fertility rates (in response to cultural influences, including education), changing patterns of mortality (in response to public-health circumstances, including AIDS), migration between countries, and population aging. Both make projections by region and by country as well as globally, and both recognize the inherent chanciness of such a predictive enterprise by offering alternate scenarios (high population, medium, relatively low) based on alternate assumptions (high fertility, medium, low) about how reproductive behavior may change. We all know that the total number of humans is still growing—and growing fast, though not quite so fast as previously—and we all assume that eventually the growth will stop. It has *got* to stop—hasn't it?—at some point in the crowded future. These reports sketch the likely shape of the continuing growth curve, and the 1998 "Projections to 2150" even addresses the questions of when, and at what final level, it may top out.

The "2000 Revision" is the less dramatic of the two. "World population reached 6.1 billion in mid-2000," it states, "and is currently growing at an annual rate of 1.2 per cent, or 77 million people per year." Of the current annual increase, India alone accounts for one-fifth (which can be taken, though the UN doesn't say so, as bad news for the Asiatic lion). Most of the planet's new burden of humans will come in less developed areas—notably, in struggling countries of the tropics. By the year 2050, the report forecasts, the total world population will be somewhere between 7.9 and 10.9 billion. According to the medium-fertility scenario, which can be considered the most likely, the total will reach 9.3 billion.

It won't stop there. The "2000 Revision," with its relatively short-term view, doesn't treat the question of when population growth might end. It says nothing about an eventual maximum. The earlier, longer-term report does, and that's what makes it so dismally interesting. "According to the medium-fertility scenario, which assumes fertility will stabilize at replacement levels of slightly above two children per woman," it says, "the world population will grow from 5.7 billion persons in 1995 to 9.4 billion in 2050, 10.4 billion in 2100, and 10.8 billion by 2150." Then the growth curve will level off, at just under eleven billion people, all of whom will continue to occupy space, drink water, burn energy, consume solid resources, produce wastes, aspire to material comfort and safety for themselves and their 2.0 children, and eat.

Call me a pessimist, but when I look into that future, I don't see any lions, tigers, or bears.

72

Sagacious biologists have warned us about this. Back in 1986, for instance, David Ehrenfeld published an essay titled "Life in the Next Millennium: Who Will Be Left in the Earth's Community?" In hazarding an answer, he stipulated one cautious assumption: that the next fifty years would unfold without any drastic interruption of recent trends.

In other words, there will be more people, there will be more industrialization, there will be more urban growth, there will be more standardization, there will be more corporate conglomeration and bigger organizations, there will be more power-oriented consumer goods, there will be more tourism, there will be bigger weapons budgets for more elaborate weapons, there will be more advertizing and image making, there will be less room for personal eccentricities, and there will be more mechanization of agriculture and chemical farming.

If those trends hold steady, Ehrenfeld claimed, "it becomes a fairly easy job to predict the fate of the animal and plant species on Earth." He then recapitulated an analysis he'd made sixteen years earlier, in a book titled *Biological Conservation*, of the general attributes of those species most likely to suffer a high jeopardy of extinction.

Biological Conservation was a small, readable volume, published originally as a text for college biology students, that became one of the foundational documents of the field now known as conservation biology. In its discussion of extinction and endangerment, it included a list of "those characteristics of animal species that can lower their survival potential"—characteristics such as large body size, restricted distribution, slow reproductive rate, reproduction in colonial aggregates, hunting pressure in the absence of effective game management, migration across international boundaries, and intolerance of human presence. Along with each fateful item on that list, of which there were ten, Ehrenfeld cited a representative species: the cougar for size, the Bahamas parrot for restricted distribution, the giant panda for slow reproductive rate, the green sea turtle for migration, and so on. Then he concocted a hypothetical composite, a "most endangered animal" so unfortunate as to possess many of the dooming characteristics. "It turns out to be a large predator," Ehrenfeld wrote,

> with a narrow habitat tolerance, long gestation period, and few young per litter. It is hunted for a natural product and/or for sport, but is not subject to efficient game management. It has a restricted distribution, but travels across international boundaries. It is intolerant of man, reproduces in aggregates, and has nonadaptive behavioral idiosyncracies. Although there probably is no such animal, this model, with one or two exceptions, comes very close to being a description of a polar bear.

That was Ehrenfeld's paradigmatic candidate for early extinction. Close behind *Ursus maritimus*, according to his profile of jeopardy, would come *Panthera tigris*, *Ursus arctos*, and *Panthera leo*.

On the other hand, he added, the typical survivor species of the

twenty-first century and beyond would embody antithetical characteristics, such as small body size, herbivorousness, fast reproductive rate,
wide distribution. Those parameters were well satisfied by certain
weedy, tenacious creatures, such as the house sparrow, the gray squirrel, the Virginia opossum, and the Norway rat. After quoting from his
1970 warning, in his later essay Ehrenfeld reaffirmed glumly, "These
are the jolly companions we can expect if the world continues much
longer on its present course." So far, the world has.

An ecologist named John Terborgh, who did his fieldwork in the
American tropics, published a similar analysis in 1974. Terborgh's
paper was titled "Preservation of Natural Diversity: The Problem of
Extinction Prone Species." The essence of the problem, he wrote, is
that human population growth, consumption, and transformation of
landscape are everywhere causing destruction to ecological communities; but the earliest and most acute effects are felt by some species
more than others. In order to plan effectively for conserving as much
of the natural world as possible, Terborgh argued, it's "important to
recognize the categories of species that are most vulnerable." He
defined six such categories, most of which overlap with Ehrenfeld's.
Among his six: the category of species with colonial nesting habits, the
category of species with restricted continental distribution, the category of species endemic to one or another island, and the category of
species that migrate internationally. Before all others on Terborgh's list
came the category of large-bodied species, especially those at the top of
a food chain, with their low reproductive rates and their high needs for
energy and territory. In other words, big predators.

Around the same time as these warnings of high extinction jeopardy
were made, ecologists had begun awakening to another important
thing about big predators: their crucial role, at least in some cases, as
keystone species. The label was new, coined to indicate a species of inordinate importance to the overall structure of an ecosystem.

A keystone, in the literal sense, is the single wedge-shaped block that absorbs and balances the unstable forces in a stone arch. A keystone species, by analogy, is one that exerts a large, stabilizing influence throughout an ecological community, despite its relatively small numerical abundance. Remove the keystone species, through extinction or experimental exclusion, and the whole community begins to fall out of whack, suffering population explosions or crises of rarity that ramify from one interrelated species to another. This idea had its roots in the work of Charles Elton (remember him? the English ecologist who, having studied foxes and lemmings in Spitsbergen, invented the concept of the food chain) and of Raymond Lindeman (remember him? the precocious grad student who, having studied sunfish and pondweed in a senescent Minnesota lake, advanced from Elton's work to the concept of trophic levels), and it sprouted from those roots decades later. A related idea that sprouted alongside it was *trophic cascades*, referring to the secondary disruptions that go cascading through a system from which a keystone species has been removed.

Neither Elton nor Lindeman went so far as to talk about keystone species or trophic cascades. Nor did three scientists who became forever immortalized in the ecological literature as the triad "Hairston, Smith, and Slobodkin" (sounds like a law firm) for a paper they co-authored in 1960. The subject addressed by Hairston and his colleagues (sometimes labeled HSS, for convenience) was how populations of species are limited—meaning, what prevents a given species from being more abundant than it is. The main mechanism of limitation but not the only one, according to HSS, is competition. That was reflected in their paper's straightforward title: "Community Structure, Population Control, and Competition."

They began with the understanding that all species interconnected in any food chain can be grouped into four trophic classes: producers (plants), decomposers (detritus-eaters and microbes), herbivores (plant-eaters), and predators (animal-eaters). Each species is subject to limiting factors that keep its population from exploding. Populations of producers, decomposers, and predators are generally limited by food supply, HSS stated. In the case of plants, food supply means water, nutrients, and sunlight; for the other classes, its meaning is self-evident. If food supply is low and demand is high, then competition among species must be fierce and significant, helping to drive evolution by forcing each competitor to adapt or disappear. This is a bottom-up sort of limitation: Food shortage impinges upward, causing competition. But competition isn't the whole story. Almost as an afterthought, HSS made a point about the trophic class that stood as exception to their rule: "Herbivores are seldom food-limited, appear most often to be predator-limited, and therefore are not likely to compete for common resources." The implication here is that predators play a critical role in structuring ecological communities by exerting pressure from the top down. Herbivores too must evolve and adapt, yes, but they adapt not so much to the nuisance of competition as to the threat of predation. Herbivore population sizes will be determined by the supply of lions, tigers, leopards, cougars, cheetahs, hyenas, and wolves, not by the supply of grass.

The next step was taken by a marine ecologist named Robert T. Paine, from the University of Washington, who had paid close attention to a certain species of starfish, *Pisaster ochraceus,* in the intertidal zone of a rocky shoreline at Mukkaw Bay, just west of Seattle. This starfish is a predator that eats barnacles, mussels, chitons, and other invertebrate prey off surf-covered rocks. Unlike a barnacle or a mussel, it can't close itself within shell walls to avoid desiccation, and so its predatory activity is mainly confined to the lower range of the intertidal zone. Even down there, it's not especially abundant, and its influence—though considerable, as Paine learned—is exerted slowly and quietly. Among its favorite prey is the mussel *Mytilus californianus*, a tasty and convenient victim that tends to colonize the middle and lower intertidal ranges. Besides being toothsome to the starfish, *Mytilus californianus* happens also to be a very effective competitor for living space against the barnacles, limpets, and other sessile creatures native to the same stretch of coast. In the middle range of the intertidal zone, from which the starfish is excluded by the pace at which it moves (not very speedy, though fast enough to make a barnacle look like it's standing still) and by the threat of desiccation during low tide, the mussel is a dominant species. In the lower range, where *Pisaster ochraceus* lives comfortably, the mussel occurs only sparsely amid a gaggle of other creatures, including limpets, barnacles, chitons, anemones, algae, sponges, and at least one species of sponge-eating sea slug. As an experiment, beginning in June 1963, Paine persistently removed every *Pisaster* starfish from a short stretch of coastline and watched for secondary effects.

Within a year, the diverse community of species in the lower intertidal range had disappeared, and the aggressive mussel was now colonizing that area. For a short while, it coexisted with two species of barnacle. Within a few more years, the barnacles too were being crowded out, the sponge-eating sea slug was gone, and diversity within the experimental area had fallen from fifteen species to eight. Eventually it would fall even farther as the mussel emerged supreme, monopolizing that patch of habitat almost totally. Paine's conclusion, reported in a 1966 paper, was that predation by *Pisaster ochraceus* interrupts the competitive ascendancy of the pushy mussel, thereby allowing a more diverse community to exist. Removing the starfish causes a trophic cascade that affects, indirectly, even the sea slug. In a second paper, published three years later, Paine described *Pisaster ochraceus* as "the keystone of the community's structure," and then explicitly, for the first time, called it a "keystone species." An important ecological principle had been recognized and named.

It caught on. Other ecologists began noticing keystone effects in ecosystems they studied. The sea otter (*Enhydra lutris*) became a famous example, identified as a keystone species along the Alaska coast because of its role in preserving subtidal kelp forests. Sea otters

eat sea urchins, and sea urchins eat kelp. Researchers saw that with sea
otters present, the kelp forests thrived, despite cropping by a modest
population of sea urchins; with the otters removed (not experimentally
but by hunting), the urchins multiplied grossly and ate all the kelp.

The phenomenon of trophic cascades, triggered by the loss of one
keystone species or of several top-level predators, appeared also in
some terrestrial ecosystems, such as Panama's Barro Colorado Island.
Until the early twentieth century, Barro Colorado had been just a hilltop within a larger expanse of mainland forest, accessible to jaguars
and cougars, among other animals. Then the hilltop became insularized within Lago Gatun, an impoundment created for the Panama
Canal, and the big cats could no longer enter and leave at will. Small
as the island was, it couldn't support a resident population of either
jaguars or cougars. In the absence of those predators, as decades
passed, Barro Colorado's ecosystem changed. Middle-sized predators
such as the collared peccary and the coatimundi, now suddenly at the
top of the food chain, became very abundant. Bird populations disappeared (notably, ten species that nested or foraged on the ground), possibly because they'd been exterminated by the middle-sized predators.
Seed-eating mammals, such as the paca and the agouti, also grew
unusually abundant. Did the proliferation of seed-eaters thwart the
reproduction of tree species producing large, edible seeds? Evidence
had only begun to trickle in, as of 1988, when John Terborgh articulated that hypothesis. If the loss of cougars and jaguars from Barro
Colorado Island had in fact led to the extinction of bird populations
and to the failure of reproduction among certain trees, it was a classic
case of trophic cascades.

Another case, described more recently by Terborgh and ten co-
authors, involves a set of forested islands in Lago Guri, an artificial
lake created for a hydroelectric project in Venezuela. On the smallest
of those islands, since their isolation in 1986, big predators have been
absent and so have armadillos; leaf-cutter ants (which armadillos eat)

have thrived, as have iguanas and howler monkeys (which eat vegetation, including young tree shoots). The inordinate populations of iguanas, howler monkeys, and leaf-cutter ants seem to be impeding the reproduction of canopy trees and changing the flora of the islands. Terborgh and his colleagues reported their findings in the journal *Science*, under a heated title: "Ecological Meltdown in Predator-Free Forest Fragments." Near the end of the paper they wrote: "These observations are warnings, because the large predators that impose top-down regulation have been extirpated from most of the continental United States and, indeed, much of Earth's terrestrial realm." Ecological meltdown, they implied, is coming soon to an ecosystem near you.

The eminent conservation biologist Michael Soulé, along with a young co-author named Kevin R. Crooks, has described a parallel situation on small patches of native scrub habitat surrounded by urban sprawl in southern California. The largest predator thereabouts is the coyote, *Canis latrans*. Coyotes are legendarily wily and opportunistic creatures that can survive in close proximity to humans, but they choose their spots and take no unnecessary risks. With continued fragmentation and shrinkage of the scrub patches, as human development presses all around, coyote presence on some of those patches has declined. In counterpoint to the decline have come increases in the activity of middle-sized predators, such as the raccoon, the domestic cat, and the opossum. The boom among middle-sized predators correlates, in turn, with a bust for scrub-nesting birds. Who would have suspected that the coyote might serve as guardian angel to the spotted towhee, the greater roadrunner, or the California gnatcatcher? An ecologist interested in keystone species, that's who.

It's no coincidence that the coyote of suburban San Diego, the jaguar of Panama, and the sea otter of coastal Alaska, like Robert Paine's original starfish, are predators occupying high trophic levels within their respective communities. Early thinking on the subject tended to associate the keystone phenomenon almost completely with high-level predation. The term "keystone predators" was even used as an alternative to "keystone species." Then came a period during which the concept grew fashionable and, like many fashions, was oversold. The terminology became stretched as ecologists talked about "keystone resources," "keystone herbivores," "keystone mutualists," "keystone modifiers," keystone foot-worms, keystone whatevers, all reflecting the undeniable truth that many species carry special significance of one sort or another. The broader it went, the less rigor it carried. More recently, several teams of ecologists have published papers critiquing the loose application of "keystone" to this and that, and trying to bring the concept back into focus. One of those teams, led by a Berkeley professor named Mary E. Power, defined "keystone species" in a simple, lean way, as "one whose impact on its community or ecosystem is large, and disproportionately large relative to its abundance." To make that definition useful, Power's group proposed a mathematical framework for gauging the "community importance" of a single species within its ecosystem. That is, they invented an equation.

Here we go, friends, on a very short plunge into the cold lake of ecological math. Your goose pimples, I promise, won't be any worse than mine. Let "community importance" be represented as CI, as Mary Power and her colleagues stipulated. Let the little letter i stand for the species whose keystone significance is in question. The symbol p will indicate the proportional abundance of that species within its ecosystem, and t is for trait, meaning any trait of the ecological community (such as its species diversity, or the relative abundances of its species) that can somehow be quantified as a measure of change to the community as a whole. Now, if species i can be totally removed from the system—exterminated by accident or malice, or plucked out experimentally, as Paine plucked his starfish—and the community trait can be measured both before and after its removal, then N (as in normal) represents the *before* status and D (deleted) represents *after*. Okay, are you ready to dive? Dry towels and summer sunshine will be available, immediately following, for anyone whose heart hasn't stopped.

Having defined their variables, Power and her co-authors (one of whom was Robert T. Paine himself) offered this simple formula:

$$CI_i = [(t_N - t_D)/t_N](1/p_i).$$

If the species in question is really a keystone, they added, the value of CI will be much greater than 1. To say that in plain English: The community importance of a keystone species will register far higher than that of an average species.

To say it more plainly still: With or without algebra, we'll never know just what we have lost, in ecological terms, until we see the damage six ways to Sunday.

The last word on this topic goes to John Terborgh, again in a chorus of multiple authorship. For a quietly convened 1997 workshop on scientific issues in conservation planning at the continental scale, he and six colleagues wrote a paper titled "The Role of Top Carnivores in Regulating Terrestrial Ecosystems." One question they addressed was whether top-down processes (such as predation) or bottom-up processes (such as food-supply limitation, resulting in competition) were most responsible for determining the structure and operation of ecological communities. At the end of their sweeping analysis, Terborgh and company concluded: "The evidence reviewed here overwhelmingly supports the strong top-down role of top carnivores in regulating prey populations—and thereby stabilizing the trophic structure of terrestrial ecosystems." Lose the big predators, and there may (will?) come an overabundance of middle-sized predators, of herbivores, of seed predators—a pestilence of minor nibblers, cropping the vegetation down to stubs, interfering with tree reproduction, jeopardizing the long-term renewal of the forest canopy, exterminating populations of ground-nesting birds and probably of other small creatures as well. The evidence suggests "a crucial and irreplaceable regulatory role" for those large, dangerous beasts at the acme of their respective food chains. "The absence of top predators," said Terborgh's group, "appears to lead inexorably to ecosystem simplification accompanied by a rush of extinctions."

Which is bad enough. And yet the toll of losses in sheer ecological terms is not the only toll to be considered. The field data from Venezuela, Panama, Mukkaw Bay, and elsewhere say nothing about that other dimension of big predators, which may also be sorely missed: their "crucial and irreplaceable regulatory role" within the human mind.

73

Planet LV-426 is a storm-blasted lump of rock, desolate and forbidding, that lies somewhere in the lonely outback of the universe, roughly 112 billion miles from Earth. It's devoid of indigenous life. But it's not altogether empty. Through an accident of interplanetary infection, it harbors the dormant eggs—they're leathery, translucent, big as conga drums—of a single extraordinary species. No one knows how long they've rested there, these eggs, awaiting a chance to be reawakened into the full pageant of their life cycle. No one knows what sort of world gave rise to the creature that laid them. In fact, no one knows that such a species exists, until humans one day arrive aboard a commercial towing starship, the *Nostromo*, which has been ordered to divert from its deep-space industrial mission and investigate a mysterious signal from LV-426. The signal suggests that, against all expectation, the planet might harbor intelligent life.

It does, in a sense, since the alien life-form lurking within the marooned eggs has its own sort of intelligence—a rapacious cunning, formidable and quick, though nothing a person would seek out for intergalactic cultural exchange. The misleading message only lures the *Nostromo*'s crew into ghastly trouble. The unknown creature turns out to be a consummate predator, and *Homo sapiens*, though strange to it, turns out to be suitable prey.

This, in case you don't recognize it, is the opening premise of the 1979 movie *Alien*. A nifty Hollywood film shot mostly in England, by an English director using an American script, *Alien* evoked a clamor of merry revulsion at the time of release. It succeeded well at the box office ($60 million gross plus $40 million for video, as return on an $11 million budget, according to one source), won a single Academy Award (for special effects), and became the founding episode of a four-movie (so far) series that collectively represents the most interesting cinematic monster saga ever made.

Later installments of the series are *Aliens* (released in 1986), *Alien³* (1992), and *Alien Resurrection* (1997). Each of them, like the original *Alien*, stars Sigourney Weaver as an indomitable character named Ellen Ripley, sometime lieutenant in the astronautical service of the vast monolithic Company that, at that point in time, runs the known universe. It's Weaver's Ripley who, in the original movie, fights her way through a juggernaut of horrors to become the last survivor of the *Nostromo*.

The other central figure throughout all four movies is the title character—which is not actually a single character, an individual being (like King Kong or Godzilla), but a lineage of creatures, prolific and terrible, to which no name other than "the alien" is ever given. The very namelessness of this life-form adds to its mystique. It's variously referred to as a "hostile organism," a "monster," a "xenomorph," a "thing," and at one moment, in blithering testimony from a man who has just seen two of his buddies slaughtered, "the dragon." Although the repeated presence of Sigourney Weaver has provided a marquee name and a bit of cleverly mixed sex appeal (Ripley as babe, Ripley looking butch, Ripley being macho), the alien—or, as it deserves to be known, the Alien—is the real star, giving the movies their main villain, their shock value, and their wonderfully awful energy. One associate producer of the first movie, Ivor Powell, commented sensibly during a twenty-year retrospective event in 1999, "I think the Alien itself is the franchise, really. It's a powerful, powerful visual." Part of what makes it so powerful, especially in *Alien*, less so as the series proceeds, is that the creature is seen only in fleeting and incomplete glimpses. Its visual impact is directly proportional to its elusiveness. By the time Ripley escapes in the *Nostromo*'s little shuttle, even an attentive viewer (unless that viewer is watching the movie on video, with a remote in hand for freezing frames and rerunning certain lurid instants) doesn't quite know what the Alien looks like.

Another part of its power, not just visually but conceptually, derives from a certain persuasive biological complexity built into it by the creators of the script, Dan O'Bannon and Ron Shusett. O'Bannon particularly, who conceived the basic idea and then invited Shusett to collaborate, deserves credit for having imagined not just a nasty extraterrestrial beast but the multistage metamorphic life cycle of a nasty extraterrestrial beast.

O'Bannon reportedly took his dark inspiration from the world of insects—specifically, from what are loosely called spider wasps, vespoid hymenopterans that prey on spiders to supply food for their young. Most of these species belong to the Pompilidae family (though O'Bannon probably didn't know that or care), a group of slender, spiny-legged, solitary wasps that are notable for the potency of their sting. The typical reproductive routine for a female spider wasp goes like this: After mating, she hunts down a succulent spider, stings it (causing paralysis), drags it back to a nest cavity she has prepared, lays an egg on it, and leaves. Spider wasps of the genus *Pepsis* are known as "tarantula hawks," because they prey on tarantulas the size of mice.

The paralyzed spider is helpless to escape, but it doesn't die, rot, or dry out in the nest cavity. Eventually, from the egg hatches a wasp larva, which feeds on the fresh meat left behind so considerately by mom. Entomologists refer to this strategy as provisioning of eggs. Besides the spider wasps, various other species of parasitic, solitary wasps provision their eggs with other sorts of victim—paralyzed beetles, paralyzed cockroaches, paralyzed cicadas. Dan O'Bannon, by analogy, dreamed up an Alien that provisions its eggs with paralyzed humans.

The gruesome relationship between the two species begins when an overly curious character named Kane (played by John Hurt) leans too close to one of those ready eggs. The egg springs open like a jack-in-the-box, releasing a demonic little hatchling that crashes through Kane's helmet and attaches itself to his face. Carried back to the spaceship, Kane lies comatose with this hideous parasite stuck onto him like a cream pie. It resembles a crab, with eight long, strong, clutchy legs that hold it in place, a bulbous body, and a short, ugly tube—a sort of ovipositor—inserted down Kane's throat. The filmmakers, during their discussions about design and realization, called this thing the Facehugger, though the characters themselves never speak that term. If the egg is stage one in the Alien life cycle, the Facehugger is stage two. It can't be removed surgically without killing the patient. But after a short time, when no one's watching, it drops away of its own accord, and it's later found lying dead, like a scuzzy crustacean tossed up on the beach. Kane awakes, seeming to be fine.

He's not fine. Unbeknownst to anybody, Kane has been impregnated with an embryonic form of stage three, which grows within his chest. For all its lengthy dormancy on LV-426, the Alien is a fast-maturing creature once it gets started, and the gestation period for stage three is brief. Kane, appetite recovered, sits down for a meal with the rest of the crew. They're all feeling relieved, eager to distance themselves from this sinister planet and skedaddle home. After a few bites, Kane is afflicted with what looks at first to be indigestion.

He coughs, sputters, and heaves, suffering some sort of paroxysmal seizure, until a large welt rises alarmingly from his sternum; the welt explodes bloodily; and a small, ferocious, very ugly varmint bursts out of his chest, killing Kane instantly and astonishing his crewmates, whose jaws fall like a dumped truckload of slag. (Even the actors, not warned what to expect from the special-effects rigging on John Hurt, were reportedly shocked.) The new creature appears to be little more than a knob of tumescent tissue, naked and blood-smeared and tipped with chattering teeth but no eyes. After a brief pause to display its seething vitality, it streaks away across the table, scattering dishes as it disappears.

Among the filmmakers, this instar of the Alien—stage three—was known as the Chestburster. Ivor Powell has remarked that if the Chestburster scene hadn't been in the script, to so horrify and fascinate those Hollywood executives who read it, the movie might never have gotten made. Ron Shusett recollects that, just after the scene was shot, the cinematographer himself went aside to vomit. Shusett thought, "My God, what have we wrought here?"

But stage three isn't the end. It's only the prelude to the Alien's final phase. Somewhere in the deep chambers and passages of the *Nostromo*, the creature metamorphoses again, by way of another biologically recognizable process—it molts. The skin splits away, a new shape emerging from the old.

We don't witness that transformation, nor does the crew. As they search the ship to exterminate this terrifying pest, an engine-room technician named Brett (played by Harry Dean Stanton) finds a limp, molted shuck. Seconds later, Brett becomes the next victim, and we get our first glimpse of stage four, the adult Alien.

Its teeth, thin as icicles, drip with clear slime. From within its jaws comes a second pair of jaws, on an extrusive appendage, which shoots out to strike like the tongue of a toad. Its circulatory system (we learn eventually) flows with acid, not blood. From who-knows-which orifice it spins a sticky, spidery silk, used for "cocooning" the semi-comatose humans with which it provisions its eggs. Its head is smooth and sleek and dark, like an eggplant, but elongated front to back, like an ice ax. It carries six clawlike fingers on each of two arms, and it stands erect on two legs, like a human, or an ostrich, or maybe *Tyrannosaurus rex*. A long tentacle protrudes forward from its belly (in the first movie, anyway; in *Alien³* this belly prong has been revised into a tail) like the arm of a giant squid. Analogies, analogies. The fact is that though its head may be "like" this, its limbs may be "like" that, the totality succeeds well at being grotesquely unique. The Alien itself isn't like *anything*.

It was designed for the film by a febrile Swiss artist named H. R. Giger, commissioned for that task based on the visionary creepiness of his paintings, which Dan O'Bannon had seen at a gallery. O'Bannon hired Giger to do some provisional drawings, and when the movie's eventual director, Ridley Scott, was shown samples of what Giger did, the artist became a principal member of the team. He was the mad genius of the production effort, dressed all in black, temperamental, working long hours in a specially built cubicle at one side of the studio. "The man has an essentially unwholesome viewpoint," O'Bannon has commented appreciatively. "His paintings are perverse, obnoxious and disgusting, but absolutely gorgeous." Yet the paintings—the reproductions of them I've seen, anyway—are spooky but static, lacking the wicked dynamism of his made-to-order movie monster. Giger's Alien is a garish visual fulfillment of O'Bannon's narrative concept: a giant, waspish predator whose zeal for procreation and whose appetite for prey, though innocently amoral in Darwinian terms, seem not just frightening but evil.

And then there's the mystery of its metabolism. Many adult wasps subsist on nectar or sap (high in carbohydrates, useful for flight) even while furnishing meat to their larval offspring. What does an adult Alien eat? We're never told. In the third movie, one creature appears to snatch a few hasty bites from a human carcass. But generally an Alien grows from the Chestburster stage (about the size of a marmot) to adulthood (eight feet tall, lanky and strong) in a sudden surge, based on unaccounted-for nutritional input or none at all. Human victims, rather than being devoured promptly, are set aside as provisions for off-spring—not quite paralyzed like a wasp-stung tarantula, but sedated and restrained within that gummy silk, each victim destined to host a Chestburster. This circumstance serves plot duty at several junctures, such as the scene in *Aliens* when Ripley manages to rescue a young girl, Newt, even after she has been grabbed and cocooned. But the spider-wasp model that inspired O'Bannon, and the absence of overt flesh-eating by Alien adults, shouldn't obscure one basic fact: In their raw impact, these are movies about predation, not about parasitism.

Ripley herself makes that point, late in *Alien³*, while sharing her hard-won knowledge of Alien behavior with the latest group of des-perate, besieged survivors, who don't know where their loathsome enemy might be hiding or skulking. "It won't go far," she assures them. "It'll nest in this area. Right around . . ."—she folds away a map they've been studying, then glances coldly at the immediate surround-ings—". . . *here*."

"How do you know that?" asks a very nervous man.

"It's like a lion," Ripley says. "It sticks close to the zebras."

74

So why are these slick, hectic space-bogey movies relevant to the sub-ject at hand? I'll give you two reasons. The first is that they capture a certain harsh sense of reality with their vivid portrayal of predation on human victims. No one who lives safe in a distant city, or in a biologically depauperate suburb, should be quick to dismiss the legitimate fear felt by indigenous people such as the Maldharis of Gir as they graze their buffalo and raise their children in a forest co-occupied by lions. Insofar as *Alien* and its sequels evoke responses roughly parallel to that fear, they offer a useful reminder, transferable from the giddy milieu of big-screen entertainment to the serious matter of landscape conservation on Earth. What reminder? That one shouldn't declaim glibly about the value of alpha predators without acknowledging also their costs; and that one can't weigh the costs apart from the crucial matter of who pays those costs. We're back again to the Muskrat Conundrum. If you can't imagine yourself driving buffalo to a waterhole in western India, or tending sheep on a Romanian mountainside, or spreading your fishnet across a brackish stream in Arnhem Land, at least try imagining yourself aboard the *Nostromo*, acutely aware that somewhere nearby, down this corridor or up that air shaft, lurks an Alien that covets your bodily meat.

The second reason for giving sober attention to these deftly made but less than profound movies is a counterpoint to the first. I believe that the success of the *Alien* series, like the durability of *Beowulf* and *Gilgamesh*, reflects not just our fear of homicidal monsters but also our need and desire for them.

Such creatures enliven our fondest nightmares. They thrill us horribly. They challenge us to transcendent fits of courage like Ripley's. They allow us to recollect our limitations. They keep us company. The universe is a very big place, but as far as we know it's mainly empty, boring, and cold. If we exterminate the last magnificently scary beasts on planet Earth, as we seem bent upon doing, then no matter where we go for the rest of our history as a species—for the rest of time—we may never encounter any others. The only thing more dreadful than arriving on LV-426 and finding a nest of Aliens, I suspect, would be to arrive there, and on the next unexplored planet, and on the next after that, and find nothing.

75

Eighteen months after my first visit to Gir, I return for another look at the lions. Again I'm with Ravi Chellam, who has come down from his institute in the hills north of Delhi. Again we base ourselves at a hotel on the edge of the sanctuary, hoping to spend as much time as possible in the forest. It's late November. The monsoon has recently ended, so instead of tawny dust piled deep along roadsides, fine as ground coriander and gently tracked with pug marks, we find dried mud. The hazy sepia light has been rinsed clear. The shriveled brown grass, the understory brush, and the bare, dust-flocked trees that we saw in May of the previous year are now leafy and verdant. The forest is a true jungle, not a drab, x-ray vision of one. In full flush, it has short sight lines and a capacity to surprise.

I notice other differences too. The antlers of the chital bucks are just coming out of velvet. The fawns are half grown. The peacocks are short-tailed, humbled with molting. At the market stalls of the village, I'm disappointed by the absence of mangoes and the meager seasonal offerings in their place: coconuts, small bananas, hard little apples. The mornings are chilly with approaching winter, obliging Ravi and me to wear jackets when we head into the forest before dawn.

From the tracker Mohammad Juma, Ravi's old compadre, we learn other cold news: Those three lion cubs, the ones we watched nursing at their mother's tits in the shade of a *Syzygium* tree, are all dead. They were killed by adult males, presumably males impatient to breed with the cubs' mother, that calm female whom I shortsightedly labeled the Happy Lioness. Ravi takes this information in stride. He's a biologist, and high cub mortality is an inevitability of demographic pressure within finite habitat, not (necessarily) a conservation tragedy. The law of the forest is birth, change, death. The law of the world is birth, change, death. Ravi is less inclined to be sentimental than to be intrigued—by the question of what this news might imply about changes of dominance and territory among the resident adults. Has a strong young male moved into that area, supplanting an older male who fathered the cubs?

He'll have no chance to learn the answer during this field visit, as it turns out, because bureaucracy and petty turf jealousy prevent us from tracking lions within the sanctuary. A new fellow has taken over as local administrator for the Forest Department, a pompous and insecure young man, who seems to have been inoculated by certain of his patrons with a truculent bias against Ravi, the well-educated interloper from Dehra Dun. To make matters worse, the son of a powerful political boss happens also to be visiting (along with his armed bodyguards) from Bombay, and the bending of rules for that princeling has made the administrator self-conscious. He dodges us. We need his signature on a piece of paper. Nothing doing, not even (especially not?) for Dr. Ravi Chellam. Although Ravi is welcome to ride through the sanctuary in a guided vehicle, like any tourist, he may not make field observations afoot—permission denied. Denying me the same privilege is merely a corollary. At one point I even receive an anonymous phone call at the hotel, warning me that "Ravi Chellam is not a welcome creature at Gir" and that "so long as you continue with him, you will have a problem." It's frustrating, given that I've sent written requests for access long in advance, then traveled ten thousand miles, arriving in company with the world's most respected authority on *Panthera leo persica*. But even more than frustrating, it's sad. If this politicized double standard symptomizes management practice for the Gir Wildlife Sanctuary and National Park, what hope can there be for the lions?

Anyway, Ravi and I get the message and turn our attention elsewhere. We call on Maldhari elders. We visit another of Ravi's old trackers, himself a Maldhari by extraction, who has recently lost his job with the Forest Department and moved to an outlying village, where he now grazes some buffalo, sells milk, and treasures a scrapbook reflecting his years as an assistant to lion research. His memories, and his yellowing letters of recommendation, go back as far as Paul Joslin and Steve Berwick. We talk with a man who was mauled by a lion, several years ago, when he tried to help shoo the cat from a mango orchard. The lion was an injured animal, sheltering there among the mangoes while nursing a bad leg, and it reacted crossly to being driven out. The man, an unskilled Forest Department employee named Ibrahim, shows me his mangled hand and the scar on his forearm.

Accepting the courtesies of our various hosts, we drink bellyfuls of sweet tea, smoky and rich with buffalo milk, served sizzling hot in flat steel saucers. We hike through a stretch of woodland outside the sanctuary, where bureaucracy and politics don't govern so strictly, and hear the roar of an unseen lion. We talk with whomever we can about lion behavior, lion attacks, the condition of the forest, the encroachments that threaten it, Forest Department policies, the treatment of Maldharis, the conflict between predators and people, the conflict between traditions and aspirations, and the serious changes in culture and landscape that are currently impinging on Gir. Has there ever been a time, I wonder, when serious changes *weren't* impinging here? I spend hours amid a tiny collection of old books at the hotel, reading back through local history as personified by the successive Nawabs of Junagadh. Best of all, Ravi and I return to the thorn-fenced ness of his pal Ismail Bapu, beside a stream called the Lakadvera, for a few days of Maldhari hospitality and conversation, a few nights of sleeping among buffalo.

Bapu, round-faced and dignified, with the sparkly eyes of a gentle ironist, has seen plenty of changes himself and has felt them deeply. In the evenings, as we sit outside the perimeter of his fence, drinking tea and savoring the twilight, he takes up that theme more than once. He cherishes Ravi like an adopted son—the son who had brilliance and advantages, who got educated, but who still understands Maldhari ways—and Ravi's presence seems to trigger Bapu's reminiscent mood. He's also keen to the opportunity of sharing his thoughts with an interested foreigner.

Fifty-six years, all his life, Bapu has lived hereabouts. His family came down from the Sind, in what is now Pakistan, maybe six or seven generations ago. His grandfather once had a ness in this very spot. Bapu himself was the youngest of four boys, with a younger sister. He's the only one still living a Maldhari life; this brother became a farmer, that one runs a shop, the other owns a couple of trucks and does hauling. When he first established a ness of his own, it housed just himself, his mother, and the sister. He started with twenty buffalo, built his herd up to eighty, sold some each year, got married, fathered children, sold more buffalo to pay for weddings as the children came of age, and always, after each reversal or expenditure, built back his herd. Who says you can't raise livestock in the presence of lions? he asks. Evidently you can. His mother taught him this: Live your life straightforwardly, and show hospitality. If people visit, give them tea and food. Do that, she said, and all else will take care of itself. Bapu seems satisfied with the degree to which she was right.

But he's not a complacent man. He's a happy, grateful man who nevertheless knows the dull aches of loss and regret. When I was twenty, he says, I could carry the biggest log out of the forest. I'd cut trees, take timber, never thinking about limits, never contemplating how many years might be required for a forest to grow back. I may have cut a thousand trees, he says. Now I have three sons; if I can teach them more wisely than I was taught, I'll save three thousand trees from destruction. Also, in those days I hated the lions, he says. Without them, things would have seemed so much easier. You could simply graze your animals, day or night, with no fear of attacks, no problems. None of that constant concern about what the lions might do. The lions . . . always the lions. I've spent so many hours of my life thinking about lions, says Bapu, that if I had spent the time meditating, God himself would have appeared to me. Slowly I came to realize, he says, that this landscape belongs to the lions if it belongs to anyone. "And if they can't stay here, where will they go? We're the intruders."

All of this comes to me translated by Ravi from Bapu's gumbo of Gujarati, Sindhi, and Hindi. I've never heard him speak a word of English, but his face is so lively, so jovial and avuncular, that I feel almost as though I can catch his meaning by some sort of visual telepathy.

The landscape is different now, Bapu says. The water table is lower, for one thing. The Lakadvera used to flow year-round. Nowadays it goes dry in summer. Much of the old forest has been converted to crop farming. Wheat, lentils, garlic, peanuts. Sugar cane. Orchards of mango. The village nearby, where he got his five years of schooling, has more than tripled in size. If trends continue as they are, Bapu predicts, within another two or three generations there will be only photographic evidence that such a creature as the lion of Gir ever lived. Although it's safe to assume that Bapu has never heard of trophic cascades, he embraces the concept without needing a fancy vocabulary. "If the lion goes, the forest goes and everything else goes."

For three days, Ravi and I are nourished by Bapu's company, and by his hot curried eggplant, his coarse lentil roti, his buffalo yogurt. Then, reluctantly, we go back to the hotel. Our little political problem with the Forest Department administrator proves insoluble, so we pack to leave. Ravi will return to his duties at the institute. I'll do some archival research in Bombay, then fly to eastern India for a look at those crocodiles in the Bhitarkanika River. Anything further I learn about *Panthera leo persica* will have to come from journal papers, dissertations, and books. On our final morning in the area, we swing back to Bapu's ness for farewells.

"Last tea?" Bapu says—in English!—surprising both Ravi and me, and smirking proudly at his venture into still another language.

Steel saucers are passed around. I hold mine in the tripod grip from below, fingertips on rim, careful to keep it level. Bapu himself pours. Steam rises. I say, *"Bhus,"* colloquial Hindi for "enough," the polite demurral. Good manners call for limiting one's drain on a host's resources. But today, shameless, I say it only when my shallow little bowl is already aflood. Truth is, I want nothing less than a full portion.

All morning I've been thinking about the juggernaut of change that moves through landscapes and cultures with such speed, causing lamentable but seemingly inevitable losses. Now I just want to embrace the here and present. I try to concentrate on this serving of Maldhari tea, aware that I won't soon get another. The gracious way to consume it, I know, is with loud, appreciative slurps, and quickly, right off the stove. While the conversation flies back and forth, all incomprehensible to me, I take a sip, nearly scalding my tongue. Ouch. The liquid is rich with that sweet, musky flavor of sugared milk from a buffalo that has grazed in a forest among lions. It's luscious. But it's too hot to drink. And then, a moment later, it's too cool.

SOURCE NOTES

7 *"Oh Lord my God, in you I take . . ."* Psalms 7:1–2 (*New Oxford Annotated Bible*).

10 eyes *"like the eyelids . . ."* Job 41:18 (King James Version).

11 *"the Lord with his cruel and great . . ."* Isaiah 27:1 (*New Oxford Annotated Bible*).

11 *"the crooked serpent"* Isaiah 27:1 (King James Version).

11 *"crushed the heads of Leviathan"* Psalm 74:14 (*New Oxford Annotated Bible*).

11 *"Who can open the doors of his face? . . ."* Job 41:14, 15, 19–21. (King James Version).

11 *"Canst thou draw out leviathan . . ."* Ibid., verses 1–2.

12 *"Will he make many supplications . . ."* Ibid., verses 3–5.

12 *"None is so fierce that dare . . ."* Ibid., verse 10.

12 *"The arrow cannot make him flee . . ."* Ibid., verses 28–34.

25 *"came down from their haunts at night"* Herodotus (1972), p. 482.

26 *"I am Ashurbanipal . . ."* Hobusch (1980), p. 34.

26 *"with a stout heart I caught 15 strong lions . . ."* Ibid., p. 35.

26 *"On the order of Ninurta, my patron . . ."* Ibid., p. 35.

26 *"100 maned lions"* Pliny the Elder (1991), p. 115.

27 *"Five hundred lions were killed . . ."* Plutarch (1999), p. 270.

27 *"ruler of the world"* Clutton-Brock (1996), p. 377.

27 *"the 'amusements' with wild animals continued . . ."* Ibid., p. 377.

27 everyone *"the right to kill lions. . ."* Ibid., p. 378.

28 *"The lion is frequently met . . ."* Quoted in Kinnear (1920), p. 35.

28 *"for fear of marauders and thieves . . ."* Quoted in ibid., p. 34.

28 *"five years ago a lion appeared . . ."* Quoted in ibid., p. 36.

28 *"a few years ago the carcass of one . . ."* Quoted in ibid., p. 36.

28 *"lions still exist along the banks . . ."* Quoted in ibid., p. 37.

28 *"I once saw a dead one floating . . ."* Quoted in ibid., p. 37.

32 *"milkmen"* Tambs-Lyche (1997), p. 148.

32 *"regarded with the mixture of suspicion . . ."* Ibid., p. 148.

32 *"pastoral caste"* Ibid., p. 148.

32 *"Migrating with flocks of a hundred animals . . ."* Ibid., pp. 148–49.

40 *"The lion is a far more noisy animal . . ."* Fenton (1909), p. 14.

40 *"It is unnecessary to look far . . ."* Wynter-Blyth (1949), p. 494.

40 *"The lion, with his comparatively bold nature . . ."* Rashid and David (1992), p. 37.

40 *"The lion's bold nature and more social way . . ."* Ibid., p. 38.

41 *"The Kathis were a brave and warlike race . . ."* Wilberforce-Bell (1916; reprinted in 1980), p. 68.

42 *"everything was chaos . . ."* Ibid., p. 179.

43 the Kathis *"began to tire . . ."* Ibid., p. 198.

44 *"wandering in a field at night . . ."* Ibid., p. 202.

44 *"not more than a dozen"* Wynter-Blyth (1950), p. 467.

46 *"Lions are much bolder . . ."* Ibid., pp. 467–68.

47 *"very bold"* Ibid., p. 468.

47 *"Lord Curzon, when he abandoned . . ."* Edwards and Fraser (1907), p. 172.

47 *"Apparently in consequence of the outcry . . ."* Ibid., p. 172.

47 *"there cannot be the slightest doubt . . ."* Fenton (1909), p. 4.

48 *"However, a most interesting fact to observe . . ."* Wynter-Blyth (1950), p. 469.

51 *"Following the drought, lions began . . ."* Saberwal et al. (1994), p. 503.

57 *"Their conduct is frequently characterized . . ."* Quoted in Daniel (1996), p. 67.

57 *"Like the tiger, the panther sometimes . . ."* Quoted in Ibid., pp. 181–82.

57 Prayag . . . meaning *"confluence"* Corbett (1991), *The Man-Eating Leopard of Rudraprayag*, p. 5.

57 *"cascades unconfined and merrily over rocks . . ."* Ibid., p. 3.

58 *"scorpion sting"* Ibid., p. 174.

58 *"Their habits bring them into closer . . ."* Quoted in Daniel (1996), p. 182.

58 *"I believe it is a female . . ."* Quoted in Ibid., pp. 185–86.

59 *"The four little girls . . ."* Quoted in Ibid., p. 186.

60 *"Leopards have lived in close association . . ."* Daniel (1996), p. 225.

60 *"Among the larger mammals, the leopard comes second . . ."* Ibid., p. 225.

60 *"fatalities caused by leopards . . ."* Ibid., p. 225.

61 *"The leopard once it takes to man-eating . . ."* Ibid., p. 225.

61 *"the perfect predator for this day and age"* Ibid., p. 225.

66 *"the science that reasons why"* Colinvaux (1978) chapter title, p. 5.

66 *"Animals come in different sizes . . ."* Ibid., p. 18.

67 *"It is an extraordinary thing but true . . ."* Ibid., p. 19.

67 *"Animals in the larger sizes . . ."* Ibid., p. 19.

67 *"And indeed it is hard to avoid the belief . . ."* Herodotus (1972), p. 248.

68 *"Darwin, himself a magnificent field naturalist . . ."* Elton (1966), p. 3.

68 *"because there are fewer species of plants . . ."* Ibid., p. 3.

69 *"Food is the burning question . . ."* Ibid., p. 56.

69 *"food-chain" . . . "food-cycle"* Ibid., p. 56.

70 *"There are very definite limits . . ."* Ibid., p. 59.

71 *"Man is the only animal which can deal . . ."* Ibid., p. 61.

71 *"the Pyramid of Numbers"* Ibid., p. 68.

72 *"Finally, a point is reached"* Ibid., p. 69.

73 *"the major contribution of one of the most creative . . ."* Addendum, by Hutchinson, to the posthumously published Lindeman (1942), p. 418.

74 *"progressive efficiency"* Lindeman (1942), p. 407.

89 *"The face of Nature may be compared . . ."* Darwin (1964), p. 67.

89 *"heavy destruction inevitably falls . . ."* Ibid., p. 66.

90 *"some 75 percent contained hair . . ."* Joslin (1984), p. 653.

92 *Harijans, "Children of God"* Mendelsohn and Vicziany (1998), p. xiii.

92 *"Large gatherings collect at animal carcasses . . ."* Ali (1996), p. 110.

96 *"including, of course, the last of the Asiatic lions"* S. Berwick (1976), p. 38.

96 *"If the Forest Department could remove the Maldharis . . ."* Ibid., p. 38.

97 *"thirteen castes"* Tambs-Lyche (1997), p. 112.

98 *"to graze" . . . "to spread"* Enthoven (1997), vol. 1, p. 271.

98 *"the dirt of her body"* Ibid., p. 273.

98 *"The lion attacked the cow . . ."* Ibid., p. 273.

99 *"a tall good-looking fair-skinned tribe . . ."* Ibid., p. 275.

99   *"strong, tall and well-made . . ."* Ibid., vol. 3, p. 254.

99   *"The men are dull and stupid . . ."* Ibid., p. 257.

100   *"it was often the women who decided . . ."* Westphal-Hellbusch quoted in Tambs-Lyche (1997), p. 158.

100   *"he who lives outside"* Enthoven (1997), vol. 3, p. 253.

100   *"goers out of the path"* Ibid., p. 253.

101   *"Data indicate that the* Maldharis . . ." M. Berwick (1990), p. 92.

102   *"there is no reason to suppose that lions . . ."* Joslin typescript of his IUCN presentation (from the files of Steve Berwick), p. 11.

103   *"forests free from fear"* Government of Gujarat (1975), p. 19.

103   *"In these protected forests . . ."* Ibid., p. 19, note††.

103   *"dangerous"* Kautilya (1992), p. 321.

104   *"apprehensive"* Singh and Kamboj (1996), p. 84.

104   *"some relocated families could not . . ."* Ibid., p. 84.

104   *"The Maldhari translocation programme has failed . . ."* Chellam and Johnsingh (1993), p. 417.

104   *"Our heart is disconcerted . . ."* Quoted in Raval (1997), p. 87.

104   *"Gir means [our] heart."* Quoted in ibid., p. 87.

105   *"If [they] throw us back into the Gir . . ."* Quoted in ibid., p. 87.

105   *"Millenniums will change . . ."* Quoted in ibid., p. 89.

105   *"even with protection the Gir will degrade . . ."* Quoted in ibid., p. 87.

105   *"If there is no jungle . . ."* Quoted in ibid., p. 89.

105   *"When the sky gets dark . . ."* Quoted in ibid., p. 87.

105   *"perceptions of resource management . . ."* Ibid., title page.

119   *"Many respondents (62%) reported . . ."* Saberwal et al. (1994), p. 504.

121   *"a live crayfish . . ."* Errington (1967), p. 241.

121   *"The game that I ate . . ."* Ibid., p. viii.

122   *"Undoubtedly, muskrat flesh . . ."* Ibid., p. 24.

123   *"wastage parts"* Errington (1963), p. 184.

123   *"overproduced young"* Ibid., p. 184.

124   *"Proper consideration of this factor . . ."* Ibid., p. 184.

128   *"fellow crocophile"* Graham and Beard (1973), p. 31.

128   *"So long as one is constantly threatened . . ."* Ibid., p. 201.

129   *"Died: William H. Olson . . ."* Time, April 22, 1966, p. 77.

130   *"We found his legs . . ."* Graham and Beard (1973), p. 200.

131   *"did not care if crocs laid hard-boiled eggs . . ."* Quoted in ibid., pp. 31–32.

131 *"peaceful in a way that only contented people . . ."* Ibid., p 96.

131 *Such a God, as Graham saw it, "was indistinguishable . . ."* Ibid., p. 66.

131 *"Theirs is not the contempt . . ."* Ibid., p. 68.

132 *"The fear of being eaten . . ."* Ibid., p. 69.

132 *"One of civilization's imperative taboos . . ."* Ibid., p. 69.

132 *"Such fears generate volcanoes . . ."* Ibid., p. 69.

132 *"It is around the matter of cannibalism . . ."* Ibid., p. 69.

133 *"They appear to be quite callous . . ."* Ibid., pp. 69–70.

134 *"in the nursery logic of our unconscious . . ."* Ibid., p. 69.

135 *"Our knowledge of crocodiles . . ."* Ibid., p. 218.

136 *"And what would our increased knowledge . . ."* Ibid., p. 223.

136 *"there is in man a cultural instinct . . ."* Ibid., p. 201.

141 *"flourishes along the shores of the Bay of Bengal"* Quoted in Bustard and Choudhury (1981), p. 204.

143 *"Crocodiles do not readily arouse public sympathy . . ."* Bustard (1969), 249.

144 *"Clearly," Bustard wrote, "there is no incentive . . ."* Ibid., p. 253.

144 *"could be applied in many areas . . ."* Ibid., p. 255.

145 *"the problem was of greater magnitude . . ."* Bustard (1974), p. 1.

145 *"Indian villagers do not fear crocodiles . . ."* Ibid., p. 11.

145 *"It is well known that man is not within . . ."* Ibid., p. 11.

146 *"The area is prime saltwater crocodile habitat . . ."* Ibid., p. 34.

147 *"In order to retain the co-operation . . ."* Bustard and Choudhury (1981), pp. 209–210.

159 *"had a very narrow escape from them . . ."* Stokes (1846), vol. 1, p. 397.

159 *"the paw of a favourite spaniel . . ."* Ibid., p. 397.

159 *"an alligator rose close by . . ."* Stokes (1846), vol. 2, p. 36.

160 *"reached the opposite shore just in time . . ."* Ibid., p. 37.

160 *were closely attuned to the imperatives* Grahame Webb reminds me that, in the ancient period following initial human immigration to Australia, the use of fire by Aboriginal people may have caused major transformations of some areas of landscape, and their hunting practices could have been a factor in some Pleistocene faunal extinctions. Populations of flightless bird, for example, may have been far more vulnerable to early hunting methods than populations of crocodiles were.

186 *"I am going out with the party to Arnhem Land . . ."* Quoted in Egan (1996), p. 23.

187 *it was "widely believed in Darwin . . ."* Egan (1996), p. 192.

187 *"for the use and benefit . . ."* Blackburn (1971), p. 8.

187 *if "any part of a native reserve . . ."* Quoted in Berndt and Berndt (1954), p. 202.

188 *"That the procedures of the excision . . ."* The full text of the Bark Petition, in both languages, is reproduced in Dean (1963), p. 6.

189 *"native title"* Blackburn (1971), p. 58.

189 *"the aboriginals have a more cogent feeling . . ."* Ibid., p. 130.

190 *"It is dangerous to attempt to express . . ."* Ibid., pp. 130–31.

190 *"We believe that we came as a crocodile . . ."* Quoted in Watson et al. (1989), p. 26.

198 *"Yesterday we went to Garrangali . . ."* Laynhapuy Schools A.S.S.P.A. Committee (1992), p. 16.

199 *"We saw three Bäru . . ."* Ibid., p. 18.

199 *"We tasted salty water . . ."* Ibid., p. 17.

212 *firearms and the right to bear them* Almost three years after my visit with him, Andrew Cappo informs me with his characteristic penetrating candor that he has "gone soft on this one," because of "the redneck killer litterbugs" who give gun ownership in Australia a bad smell.

219 *try telling that to Hell's Angels* Yet, according to his later update, Andrew did voice these misgivings to the Angels, diplomatically, after which "they agreed to the skull option and it turned out great!"

224 *"native intelligence, phenomenal memory . . ."* Quoted in Cullen (1990), p. 96.

225 *"among the world's leading proponents . . ."* Quoted in Judt (2001), p. 44.

225 *"one of Europe's good Communists"* Quoted in ibid., p. 44.

227 *"Stop it, Nicu . . ."* Quoted in Behr (1991), p. 26.

227 *"I can't believe this . . ."* Quoted in ibid., p. 26.

238 *"Sometimes more stags were 'found' than . . ."* Crişan (1994), typescript translation by Eduard Érsek, pp. 19–20.

238 *"to the limit of its resistance"* Ibid., p. 24.

238 *"It was an honor for them . . ."* Ibid., p. 25.

240 *"The bears were running . . ."* Ibid., p. 46.

240 *"We, the foresters, gathered . . ."* Ibid., p. 46.

250 *in the range of 3.1 to 7.7 million hectares* A recent estimate putting the habitat area at 7.67 million hectares is probably accurate, according to Annette Mertens.

255 Sha naqba īmuru, *or "He Who Saw the Deep"* George edition (1999), p. xxv.

257 *"I knew him, my friend, in the uplands . . ."* Ibid., p. 18.

257 *"My lord, you have not set eyes . . ."* Ibid., p. 154.

257 *"With your spineless words . . ."* Ibid., p. 19.

258 *"fearsome ogre" cloaked in "seven . . ."* Ibid., p. xxxii.

258 *"monster" with a face like coiled intestines* Dalley edition (1998), pp. 42, 323.

258 *"Humbaba is 'Evil' . . ."* Sandars edition (1972), p. 33.

258 *"a strange face and long hair . . ."* Lambert (1987), p. 43.

258 *"So I shall bite through your windpipe . . ."* Dalley edition (1998), p. 72.

259 *"the tusks"* George edition (1999), p. 44.

259 *"to trample the forest"* Ibid., p. 46.

259 *"Gilgamesh was cutting down the trees . . ."* Dalley edition (1998), p. 76.

259 *"Lord Wind"* George edition (1999), p. 223.

259 *"door of the woodland"* Ibid., p. 55.

260 *"the most evil serpent"* Byock edition (1990), p. 59.

261 *"a beast no weapons could wound"* Hamilton (1961), p. 164.

262 meaning of the word labbu, *by the way, was "lion"* Heidel edition (1963), p. 141.

262 *"the Babylonian Genesis"* Ibid., title page.

262 *"the glistening one"* King edition (1902), p. 9.

262 *"monster-serpents"* Ibid., p. 17.

262 *"With poison instead of blood . . ."* Ibid., p. 17.

263 *"wisest of the gods"* Heidel edition (1963), p. 5.

263 *"like a flat fish into two halves"* King edition (1902), p. 77.

264 *"Beowulf's life, though full of away wins . . ."* Alexander edition (1973), p 17.

264 *"a powerful demon, a prowler . . ."* Heaney edition (2000), p. 9.

264 *"the banished monsters, / Cain's clan . . ."* Ibid., p. 9.

264 *"when dawn broke and day crept in . . ."* Ibid., p. 33.

265 *"Nor did the creature keep . . ."* Ibid., pp. 49, 51.

265 *"Venturing closer, / his talon was raised . . ."* Ibid., p. 51.

266 *"grief-racked and ravenous . . ."* Ibid., p. 89.

266 brim-wylf (*"wolf of the deep" . . .*) Ibid., pp. 110, 111.

266 *"savage talons"* Ibid., p. 105.

266 *"and a bewildering horde / came at him . . ."* Ibid., p. 105.

267 *"Do not give way to pride . . ."* Ibid., p. 121.

268 *"The great man / had breathed his last . . ."* Ibid., p. 205.

268   *a "banished" monster* Ibid., p. 9.

268   *"God-cursed"* Ibid., p. 11.

268   *"lonely war"* Ibid., p. 13.

268   *"brooded on her wrongs"* Ibid., p. 89.

268   *"the trespasser who had troubled . . ."* Ibid., p. 157.

268   *"writhing in anger"* Ibid., p. 155.

269   *"Psychologically, the dragon is one's own binding . . ."* Campbell (1998), p. 149.

269   *"When Siegfried has killed the dragon . . ."* Ibid., p. 146.

270   *"A dragon is no idle fancy . . ."* Tolkien (1936), pp. 15–16.

270   *"the undiscriminating cruelty of fortune"* Ibid., p. 17.

271   *"man at war with the hostile world . . ."* Ibid., p. 18.

271   *"lif is læne . . ."* Ibid., p. 18. Jerry Coffey, of Montana State University, provided me with the translation of this line, which Tolkien omitted.

271   *"hell-serf"* Heaney edition (2000), p. 53.

271   "Godes yrre bær" Tolkien (1936), p. 25.

271   *"a monstrous hell-bride"* Heaney edition (2000), p. 89.

271   *"hell-dam"* Ibid., p. 91.

271   *"force for evil"* Ibid., p. 93.

271   *"the evil one"* Ibid., p. 171.

271   *"the old monsters became images . . ."* Tolkien (1936), p. 23.

272   *"Cain's clan"* Heaney edition (2000), p. 9.

272   *"the Lord of Life"* Ibid., p. 3.

272   *"but for the nearness of a pagan time . . ."* Tolkien (1936), p. 23.

272   *"a wild litany / of nightmare and lament . . ."* Heaney edition (2000), p. 211.

273   *"Culture teems with animals who have no . . ."* Shepard (1997), p. 175.

273   *"a fire-breathing old grump . . ."* Ibid., p. 176.

274   *"Our fear of monsters in the night . . ."* Ibid., p. 29.

276   *"silent hounds with cruel sharp beaks"* Mayor (2000), p. 29.

276   *"a quadruped like a lion . . ."* Aelian (1971), vol. 1, book 4, p. 241.

277   *"Bee-Wolf"—that is, bear* Alexander edition (1973), p. 8.

289   *"there was not enough parking space . . ."* Schullery (1992), p. 106.

304   *"a piece of lip and scalp . . ."* Herrero (1985), p. 72.

304   *"Quite a bit of her soft tissue . . ."* Ibid., p. 73.

305   *"injuries inflicted by a bear"* Quoted in McMillion (1998), p. 92.

305   *"Then, once he was dead . . ."* Quoted in ibid., p. 92.

305   *"just like she would have treated . . ."* McMillion (1998), p. 85.

306 *"A grizzly bear had killed and eaten . . ."* Ibid., p. 13–14.

308 *"bite-and-spit paradox"* Ellis and McCosker (1991), p. 110.

310 *"The vine stopped the youngster . . ."* Auffenberg (1981), p. 320.

310 *"bear a greater resemblance . . ."* Ibid., p. 209.

311 *"Few of those who have experienced . . ."* Plumwood (1996), p. 35.

311 *"beyond words"* Ibid., p. 35.

313 *"Again it came . . ."* Ibid., p. 34.

313 *"between the legs"* Ibid., p. 35.

313 *"a centrifuge of whirling . . ."* Ibid., p. 35.

314 *"I could see the crocodile taking . . ."* Ibid., p. 36.

315 *"the masculinist monster myth"* Ibid., p. 40.

315 *"an implacable monster"* Ibid., p. 40.

315 *"boundary breakdowns"* Ibid., p. 42.

315 *"If ordinary death is a horror . . ."* Ibid., p. 42.

318 *"Strange though it may sound . . ."* Ibid., p. 58.

323 *"shear-bite"* Akersten (1985), p. 18.

324 *"The pride regroups at a distance . . ."* Ibid., p. 18.

324 *"massive and, more importantly, rapid damage . . ."* Turner (1997), p. 124.

326 *"Another factor . . . was almost certainly the rise . . ."* Brain (1981), p. 158.

328 *"more than one species of cat . . ."* Ibid., p. 273.

342 *"Races of warm-blooded vertebrates . . ."* Quoted in Mayr (1956), which is reprinted in Mayr (1976), p. 211.

343 *"very poor scientific basis for the eight . . ."* Kitchener (1999), p. 22.

345 *"appears to vary clinally . . ."* Ibid., p. 35.

345 *"While it is true that in its winter pelage . . ."* Matthiessen (2000), p. 47.

354 *"ideological perestroika"* Stephan (1994), p. 219.

356 *"the Manchurian tiger"* Baikov (1925), typescript translation by Misha Jones, p. 1.

356 *"The whiskers, heart, blood, bones . . ."* Ibid., typescript translation, p. 17.

356 *"It is rather tender, tasty . . ."* Ibid., typescript translation, p. 17.

356 *"semi-barbarian inhabitants"* Ibid., typescript translation, p. 19.

357 *"under the spell of the threatening 'tsar of the taiga' . . ."* Ibid., typescript translation, p. 19.

357 *"If a tiger eats a human . . ."* Ibid., typescript translation, p. 19.

358 *"A million people (not including forced laborers) . . ."* Stephan (1994), p. 185.

359 *"should develop an independent . . ."* Ibid., p. 185.

359 *"hammered hard by Communist Party nimrods"* Matthiessen (2000), p. 10.

359 *"the tigress is madly courageous . . ."* Baikov (1925), quoted in Stroganov (1969), p. 495.

363 *"Preservation of the Amur tiger, the largest . . ."* Pikunov (1988a), p. 179.

363 *"old-fashioned attitude to and protection of . . ."* Bragin and Gaponov (1989), typescript translation by Bragin, p. 1.

363 *"Similar to all other fields of national economy . . ."* Ibid., typescript translation, p. 9.

364 *"a special nature management regime"* Ibid., typescript translation, p. 9.

364 *"Without this minimum area . . ."* Ibid., typescript translation, p. 9.

365 *"zones of a special economic regime"* Ibid., typescript translation, p. 9.

365 *"it would be most reasonable to give the tiger . . ."* Ibid., typescript translation, p. 10. Actually, the typescript reads "it would be most resonable [*sic*] to give the tiger . . ."

366 *"For the purpose of obtaining the maximum level . . ."* Ibid., typescript translation, p. 10.

366 *"If in the present conditions the leaseholder . . ."* Ibid., typescript translation, p. 10.

367 *"sent the Far East into an economic free fall . . ."* Stephan (1994), p. 290.

367 *"the region was an economic basket case"* Ibid., p. 288.

367 *"Yet most Far Easterners showed . . ."* Ibid., p. 290.

369 *"He had full access to the taiga . . ."* Galster and Eliot (1999), pp. 235–36.

370 *"On Saving the Amur Tiger and Other . . ."* Ibid., p. 237.

381 *"Dogs not trained for tigers are useless . . ."* Baikov (1925), typescript translation, p. 14.

381 *"when some cowardly dog near a hunter . . ."* Abramov (1965), typescript translation by Misha Jones, p. 6.

382 *"a young tiger regularly hunted dogs . . ."* Matyushkin et al. (1980), p. 35.

383 *"Cases of uninjured tigers attacking . . ."* Baikov (1925), typescript translation, p. 14.

383 *"The attack, by a tigress accompanied . . ."* Smirnov (1992) abstract, in Matyushkin (1998), p. 337.

383 *"The man was killed in the tiger's fourth bound . . ."* Nikolaev and Yudin (n.d.), typescript in English, p. 6.

384 *"would have been eaten entirely . . ."* Ibid., p. 6.

384 *"Tiger numbers have increased substantially . . ."* Ibid., p. 1.

384   *"Not only are such animals likely . . ."* Ibid., p. 3.

384   *"any deviations in a tiger's normal behavior . . ."* Ibid., p. 7.

385   *"With this knowledge it can be concluded . . ."* Ibid., p. 7.

385   *"The normal tiger exhibits a deep-rooted aversion . . ."* McDougal (1987), p. 435.

386   *"Sixty persons were killed . . ."* Ibid., p. 437.

386   *"man-eaters entered huts to carry off . . ."* Ibid., p. 438.

387   *"either due to the disturbance of tiger habitat . . ."* Ibid., p. 443.

388   *"Tigers do not kill human beings in numbers proportionate . . ."* Seiden-sticker and McDougal (1993), p. 105.

388   *"One of the puzzling aspects . . ."* Ibid., p. 121.

388   *"Walking in a normal upright posture . . ."* Ibid., p. 122.

388   *"rubber tappers who go out . . ."* Ibid., p. 122.

389   *"When a tiger openly attacks a person . . ."* Baikov (1925), typescript translation, p. 15.

389   *"as a rule, a tiger avoids attacking humans . . ."* Abramov (1965), typescript translation, p. 7.

389   *"But then how do you explain that the Udege . . ."* Ibid., typescript translation, p. 7.

401   *"In the mountains it used caves . . ."* Vereshchagin and Baryshnikov (1984), p. 497.

402   *"gigantic" even possibly "the largest felid . . ."* Kurtén (1968), p. 85.

403   *"To measure the resonance of the echo . . ."* Chauvet et al. (1996), p. 35.

404   *"We were overwhelmed . . ."* Ibid., pp. 40–41.

405   *"was very strange and seemed a bit . . ."* Ibid., p. 48.

405   *"Suddenly our lamps lit up . . ."* Ibid., p. 58.

407   *"Exact dates for the paintings . . ."* Clottes (1995), p. 34.

407   *"some details point to a pre-Magdalenian period . . ."* Ibid., p. 34.

407   *"powerful originality"* Clottes (1996), p. 121.

407   *"That changed our whole conception . . ."* Jean Clottes lecture in Bozeman, Montana; June 1, 2001.

409   *"healing the sick, foretelling the future . . ."* Clottes and Lewis-Williams (1998), p. 19.

409   *"Chauvet Cave reveals that big cats . . ."* Clottes (1996), p. 127.

409   *"All speculations about this problem . . ."* Ibid., p. 127.

413   *"World population reached 6.1 billion . . ."* United Nations Population Division (2001), p. v.

413 *"According to the medium-fertility scenario . . ."* United Nations Population Division (1998), p. ix.

414 *"In other words, there will be more people . . ."* Ehrenfeld (1986), pp. 176–77.

414 *"it becomes a fairly easy job . . ."* Ibid., p. 177.

415 *"those characteristics of animal species . . ."* Ehrenfeld (1970), p. 129.

415 *"most endangered animal"* Ibid., p. 130.

415 *"It turns out to be a large predator . . ."* Ibid., p. 130.

416 *"These are the jolly companions . . ."* Ehrenfeld (1986), p. 178.

416 *"important to recognize the categories of species . . ."* Terborgh (1974), p. 719.

418 *"Herbivores are seldom food-limited . . ."* Hairston, Smith, and Slobodkin (1960), p. 424.

419 *"the keystone of the community's structure"* Paine (1969), p. 92.

419 *"keystone species"* Ibid., p. 92.

421 *"These observations are warnings . . ."* Terborgh et al. (2001), p. 1925.

422 *"one whose impact on its community or ecosystem . . ."* Power et al. (1996), p. 609.

423 *offered this simple formula* The equation is in Power et al. (1996), p. 610.

423 *"The evidence reviewed here overwhelmingly supports . . ."* Terborgh et al. (1999), pp. 57–58.

424 *"a crucial and irreplaceable regulatory role"*. . . Ibid., p. 58.

424 *"The absence of top predators"* . . . *"appears to lead. . ."* Ibid., p. 58.

426 *"I think the Alien itself is the franchise . . ."* Ivor Powell's statement appears in a documentary film, *The Alien Legacy*, made in conjunction with a celebration of the twentieth anniversary of the release of the original *Alien*.

428 *"My God, what have we wrought . . ."* Ron Shusett, like Ivor Powell, speaks as an interviewee in *The Alien Legacy*.

429 *"The man has an essentially unwholesome viewpoint . . ."* Dan O'Bannon's comments are printed as a blurb on Giger (1996).

BIBLIOGRAPHY

A small disclaimer to those of you who read the fine print, love books and book history, and notice dates: I cite here whatever edition of each book happened to be available to me for research, which in many cases is not the first or most famous edition. This leads to some apparent anachronisms, such as Darwin (1964) and Elton (1966). I trust you to know that Charles Darwin didn't publish *The Origin of Species* in 1964, and to appreciate that other works such as Charles Elton's pioneering 1927 volume are now more readily found in later editions.

Abdi, Rupa Desai. 1993. *Maldharis of Saurashtra: A Glimpse into Their Past and Present.* Bhavnagar, India: Suchitra Offset.

Abramov, K. G. 1965. "Tigr Amursky—relikt fauny Dal'nego Vostoka" ("The Amur Tiger: Relict Fauna of the Far East"). *Zapiski Primorskogo filiala Geograficheskogo obshchestva SSSR (Notes of the Primoriskiy Branch of the USSR Geographic Society)*, Vol. 1, No. 24. Vladivostok: Dal'nevostochnogo knizhnogo izdatal'estva (Far East Publishing House). Typescript translation by Misha Jones.

Aelian. 1971. *On the Characteristics of Animals*, trans. A. F. Scholfield. London: William Heinemann.

Akersten, William A. 1985. "Canine Function in Smilodon (Mammalia; Felidae; Machairodontinae)." *Contributions in Science*, no. 356. Los Angeles: Los Angeles County Museum.

Alexander, Michael, ed. and trans. 1973. *Beowulf.* Harmondsworth, Middlesex, Eng.: Penguin Books.

Ali, Salim. 1996. *The Book of Indian Birds*. Bombay: Bombay Natural History Society and Oxford University Press.

Allen, Judy, and Jeanne Griffiths. 1979. *The Book of the Dragon*. Secaucus, N.J.: Chartwell Books.

Allen, Thomas B. 1999. *The Shark Almanac*. New York: Lyons Press.

Almăşan, Horia. N.d. *Bonitatea Fondurilor de Vînătoare şi Efectivele Optime la Principalele Specii de Vînat din R. S. România.* (Privately translated by Eduard Érsek as "Carrying Capacity of Hunting Areas and Optimal Population Numbers of Game in Romania.") Bucharest: Institutul de Cercetărişi Amenajări Silvice.

Anderson, Elaine. 1984. "Who's Who in the Pleistocene: A Mammalian Bestiary." In Martin and Klein (1984).

Arseniev, V. K. 1996. *Dersu the Trapper*, trans. Malcolm Burr, preface by Jaimy Gordon. Kingston, N.Y.: McPherson.

Auffenberg, Walter. 1981. *The Behavioral Ecology of the Komodo Monitor*. Gainesville: University Presses of Florida.

Baikov, N. A. 1925. "Man'chzhurskii tigr" ("The Manchurian Tiger"). Harbin, China: Obshchestvo izucheniya Man'chzhurskogo Kraya (Society for the Study of Manchuria). Typescript translation by Misha Jones.

Bailey, Theodore N. 1993. *The African Leopard: Ecology and Behavior of a Solitary Felid*. New York: Columbia University Press.

Bakels, Jet. 1992. "Tiger by the Tail: On Tigers, Ancestors, and Nature Spirits in Kerinci." Typescript of a paper written for the Sumatran Tiger PHVA Workshop, November 22–29, 1992. Padang, Sumatra.

Behr, Edward. 1991. *Kiss the Hand You Cannot Bite: The Rise and Fall of the Ceauşescus*. New York: Villard Books.

Bellow, Saul. 1998. *The Dean's December*. New York: Penguin Books.

Berndt, Ronald M. 1964. "The Gove Dispute: The Question of Australian Aboriginal Land and the Preservation of Sacred Sites." *Anthropological Forum*, Vol. 1, No. 2.

Berndt, Ronald M., and Catherine H. Berndt. 1954. *Arnhem Land: Its History and Its People*. Melbourne: F. W. Cheshire.

Berndt, Ronald M., and Catherine H. Berndt, with John E. Stanton. 1998. *Aboriginal Australian Art*. Sydney: New Holland Publishers.

Berwick, Marianne. 1990. "The Ecology of the *Maldhari* Graziers in the Gir Forest, India." In *Conservation in Developing Countries: Problems and*

Prospects, ed. J. C. Daniel and J. S. Serrao. Bombay: Bombay Natural History Society.

Berwick, Stephen. 1971. "The Gir Forest: Its Wildlife and Ecology." *Span.* December 1971.

————. 1974. "The Community of Wild Ruminants in the Gir Forest Ecosystem, India." Ph.D. dissertation, Yale University.

————. 1976. "The Gir Forest: An Endangered Ecosystem." *American Scientist*, vol. 64.

Bhaskarananda, Swami. 1994. *The Essentials of Hinduism: A Comprehensive Overview of the World's Oldest Religion*. Seattle: Viveka Press.

Biknevicius, A. R., B. Van Valkenburgh, and J. Walker. 1996. "Incisor Size and Shape: Implications for Feeding Behaviors in Saber-Toothed 'Cats.'" *Journal of Vertebrate Paleontology*, vol. 16, no. 3.

Blackburn, Mr. Justice. 1971. *Milirrpum v. Nabalco Pty. Ltd. and the Commonwealth of Australia*. Decision in a case before the Supreme Court of the Northern Territory. Sydney: Law Book Company.

Blainey, Geoffrey. 1993. *The Rush That Never Ended: A History of Australian Mining*. Melbourne: Melbourne University Press.

Blank, Jonah. 1992. *Arrow of the Blue-Skinned God: Retracing the Ramayana Through India*. New York: Doubleday/Image Books.

Bleakley, J. W. 1929. "The Aboriginals and Half-Castes of Central Australia and North Australia." Report to the Parliament of the Commonwealth of Australia.

Bloch, Maurice. 1997. *Prey into Hunter: The Politics of Religious Experience*. Cambridge: Cambridge University Press.

Bomford, Mary, and Judy Caughley, eds. 1996. *Sustainable Use of Wildlife by Aboriginal People and Torres Strait Islanders*. Canberra: Australian Government Publishing Service.

Booth, Martin. 1990. *Carpet Sahib: A Life of Jim Corbett*. New Delhi: Oxford University Press.

Bowlby, John. 1992. *Charles Darwin: A New Life*. New York: W. W. Norton.

Bragin, A. P. 1986. "Population Characteristics and Social-Spatial Patterns of the Tiger (*Panthera tigris*) on the Eastern Macroslope of the Sikhote-Alin Mountain Range, USSR." Vladivostok: VINITI. Typescript translation by Bragin.

Bragin, Anatoly P., and Victor V. Gaponov. 1989. "Problems of the Amur

Tiger." *Okhota i okhotnichie khozyaistvo* (*Hunters and Hunting*), no. 10. Typescript translation by Bragin.

Brain, C. K. 1970. "New Finds at the Swartkrans Australopithecine Site." *Nature*, vol. 225, March 21, 1970.

———. 1981. *The Hunters or the Hunted? An Introduction to African Cave Taphonomy*. Chicago: University of Chicago Press.

Braun, Clait E., ed. 1991. "Mountain Lion-Human Interaction." Proceedings from a symposium and workshop, April 24–26, 1991. Denver: Colorado Division of Wildlife.

Brazaitis, Peter, Myrna E. Watanabe, and George Amato. 1998. "The Caiman Trade." *Scientific American*, vol. 278, no. 3.

Breeden, Stanley, and Belinda Wright. 1998. *Kakadu: Looking After the Country—the Gagudju Way*. Marleston, South Australia: J. B. Books.

Brown, David E. 1985. *The Grizzly in the Southwest: Documentary of an Extinction*. Norman: University of Oklahoma Press.

Brown, Gary. 1993. *The Great Bear Almanac*. New York: Lyons and Burford.

Burford, Tim, and Dan Richardson. 1998. *Romania: The Rough Guide*. London: Rough Guides.

Burkert, Walter. 1998. *Creation of the Sacred: Tracks of Biology in Early Religions*. Cambridge, Mass.: Harvard University Press.

Busch, Robert H. 2000. *The Grizzly Almanac*. New York: Lyons Press.

Bustard, H. Robert. 1969. "A Future for Crocodiles." *Oryx*, vol. 10, no. 4.

———. 1974. "A Preliminary Survey of the Prospects for Crocodile Farming (India)." Report FO: IND/71/033. Rome: FAO.

Bustard, H. R., and B. C. Choudhury. 1980a. "Long Distance Movement by a Saltwater Crocodile (*Crocodylus porosus*)." *British Journal of Herpetology*, vol. 6.

———. 1980b. "Conservation Future of the Saltwater Crocodile (*Crocodylus porosus* Schneider) in India." *Journal of the Bombay Natural History Society*, vol. 77, no. 2.

Butler, W. Harry. 1987. "'Living with Crocodiles' in the Northern Territory of Australia." In Webb, Manolis, and Whitehead (1987).

Byock, Jesse L., ed. and trans. 1990. *The Saga of the Volsungs*. New York: Penguin Books.

Campbell, Joseph. 1972. *The Hero with a Thousand Faces*. Princeton, N.J.: Bollingen Series/Princeton University Press.

————. 1988. *The Power of Myth*. With Bill Moyers. New York: Doubleday.

————. 1991. *Primitive Mythology: The Masks of God*. New York: Penguin Books.

Capstick, Peter Hathaway. 1998. *Maneaters*. Long Beach, Calif.: Safari Press.

Caputo, Philip. 2002. *Ghosts of Tsavo: Stalking the Mystery Lions of East Africa*. Washington, D.C.: National Geographic Adventure Press.

Caras, Roger A. 1977. *Dangerous to Man: The Definitive Story of Wildlife's Reputed Dangers*. South Hackensack, N. J.: Stoeger.

Carment, David. 1996. *Looking at Darwin's Past*. Darwin, Northern Territory, Aus.: North Australia Research Unit.

Carpenter, Stephen R., James F. Kitchell, and James R. Hodgson. 1985. "Cascading Trophic Interactions and Lake Productivity." *BioScience*, vol. 35, no. 10.

Cartmill, Matt. 1993. *A View to a Death in the Morning: Hunting and Nature Through History*. Cambridge, Mass.: Harvard University Press.

Carver, Robert. 1999. *The Accursed Mountains: Journeys in Albania*. London: HarperCollins.

Chaloupka, George. N.d. *Burrunguy: Nourlangie Rock*. Australia (no city given): Northart.

Champion-Jones, R. N. 1945. "Occurrence of the Lion in Persia." *Journal of the Bombay Natural History Society*, vol. 45.

Charlesworth, Max. 1984. *The Aboriginal Land Rights Movement*. Richmond, Victoria, Aus.: Hodja Educational Resources Cooperative.

Chatterjee, Nilanjana. 1992. "Midnight's Unwanted Children: East Bengali Refugees and the Politics of Rehabilitation." Ph.D. dissertation, Brown University.

Chauvet, Jean-Marie, Eliette Brunel Deschamps, and Christian Hillaire. 1996. *Dawn of Art: The Chauvet Cave*. New York: Harry N. Abrams.

Chellam, Ravi, 1993. "Ecology of the Asiatic Lion (*Panthera leo persica*)." Ph.D. dissertation. Saurashtra University, Rajkot.

————. 1996. "Lions of the Gir Forest." *Wildlife Conservation*. May–June 1996.

————. 1997. "Asia's Envy, India's Pride." Srishti, vol. 2.

Chellam, Ravi, and A. J. T. Johnsingh. 1993. "Management of Asiatic Lions in the Gir Forest, India." *Symposium of the Zoological Society of London*, no. 65.

Chellam, Ravi, and Vasant Saberwal. 2000. "Asiatic Lion." In *Endangered*

Animals: A Reference Guide to Conflicting Issues. Richard P. Reading and Brian Miller, eds. Westport, Conn.: Greenwood Press.

Cherry, John, ed. 1995. *Mythical Beasts*. London: British Museum Press.

Choudhury, B. C., and H. R. Bustard. 1980. "Predation on Natural Nests of the Saltwater Crocodile (*Crocodylus porosus* Schneider) on North Andaman Island with Notes on the Crocodile Population." *Journal of the Bombay Natural History Society*, vol. 76, no. 2.

Choudhury, B. C., and S. Choudhury. 1986. "Lessons from Crocodile Reintroduction Projects in India." *Indian Forester*, vol. 112, no. 10.

Clark, Tim W., A. Peyton Curlee, Steven C. Minta, and Peter M. Kareiva, eds. 1999. *Carnivores in Ecosystems: The Yellowstone Experience*. New Haven, Conn.: Yale University Press.

Clarke, James. 1969. *Man Is the Prey*. New York: Stein and Day.

Clottes, Jean. 1995. "Rhinos and Lions and Bears (Oh, My!)." *Natural History*, vol. 104, May 1995.

———. 1996. "Epilogue: Chauvet Cave Today." In Chauvet, Deschamps, and Hillaire (1996).

———. 2001. "Chauvet Cave: France's Magical Ice Age Art." *National Geographic*, vol. 200, August 2001.

Clottes, Jean, and David Lewis-Williams. 1998. *The Shamans of Prehistory: Trance and Magic in the Painted Caves*. Text by Jean Clottes, trans. Sophie Hawkes. New York: Harry N. Abrams.

Clutton-Brock, Juliet. 1996. "Competitors, Companions, Status Symbols, or Pests: A Review of Human Associations with Other Carnivores." In Gittleman (1996).

Codrescu, Andrei. 1991. *The Hole in the Flag: A Romanian Exile's Story of Return and Revolution*. New York: William Morrow.

Cohen, Joel E. 1995. *How Many People Can the Earth Support?* New York: W. W. Norton.

Cole, Keith. 1979. *The Aborigines of Arnhem Land*. Adelaide, Aus.: Rigby.

———. 1980. *Arnhem Land: Places and People*. Adelaide, Aus.: Rigby.

Colinvaux, Paul. 1978. *Why Big Fierce Animals Are Rare: An Ecologist's Perspective*. Princeton, N.J.: Princeton University Press.

Corbett, Jim. 1991. *The Jim Corbett Omnibus*. (Consisting of: *Man-Eaters of Kumaon, The Temple Tiger and More Man-Eaters of Kumaon*, and *The Man-Eating Leopard of Rudraprayag*.) Delhi: Oxford University Press.

Cott, Hugh B. 1961. "Scientific Results of an Inquiry into the Ecology and

Economic Status of the Nile Crocodile (*Crocodylus niloticus*) in Uganda and Northern Rhodesia." *Transactions of the Zoological Society of London*, vol. 29, part 4.

Courtney, Nicholas. 1980. *The Tiger: Symbol of Freedom*. London: Quartet Books.

Craighead, Frank C., Jr. 1979. *Track of the Grizzly*. San Francisco: Sierra Club Books.

Craighead, John J., J. S. Summer, and G. B. Scaggs. 1982. *A Definitive System for Analysis of Grizzly Bear Habitat and Other Wilderness Resources*. Monograph No. 1. Missoula, Mont.: Wildlife-Wildlands.

Crawley, Michael J., ed. 1992. *Natural Enemies: The Population Biology of Predators, Parasites and Diseases*. Oxford: Blackwell Scientific Publications.

Crişan, Vasile. 1994. *Jäger? Schlächter: Ceauşescu*. (Privately translated by Eduard Érsek as "Ceauşescu: Hunter or Butcher?") Mainz, Ger.: Verlag Dieter Hoffmann.

Cronon, William. 1983. *Changes in the Land: Indians, Colonists, and the Ecology of New England*. New York: Hill and Wang.

Crooks, Kevin R., and Michael E. Soulé. 1999. "Mesopredator Release and Avifaunal Extinctions in a Fragmented System." *Nature*, vol. 400, August 5, 1999.

Crowe, David M. 1994. *A History of the Gypsies of Eastern Europe and Russia*. New York: St. Martin's.

Cullen, Robert. 1990. "Report from Romania." *The New Yorker*, vol. 66, April 2, 1990.

Dalley, Stephanie, ed. and trans. 1998. *Myths from Mesopotamia: Creation, the Flood, Gilgamesh, and Others*. Oxford: Oxford University Press.

Dalvi, M. K. 1969. "Gir Lion Census 1968." *Indian Forester*, vol. 95, no. 11.

Daniel, J. C. 1980. "An Island of Hope." *Animal Kingdom*, vol. 83, no. 5.

————. 1996. *The Leopard in India: A Natural History*. Dehra Dun, India: Natraj Publishers.

Darwin, Charles. 1964. *On the Origin of Species*. A facsimile of the first (1859) edition. Cambridge, Mass.: Harvard University Press.

Day, David. 1981. *The Encyclopedia of Vanished Species*. London: McLaren.

Dean, R. L., et al. 1963. *Report from the Select Committee on Grievances of Yirrkala Aborigines, Arnhem Land Reserve*. Canberra: Commonwealth Government Printer.

Deletant, Dennis. 1995. *Ceauşescu and the Securitate: Coercion and Dissent in Romania, 1965–1989*. London: Hurst.

Desai, J. R. 1974. "The Gir Forest Reserve: Its Habitats, Faunal and Social Problems." In *Second World Conference on National Parks*, ed. Sir Hugh Elliott. Morges, Switz.: IUCN.

Desai, Bharat. 2001. "Gir Lion May Be Cramped for Space." *Times of India*, May 28, 2001.

Deurbrouck, Jo, and Dean Miller. 2001. *Cat Attacks: True Stories and Hard Lessons from Cougar Country*. Seattle: Sasquatch Books.

Dharmakumarsinhji, K. S., and M. A. Wynter-Blyth. 1950. "The Gir Forest and Its Lions, Part III." (See under Wynter-Blyth for parts I and II.) *Journal of the Bombay Natural History Society*, Vol. 49.

————. 1998. *Reminiscences of Indian Wildlife*. Delhi: Oxford University Press.

Dharmakumarsinhji, R. S. 1968. "The Gir Lion." *Cheetal*, vol. 10, no. 2.

Diamond, Jared M. 1986. "How Great White Sharks, Sabre-Toothed Cats and Soldiers Kill." *Nature*, vol. 322, August 28, 1986.

Dilks, David. 1969. *Curzon in India*. 2 vols. New York: Taplinger.

Divyabhanusinh. 1995. *The End of a Trail: The Cheetah in India*. New Delhi: Banyan Books.

————. 1998. "A Princely Bequest." *Seminar*, no. 466, June 1998.

Dresser, B. L., R. W. Reese, and E. J. Maruska, eds. 1988. *Proceedings of the 5th World Conference on Breeding Endangered Species in Capitivity*. Cincinnati, Ohio: Cincinnati Zoo.

Duggins, David O. 1980. "Kelp Beds and Sea Otters: An Experimental Approach." *Ecology*, vol. 61, no. 3.

Dunstone, N., and M. L. Gorman, eds. 1993. *Mammals as Predators*. Proceedings of a symposium held by the Zoological Society of London and the Mammal Society, London, November 22–23, 1991. Symposia of the Zoological Society of London, number 65. Oxford: Clarendon Press.

Edgaonkar, Advait, and Ravi Chellam. 1998. "A Preliminary Study on the Ecology of the Leopard, *Panthera pardus fusca*, in the Sanjay Gandhi National Park, Maharashtra." Dehra Dun, India: Wildlife Institute of India.

Edwards, Hugh. 1989. *Crocodile Attack*. New York: Harper and Row.

Edwards, S. M., and L. G. Fraser. 1907. *Ruling Princes of India: Junagadh*. Bombay: "Times of India" Press.

Egan, Ted. 1996. *Justice All Their Own: The Caledon Bay and Woodah Island Killings*, 1932–1933. Melbourne: Melbourne University Press.

Egerton, Frank N. 1973. "Changing Concepts of the Balance of Nature." *Quarterly Review of Biology*, vol. 48.

Ehrenfeld, David W. 1970. *Biological Conservation*. New York: Holt, Rinehart and Winston.

———. 1986. "Life in the Next Millennium: Who Will Be Left in the Earth's Community?" In Kaufman and Mallory (1986).

Ehrenreich, Barbara. 1997. *Blood Rites: Origins and History of the Passions of War*. New York: Henry Holt.

Ellis, Richard. 1999. *The Search for the Giant Squid*. New York: Penguin Books.

———. 1996. *Monsters of the Sea*. New York: Doubleday/Main Street Books.

Ellis, Richard, and John E. McCosker. 1991. *Great White Shark*. Stanford, Calif.: Stanford University Press.

Elton, Charles. 1966. *Animal Ecology*. New York: October House.

Emerson, Sharon B., and Leonard Radinsky. 1980. "Functional Analysis of Sabertooth Cranial Morphology." *Paleobiology*, vol. 6, no. 3.

Enthoven, R. E. 1997. *The Tribes and Castes of Bombay*. 3 vols. Delhi: Low Price Publications.

Errington, Paul. 1946a. "Predation and Vertebrate Populations." *Quarterly Review of Biology*, vol. 21, no. 2.

———. 1946b. "Predation and Vertebrate Populations (Concluded)." *Quarterly Review of Biology*, vol. 21, no. 3.

———. 1956. "Factors Limiting Higher Vertebrate Populations." *Science*, vol. 124, August 16, 1956.

———. 1963. "The Phenomenon of Predation." *American Scientist*, vol. 51, no. 2.

———. 1967. *Of Predation and Life*. Ames: Iowa State University Press.

Estes, James, Kevin Crooks, and Robert Holt. 2001. "Predators, Ecological Role of." *Encyclopedia of Biodiversity*, vol. 4. Edited by Simon Asher Levin. San Diego, Calif.: Academic Press.

Ewer, R. F. 1998. *The Carnivores*. With a new foreword by Devra Kleiman. Ithaca, N.Y.: Cornell University Press.

Ewing, Susan, and Elizabeth Grossman, eds. 1999. *Shadow Cat: Encountering the American Mountain Lion*. Seattle: Sasquatch Books.

Farkas, Ann E., Prudence O. Harper, and Evelyn B. Harrison, eds. 1987. *Monsters and Demons in the Ancient and Medieval Worlds*. Mainz on Rhine, Ger.: Verlag Philipp von Zabern.

Feazel, Charles T. 1992. *White Bear: Encounters with the Master of the Arctic Ice*. New York: Ballantine Books.

Feder, Martin E., and George V. Lauder, eds. 1986. *Predator-Prey Relationships: Perspectives and Approaches from the Study of Lower Vertebrates*. Chicago: University of Chicago Press.

Fenton, L. L. 1909. "The Kathiawar Lion." *Journal of the Bombay Natural History Society*, vol. 19.

Ferry, David, ed. and trans. 1994. *Gilgamesh: A New Rendering in English Verse*. New York: Noonday Press/Farrar, Straus and Giroux.

Finlayson, Max, Dean Yibarbuk, Lisa Thurtell, Michael Storrs, and Peter Cooke. 1999. "Local Community Management of the Blyth/Liverpool Wetlands, Arnhem Land, Northern Territory, Australia." Canberra: Supervising Scientist, Environment Australia.

Fischer, Hank. 1995. *Wolf Wars: The Remarkable Inside Story of the Restoration of Wolves to Yellowstone*. Helena, Mont.: Falcon Press.

Fischer, Henry G. 1987. "The Ancient Egyptian Attitude Toward the Monstrous." In Farkas, Harper, and Harrison (1987).

Fischer-Galaţi, Stephen. 1970. *Twentieth Century Romania*. New York: Columbia University Press.

Flannery, Tim. 1995. *The Future Eaters*. New York: George Braziller.

Gadgil, Madhav, and Ramachandra Guha. 1993. *This Fissured Land: An Ecological History of India*. Delhi: Oxford University Press.

————. 1995. *Ecology and Equity: The Use and Abuse of Nature in Contemporary India*. New York: Routledge.

Galster, Steven Russell, and Karin Vaud Eliot. 1999. "Roaring Back: Antipoaching Strategies for the Russian Far East and the Comeback of the Amur Tiger." In Seidensticker et al. (1999).

Gans, Carl. 1986. "Functional Morphology of Predator-Prey Relationships." In Feder and Lauder (1986).

Gardner, John. 1989. *Grendel*. New York: Vintage Books.

Gee, E. P. 1964. *The Wild Life of India*. London: Collins.

Geist, Valerius. 1989. "Did Large Predators Keep Humans out of North America?" In *The Walking Larder: Patterns of Domestication, Pastoralism, and Predation*, ed. Juliet Clutton-Brock. London: Hyman Unwin.

George, Andrew, ed. and trans. 1999. *The Epic of Gilgamesh*. New York: Penguin Books.

Georgescu, Vlad. 1991. *The Romanians: A History*, ed. Matei Calinescu, trans. Alexander Bley-Vroman. Columbus: Ohio State University Press.

Gilbert, D. A., C. Packer, A. E. Pusey, J. C. Stephens, and S. J. O'Brien. 1991. "Analytical DNA Fingerprinting on Lions: Parentage, Genetic Diversity, and Kinship." *Journal of Heredity*, vol. 82.

Gillespie, Dan, Peter Cooke, and John Taylor. N.d. "Improving the Capacity of Indigenous People to Contribute to the Conservation of Biodiversity in Australia." A report to the Biological Diversity Advisory Council, Environment Australia. Darwin, Northern Territory, Aus.: Tallegalla Consultants et al.

Gittleman, John L., ed. 1989. *Carnivore Behavior, Ecology, and Evolution*. Ithaca, N.Y.: Cornell University Press.

———. 1996. *Carnivore Behavior, Ecology, and Evolution*. Vol. 2. Ithaca: Cornell University Press.

Gittleman, J. L., S. Funk, D. W. Macdonald, and R. K. Wayne, eds. 2001. *Carnivore Conservation*. Cambridge: Cambridge University Press.

Gonyea, William J. 1976. "Behavioral Implications of Saber-Toothed Felid Morhpology." *Paleobiology*, vol. 2.

Government of Gujarat. 1975. *The Gir Lion Sanctuary Project*. Revised (from 1972 edition) by K. P. Karamchandani. Gandhinagar: Government of Gujarat.

Graham, Alistair, and Peter Beard. 1973. *Eyelids of Morning: The Mingled Destinies of Crocodiles and Men*. New York: A & W Visual Library.

Groger-Wurm, Helen M. 1973. "Australian Aboriginal Bark Paintings and Their Mythological Interpretation." Australian Aboriginal Studies, no. 30. Canberra: Australian Institute of Aboriginal Studies.

Grumbine, R. Edward. 1992. *Ghost Bears: Exploring the Biodiversity Crisis*. Washington, D.C.: Island Press.

Grun, Bernard. 1982. *The Timetables of History: A Horizontal Linkage of People and Events*. New York: Touchstone.

Grzelewski, Derek. 2001. "Risky Business." *Smithsonian*, vol. 32, no. 8.

Guggisberg, C. A. W. 1963. *Simba: The Life of the Lion*. Philadelphia: Chilton Books.

———. 1972. *Crocodiles: Their Natural History, Folklore and Conservation*. Harrisburg, Pa.: Stackpole Books.

Hairston, Nelson G., Frederick E. Smith, and Lawrence B. Slobodkin. 1960.

"Community Structure, Population Control, and Competition." *American Naturalist*, vol. 94, no. 879.

Hale, Julian. 1971. *Ceaușescu's Romania: A Political Documentary*. London: George G. Harrap.

Hamilton, Edith. 1961. *Mythology*. New York: Mentor Books/New American Library.

Hansen, Kevin. 1992. *Cougar: The American Lion*. Flagstaff, Ariz.: Northland Publishing.

Harrison, Paul. 1992. *The Third Revolution: Environment, Population and a Sustainable World*. London: I. B. Tauris.

Hasluck, Paul. 1988. *Shades of Darkness: Aboriginal Affairs 1925–1965*. Melbourne: Melbourne University Press.

Heaney, G. F. 1944. "Occurrence of the Lion in Persia." *Journal of the Bombay Natural History Society*, vol. 44.

Heaney, Seamus, ed. and trans. 2000. *Beowulf: A New Verse Translation*. New York: Farrar, Straus and Giroux.

Heidel, Alexander, ed. and trans. 1963. *The Babylonian Genesis*. Chicago: University of Chicago Press.

Herodotus. 1972. *The Histories*, trans. Aubrey de Sélincourt; revised, with an introduction and notes by A. R. Burn. Harmondsworth, Middlesex, Eng.: Penguin Books.

Herrero, Stephen. 1985. *Bear Attacks: Their Causes and Avoidance*. New York: Nick Lyons Books.

Hiatt, L. R. 1965. *Kinship and Conflict: A Study of an Aboriginal Community in Northern Arnhem Land*. Canberra: Australian National University Press.

Hobbes, Thomas. 1985. *Leviathan*, ed. with an introduction by C. B. Macpherson. New York: Penguin Classics.

Hodges-Hill, Edward. 1992. *Man-Eater: Tales of Lion and Tiger Encounters*. Heathfield, East Sussex, Eng.: Cockbird Press.

Hobusch, Erich. 1980. *Fair Game: A History of Hunting, Shooting and Animal Conservation*. English version by Ruth Michaelis-Jena and Patrick Murray. New York: Arco Publishing.

Horner, John R., and Don Lessem. 1993. *The Complete T. rex*. New York: Simon and Schuster.

Hoult, Janet. 1987. *Dragons: Their History and Symbolism*. Glastonbury, Somerset, Eng.: Gothic Image Publications.

Hoogesteijn, Rafael, and Edgardo Mondolfi. 1996. "Body Mass and Skull Measurements in Four Jaguar Populations and Observations on Their Prey Base." *Bulletin of the Florida Museum of Natural History*, vol. 39, no. 6.

Hughes, Robert. 1988. *The Fatal Shore*. New York: Vintage Books.

Hummel, Monte, and Sherry Pettigrew, with John Murray. 1991. *Wild Hunters: Predators in Peril*. Niwot, Colo.: Roberts Rinehart.

Hutcherson, Gillian. 1995. *Djalkiri Wanga, The Land Is My Foundation: 50 Years of Aboriginal Art from Yirrkala, Northeast Arnhem Land*. Occasional Paper, no. 4. Perth: University of Western Australia/Berndt Museum of Anthropology.

Hutton, Jon M., and Grahame J. W. Webb. 1992. "An Introduction to the Farming of Crocodiles." From a workshop held at the 10th Working Meeting of the IUCN/SSC Crocodile Specialist Group, Gainesville, Fl., April 21–27, 1990.

———. 1994. "The Principles of Farming Crocodiles." In *Proceedings of the Second Regional Meeting (Eastern Asia, Oceania, Australasia) of the Crocodile Specialist Group of the Species Survival Commission of IUCN Convened at Darwin, Northern Territory, Australia*. Gland, Switz.: IUCN.

Ioanid, Radu. 2000. *The Holocaust in Romania: The Destruction of Jews and Gypsies Under the Antonescu Regime, 1940–1944*. Chicago: Ivan R. Dee.

Ionescu, Ovidiu. 1999. "Status and Management of the Brown Bear in Romania." In Servheen et al. (1999).

Irving, Laurence. 1972. *Arctic Life of Birds and Mammals, Including Man*. New York: Springer-Verlag.

Jagannathan, Shakunthala. 1984. *Hinduism: An Introduction*. Mumbai: Vakils, Feffer and Simons.

Janis, Christine. 1994. "The Sabertooth's Repeat Performances." *Natural History*, vol. 103, no. 4.

Janis, Christine M., Kathleen M. Scott, and Louis L. Jacobs, eds. 1998. *Evolution of Tertiary Mammals of North America*. Vol. 1, *Terrestrial Carnivores, Ungulates, and Ungulatelike Mammals*. Cambridge: Cambridge University Press.

Johanson, Donald, and Blake Edgar. 1996. *From Lucy to Language*. New York: Simon and Schuster.

Johnsingh, A. J. T., and Ravi Chellam. 1991. "India's Last Lions." *Zoogoer*, vol. 20, no. 5.

Johnsingh, A. J. T., Ravi Chellam, and G. S. Rawat. 1995. "Prospects for Ecotourism in India." In *Integrating People and Wildlife for a Sustainable Future*, ed. John A. Bissonette and Paul R. Krausman. Bethesda, Md.: Wildlife Society.

Joslin, Paul. 1973. "The Asiatic Lion: A Study of Ecology and Behaviour." Ph.D. dissertation, University of Edinburgh.

————. 1980. "The Lion's Share Is Very Small." *Animal Kingdom*, vol. 83, no. 5.

————. 1984. "The Environmental Limitations and Future of the Asiatic Lion." *Journal of the Bombay Natural History Society*, vol. 81.

————. 1985. "Lions of India." *Bison*. Spring 1985.

Judt, Tony. 2001. "Romania: The Bottom of the Heap." *New York Review of Books*, November 1, 2001.

Kanvinde, Hemal S. 1995. "Bhitarkanika Wildlife Sanctuary." In *Protecting Endangered National Parks*. Delhi: Rajiv Gandhi Institute for Contemporary Studies.

Kaplan, Robert. D. 1994. *Balkan Ghosts: A Journey Through History*. New York: Vintage Books.

Kar, S. K. 1993. "Post Hatching Dispersal and Growth of the Saltwater Crocodile, *Crocodylus porosus* Schneider, in Orissa, India." *Journal of the Bombay Natural History Society*, vol. 90.

Kar, S. K., and H. R. Bustard. 1989. "Status of the Saltwater Crocodile (*Crocodylus porosus* Schneider) in the Bhitarkanika Wildlife Sanctuary, Orissa, India." *Journal of the Bombay Natural History Society*, vol. 86.

————. 1990. "Results of a Pilot Saltwater Crocodile *Crocodylus porosus* Schneider Restocking in Bhitarkanika Wildlife Sanctuary, Orissa." *Journal of the Bombay Natural History Society*, vol. 87.

Kaufman, Les, and Kenneth Mallory, eds. 1986. *The Last Extinction*. Cambridge, Mass.: MIT Press.

Kautilya. 1992. *The Arthashastra*, ed. and trans. L. N. Rangarajan. New Delhi: Penguin Books.

Keen, Ian. 1994. *Knowledge and Secrecy in an Aboriginal Religion*. Oxford: Clarendon Press.

Keenan, Sheila. 2000. *Gods, Goddesses, and Monsters: An Encyclopedia of World Mythology*. New York: Scholastic.

Keilman, Nico. 2001. "Uncertain Population Forecasts." *Nature*, vol. 412. August 2, 2001.

Keiter, Robert B., and Mark S. Boyce, eds. 1991. *The Greater Yellowstone Ecosystem: Redefining America's Wilderness Heritage*. New Haven, Conn.: Yale University Press.

Kellert, Stephen R. 1997. *Kinship to Mastery: Biophilia in Human Evolution and Development*. Washington, D.C.: Island Press/Shearwater Books.

King, L. W., ed. and trans. 1902. *Enuma Elish: The Seven Tablets of Creation*. Vol. 1. London: Luzac. (Facsimile reprint, Escondido, Calif.: Book Tree, 1999.)

Kingsland, Sharon E. 1988. *Modeling Nature: Episodes in the History of Population Ecology*. Chicago: University of Chicago Press.

Kinnear, N. B. 1920. "The Past and Present Distribution of the Lion in South Eastern Asia." *Journal of the Bombay Natural History Society*, vol. 27.

Kitchener, Andrew. 1991. *The Natural History of the Wild Cats*. Ithaca, N.Y.: Cornell University Press.

———. 1999. "Tiger Distribution, Phenotypic Variation and Conservation Issues." In Seidensticker et al. (1999).

Kowalski, Kazimierz. 1967. "The Pleistocene Extinction of Mammals in Europe." In Martin and Wright (1967).

Kruuk, Hans. 2002. *Hunter and Hunted: Relationships Between Carnivores and People*. Cambridge: Cambridge University Press.

Kurtén, Björn. 1968. *Pleistocene Mammals of Europe*. Chicago: Aldine Publishing.

———. 1972. *The Age of Mammals*. New York: Columbia University Press.

———. 1995. *The Cave Bear Story: Life and Death of a Vanished Animal*. New York: Columbia University Press.

Lambert, David, and the Diagram Group. 1990. *The Dinosaur Data Book*. New York: Avon Books.

Lambert, Wilfred G. 1987. "Gilgamesh in Literature and Art: The Second and First Millennia." In Farkas, Harper, and Harrison (1987).

Laurance, William F. and Richard O. Bierregaard, Jr., eds. 1997. *Tropical Forest Remnants: Ecology, Management, and Conservation of Fragmented Communities*. Chicago: University of Chicago Press.

Laynhapuy Schools A.S.S.P.A. Committee. 1992. "Baniyala Garrangali Galtha Rom Workshop." Nhulunbuy, Northern Territory, Aus.: Yirrkala Literature Production Centre.

Leopold, Aldo. 1933. *Game Management*. New York: Charles Scribner's Sons.

Leyhausen, Paul. 1979. *Cat Behavior: The Predatory and Social Behavior of Domestic and Wild Cats*, trans. Barbara A. Tonkin. New York: Garland STPM Press.

Lindeman, Raymond L. 1942. "The Trophic-Dynamic Aspect of Ecology." *Ecology*, vol. 23, no. 4.

Lindskog, Birger. 1954. *African Leopard Men*. Studia Ethnographica Upsaliensia, 7. Uppsala, Swe.: Almqvist & Wiksells Boktryckeri.

Lineweaver, Thomas H., III, and Richard H. Backus. 1984. *The Natural History of Sharks*. New York: Nick Lyons Books/Schocken Books.

Long, John, ed. 1998. *Attacked! By Beasts of Prey and Other Deadly Creatures*. Camden, Me.: Ragged Mountain Press.

Loos, Noel, and Koiki Mabo. 1996. *Edward Koiki Mabo: His Life and Struggle for Land Rights*. St. Lucia, Queensland, Aus.: University of Queensland Press.

Lutz, Wolfgang, Warren Sanderson, and Sergei Scherbov. 2001. "The End of World Population Growth." *Nature*, vol. 412, August 2, 2001.

Lydekker, R. 1900. *The Great and Small Game of India, Burma, and Tibet*. London: Rowland Ward.

McCulloch, Dale R., and Reginald H. Barrett. 1991. *Wildlife 2001: Populations*. Proceedings of an International Conference on Population Dynamics and Management of Vertebrates (Exclusive of Primates and Fish). London: Elsevier Applied Science.

McDougal, Charles. 1977. *The Face of the Tiger*. London: Rivington Books and André Deutsch.

———. 1987. "The Man-Eating Tiger in Geographical and Historical Perspective." In Tilson and Seal (1987).

McGowan, Chistopher. 1997. *The Raptor and the Lamb: Predators and Prey in the Living World*. New York: Henry Holt.

McMillion, Scott. 1998. *Mark of the Grizzly: True Stories of Recent Bear Attacks and the Hard Lessons Learned*. Helena: Falcon Publishing.

McNamee, Thomas. 1990. *The Grizzly Bear*. New York: Penguin Books.

———. 1998. *The Return of the Wolf to Yellowstone*. New York: Henry Holt/Owl Books.

McNeely, Jeffrey A., and Paul Spencer Wachtel. 1988. *The Soul of the Tiger: Searching for Nature's Answers in Exotic Southeast Asia*. New York: Doubleday.

McVedy, Colin. 1972. *The Penguin Atlas of Ancient History*. Harmondsworth, Middlesex, Eng.: Penguin Books.

Mallick, Ross. 1993. *Development Policy of a Communist Government: West Bengal Since 1997.* Cambridge: Cambridge University Press.

————. 1998. *Development, Ethnicity and Human Rights in South Asia.* New Delhi: Sage Publications.

Manfredi Paola, ed. 1997. *In Danger: Habitats, Species and People.* New Delhi: Ranthambhore Foundation.

Maniguet, Xavier. 1991. *The Jaws of Death: Shark as Predator, Man as Prey,* trans. David A. Christie. Dobbs Ferry, N.Y.: Sheridan House.

Marean, Curtis W. 1989. "Sabertooth Cats and Their Relevance for Early Hominid Diet and Evolution." *Journal of Human Evolution,* vol. 18.

Marean, Curtis W. and Celeste L. Ehrhardt. 1995. "Paleoanthropological and Paleoecological Implications of the Taphonomy of a Sabertooth's Den." *Journal of Human Evolution,* vol. 29.

Marika, Wandjuk. 1995. *Wandjuk Marika, Life Story.* As told to Jennifer Isaacs. St. Lucia, Queensland, Aus.: University of Queensland Press.

Martin, P. S., and H. E. Wright, Jr., eds. 1967. *Pleistocene Extinctions: The Search for a Cause.* New Haven: Yale University Press.

Martin, Paul S., and Richard G. Klein, eds. 1984. *Quaternary Extinctions: A Prehistoric Revolution.* Tucson: University of Arizona Press.

Matthiessen, Peter. 2000. *Tigers in the Snow.* New York: North Point Press/Farrar, Straus and Giroux.

Matyushkin, E. N. 1998. *The Amur Tiger in Russia: An Annotated Bibliography, 1925–1997.* Moscow: World Wide Fund for Nature.

Matyushkin, E. N., V. I. Zhivotchenko, and E. N. Smirnov. 1980. *The Amur Tiger in the USSR.* English translation, uncredited. Gland, Switz.: IUCN.

Mayor, Adrienne. 2000. *The First Fossil Hunters: Paleontology in Greek and Roman Times.* Princeton, N.J.: Princeton University Press.

Mayr, Ernst. 1956. "Geographical Character Gradients and Climatic Adaptation." *Evolution,* vol. 10, no. 1. Revised and reprinted in Mayr's *Evolution and the Diversity of Life: Selected Essays.* Cambridge, Mass.: Belknap Press of Harvard University Press, 1976.

Mazur, Laurie Ann. 1994. *Beyond the Numbers: A Reader on Population, Consumption, and the Environment.* Washington, D.C.: Island Press.

Mech, L. David. 1981. *The Wolf: The Ecology and Behavior of an Endangered Species.* Minneapolis: University of Minnesota Press.

Meena, R. L., R. D. Kambol, and Mahesh Singh. N.d. "The Gir." Sasan, India: Gir Welfare Fund, Wildlife Division.

Mehta, Gita. 1997. *Snakes and Ladders: A View of Modern India*. London: Secker and Warburg.

Mendelsohn, Oliver, and Marika Vicziany. 1998. *The Untouchables: Subordination, Poverty and the State in Modern India*. Cambridge: Cambridge University Press.

Menge, Bruce A., and Tess L. Freidenburg. 2001. "Keystone Species." *Encyclopedia of Biodiversity*, vol. 3, ed. Simon Asher Levin. San Diego, Calif.: Academic Press.

Mertens, Annette, and Christoph Promberger. N.d. "Economic Aspects of Large Carnivore-Livestock Conflicts in Romania." (Draft/private.)

Messel, Harry. 1977. "The Crocodile Programme in Northern Australia: Population Surveys and Numbers." In *A Study of Crocodylus Porosus in Northern Australia: A Series of Five Lectures*, ed. Messel and Butler. Sydney: Shakespeare Head Press.

———. 1991. "Sustainable Utilization: A Program That Conserves Many Crocodilians." *Species*, vol. 16.

Metzger, Bruce M., and Roland E. Murphy, eds. 1994. *The New Oxford Annotated Bible*. New York: Oxford University Press.

Micu, Ion. 1998. *Ursul Brun: Aspecte Eco-Etologice*. (Privately translated by Eduard Érsek.) Bucharest: Editura Ceres.

Mills, L. Scott, Michael E. Soulé, and Daniel F. Doak. 1993. "The Keystone-Species Concept in Ecology and Conservation." *BioScience*, vol. 43, no. 4.

Minta, Steven C., Peter M. Kareiva, and A. Peyton Curlee. 1999. "Carnivore Research and Conservation: Learning from History and Theory." In Clark et al. (1999).

Mitchell, Stephen, ed. and trans. 1992. *The Book of Job*. New York: Harper Perennial.

Montgomery, Sy. 1995. *Spell of the Tiger: The Man-Eaters of Sundarbans*. Boston: Houghton Mifflin.

Moorehead, Alan. 1987. *The Fatal Impact: The Invasion of the South Pacific 1767–1840*. Sydney: Mead and Beckett Publishing.

Moraes, Frank. 1956. *Jawaharlal Nehru*. New York: Macmillan.

Morell, Virginia. 1994. "Serengeti's Big Cats Going to the Dogs." *Science*, vol. 264, June 17, 1994.

———. 1996. "New Virus Variant Killed Serengeti Cats." *Science*, vol. 271, February 2, 1996.

Morphy, Howard. 1984. *Journey to the Crocodile's Nest*. Canberra: Australian Institute of Aboriginal Studies.

————. 1991. *Ancestral Connections: Art and an Aboriginal System of Knowledge*. Chicago: University of Chicago Press.

Murphy, John C., and Robert W. Henderson. 1997. *Tales of Giant Snakes: A Historical Natural History of Anacondas and Pythons*. Malabar, Fla.: Krieger.

Naipaul, V. S. 1979. *India: A Wounded Civilization*. New York: Penguin Books.

————. 1992. *India: A Million Mutinies Now*. New York: Penguin Books.

Narayan, R. K. 1977. *The Ramayana*. New York: Penguin Books.

————. 1984. *A Tiger for Malgudi*. New York: Penguin Books.

————. 1993. *Gods, Demons, and Others*. Chicago: University of Chicago Press.

Narayan, Shankar. 1996. "Joint Management of Gir National Park." In *People and Protected Areas: Towards Participatory Conservation in India*, ed. Ashish Kothari, Neena Singh, and Saloni Suri. New Delhi: Sage Publications.

Negi, S. S. 1969. "Transplanting of Indian Lion in Uttar Pradesh State." *Cheetal*, vol. 12, no. 1.

Newell, Josh, and Emma Wilson. 1996. *The Russian Far East: Forests, Biodiversity Hotspots, and Industrial Developments*. Tokyo: Friends of the Earth—Japan.

Nichols, Michael, and Geoffrey C. Ward. 1998. *The Year of the Tiger*. Washington, D.C.: National Geographic Society.

Nikolaev, Igor G., and Victor G. Yudin. 1993. "Tiger and Man in Conflict Situations." *Bull. Mosk. Obschestva Ispytateley Prirody. Otd. Biol.*, vol. 98, no. 3.

————. N.d. "Conflicts between Man and Tiger in the Russian Far East." Typescript in English of an (unpublished?) update, circa 1999, of Nikolaev and Yudin (1993).

Nitecki, Matthew H., ed. 1984. *Extinctions*. Chicago: University of Chicago Press.

Nowell, Kristin, and Peter Jackson. 1996. *Wild Cats: Status Survey and Conservation Action Plan*. Gland, Switz.: IUCN.

Nuttall-Smith, Chris. 2002. "Symbol of Asian Pride Extinct: Report." *Vancouver Sun*, November 2, 2002.

O'Brien, Stephen J. 1994. "Genetic and Phylogenetic Analyses of Endangered Species." *Annual Review of Genetics*, vol. 28.

O'Brien, S. J., P. Joslin, G. L. Smith III, R. Wolfe, N. Schaffer, E. Heath, J. Ott-Joslin, P. P. Rawal, K. K. Bhattacharjee, and J. S. Martenson. 1987. "Evidence for African Origins of Founders of the Asiatic Lion Species Survival Plan." *Zoo Biology*, vol. 6.

O'Brien, Stephen J., Janice S. Martenson, Craig Packer, Lawrence Herbst, Valerius de Vos, Paul Joslin, Janis Ott-Joslin, David E. Wildt, and Mitchell Bush. 1987. "Biochemical Genetic Variation in Geographic Isolates of African and Asiatic Lions." *National Geographic Research*, vol. 3, no. 1.

Odum, Eugene P. 1971. *Fundamentals of Ecology*. Philadelphia: W. B. Saunders.

Okarma, Henryk, Yaroslav Dovchanych, Slavomir Findo, Ovidiu Ionescu, Petr Koubek, and Laszlo Szemethy. N.d. "Status of Large Carnivores (Brown Bear, Wolf, Lynx, Otter) in the Carpathians." (Draft/private.)

Oza, G. M. 1974. "Conservation of the Asiatic Lion: Now Limited to Gujarat State, India." *Biological Conservation*, vol. 6, no. 3.

Pacepa, Lt. Gen. Ion Mihai. 1987. *Red Horizons: The True Story of Nicolae and Elena Ceausescu's Crimes, Lifestyle, and Corruption*. Washington, D.C.: Regnery Gateway.

Packer, Craig, and Jean Clottes. 2000. "When Lions Ruled France." *Natural History*, vol. 109, November 2000.

Paine, Robert T. 1963. "Trophic Relationships of Sympatric Predatory Gastropods." *Ecology*, vol. 44, no. 1.

———. 1966. "Food Web Complexity and Species Diversity." *American Naturalist*, vol. 100, no. 910.

———. 1969. "A Note on Trophic Complexity and Community Stability." *American Naturalist*, vol. 103, no. 929.

Pakula, Hannah. 1984. *The Last Romantic: A Biography of Queen Marie of Roumania*. New York: Simon and Schuster.

Pandav, Bivash. 1996. "Birds of the Bhitarkanika Mangroves, Eastern India." *Forktail*, vol. 12.

Pandav, Bivash, and Binod C. Choudhury. 1996. "Diurnal and Seasonal Activity Patterns of Water Monitor (*Varanus salvator*) in the Bhitarkanika Mangroves, Orissa, India." *Hamadryad*, vol. 21.

Panwar, H. S. 1995. "Management of Wildlife Habitats in India." In *The Development of International Principles and Practices of Wildlife Research and Management: Asian and American Approaches*, ed. Stephen H. Berwick and V. B. Saharia. Delhi: Oxford University Press.

Patterson, Lt. Col. J. H. 1996. *The Man-Eaters of Tsavo*. New York: Pocket Books.

Penny, Malcolm. 1991. *Alligators and Crocodiles*. New York: Crescent Books.

Pericot-Garcia, Luis, John Galloway, and Andreas Lommel. 1967. *Prehistoric and Primitive Art*. New York: Harry N. Abrams.

Perry, Richard. 1965. *The World of the Tiger*. New York: Atheneum.

Peterson, Nicolas, and Marcia Langton. 1983. *Aborigines, Land and Land Rights*. Canberra: Australian Institute of Aboriginal Studies.

Pfeffer, Pierre, ed. 1989. *Predators and Predation: The Struggle for Life in the Animal World*. New York: Facts on File.

Philip, Neil. 1999. *Myths and Legends*. New York: DK Publishing.

Pikunov, D. G. 1988a. "Amur Tiger (*Panthera Tigris Altaika*): Present Situation and Perspectives for Preservation of Its Population in the Soviet Far East." In Dresser et al. (1988).

———. 1988b. "Eating Habits of the Amur Tiger (*Panthera Tigris Altaika*) in the Wild." In Dresser et al. (1988).

Pliny the Elder. 1991. *Natural History: A Selection*, trans. with intro. and notes by John F. Healey. New York: Penguin Books.

Plumwood, Val. 1993. *Feminism and the Mastery of Nature*. London: Routledge.

———. 1996. "Being Prey." *Terra Nova*, vol. 1, no. 3.

Plutarch. 1999. *Roman Lives: A Selection of Eight Roman Lives*, trans. Robin Waterfield; intro. and notes by Philip A. Stadter. Oxford: Oxford University Press.

Pocock, R. I. 1930. "The Lions of Asia." *Journal of the Bombay Natural History Society*, vol. 34.

———. 1936. "The Lion in Baluchistan." *Journal of the Bombay Natural History Society*, vol. 38.

Pope, Clifford H. 1962. *The Giant Snakes: The Natural History of the Boa Constrictor, the Anaconda, and the Largest Pythons*. London: Routledge & Kegan Paul.

Powell, Alan. 1996. *Far Country: A Short History of the Northern Territory*. Carlton, Victoria, Aus.: Melbourne University Press.

Power, Mary E., David Tilman, James A. Estes, Bruce A. Menge, William J. Bond, L. Scott Mills, Gretchen Daily, Juan Carlos Castilla, Jane Lubchenco, and Robert T. Paine. 1996. "Challenges in the Quest for Keystones." *BioScience*, vol. 46, no. 8.

Prater, S. H. 1980. *The Book of Indian Animals*. Bombay: Bombay Natural History Society and Oxford University Press.

Press, Tony, David Lea, Ann Webb, and Alistair Graham, eds. 1995. *Kakadu: Natural and Cultural Heritage and Management*. Darwin, Northern Territory, Aus.: Australia Nature Conservation Agency.

Quigley, Howard B. 1993. "Saving Siberia's Tigers." *National Geographic*. July 1993.

Quirk, Susan, and Michael Archer, eds. N.d. *Prehistoric Animals of Australia*. Drawings by Peter Shouten. Sydney: Australian Museum.

Rashid, M. A. 1978. "The Gir Wild Life Sanctuary and National Park in Gujarat State." Gujarat Forest Department, Gujarat, India.

———. 1983. "The Gir (Asiatic) Lion." *Cheetal*, vol. 24, no. 3.

Rashid, M. A., and Reuben David. 1992. *The Asiatic Lion*. Department of Environment, Government of India. Baroda, India: Vishal Offset.

Raval, Shishir R. 1991. "The Gir National Park and the Maldharis: Beyond 'Setting Aside.'" In *Resident Peoples and National Parks*, ed. Patrick C. West and Steven R. Brechin. Tucson: University of Arizona Press.

———. 1994. "Wheel of Life: Perceptions and Concerns of the Resident Peoples for Gir National Park in India." *Society and Natural Resources*, vol. 7.

———. 1997. "Perceptions of Resource Management and Landscape Quality of the Gir Wildlife Sanctuary and National Park in India." Ph.D. dissertation, University of Michigan.

Reed, A. W. 1978. *Aboriginal Legends: Animal Tales*. Kew, Victoria, Aus.: Reed Books.

Reynolds, Henry. 1992. *The Law of the Land*. Ringwood, Victoria, Aus.: Penguin Books.

Reynolds, John. 1974. *Men and Mines: A History of Australian Mining 1788–1971*. Melbourne: Sun Books.

Ridley, Mark. 1993. *Evolution*. Boston: Blackwell Scientific Publications.

Roelke-Parker, Melody, Linda Munson, Craig Packer, Richard Kock, Sarah Cleaveland, Margaret Carpenter, Stephen J. O'Brien, Andreas Pospischil, Regina Hofmann-Lehmann, Hans Lutz, George L. M. Mwamengele, M. N. Mgasa, G. A. Machange, Brian A. Summers, and Max J. G. Appel. 1996. "A Canine Distemper Virus Epidemic in Serengeti Lions (*Panthera leo*)." *Nature*, vol. 379, February 1, 1996.

Rose, Carol. 2000. *Giants, Monsters, and Dragons: An Encyclopedia of Folklore, Legend, and Myth*. New York: W. W. Norton.

Rose, Kenneth. 1969. *Superior Person: A Portrait of Curzon and His Circle in Late Victorian England*. New York: Weybright and Talley.

Ross, Charles A., ed. 1989. *Crocodiles and Alligators*. Silverwater, New South Wales, Aus.: Golden Press.

Saberwal, Vasant. 1990. "Lion-Human Conflict Around Gir." *WII Newsletter*, vol. 5, no. 3.

Saberwal, Vasant K., James P. Gibbs, Ravi Chellam, and A. J. T. Johnsingh. 1994. "Lion-Human Conflict in the Gir Forest, India." *Conservation Biology*, vol. 8, no. 2.

Saharia, V. B. 1982. *Wildlife in India*. Dehra Dun: Natraj Publishers.

Saitoti, Tepilit Ole, and Carol Beckwith. *Maasai*. New York: Abradale Press/Harry N. Abrams.

Salvatori, V., O. Ionescu, H. Okarma, and S. Findo. N.d. "The Hunting Legislation in the Carpathian Mountains: Implications for the Conservation and Management of Large Carnivores." (Draft/private.)

Sandars, N. K., ed. and trans. 1972. *The Epic of Gilgamesh*. Harmondsworth, Middlesex, Eng.: Penguin Books.

Savage, R. J. G., and M. R. Long. 1986. *Mammal Evolution: An Illustrated Guide*. New York: Facts on File.

Savill, Sheila. 1977. *Pears Encyclopedia of Myths and Legends: Western and Northern Europe, Central and Southern Africa*. London: Pelham Books.

Savill, Sheila, and Elizabeth Locke. 1976. *Pears Encyclopedia of Myths and Legends: Ancient Near and Middle East, Ancient Greece and Rome*. London: Pelham Books.

Schaller, George B. 1967. *The Deer and the Tiger: A Study of Wildlife in India*. Chicago: University of Chicago Press.

———. 1972. *The Serengeti Lion: A Study of Predator-Prey Relations*. Chicago: University of Chicago Press.

———. 1993. *The Last Panda*. Chicago: University of Chicago Press.

Scholander, P. F. 1955. "Evolution of Climatic Adaptation in Homeotherms." *Evolution*, vol. 9.

Schullery, Paul. 1992. *The Bears of Yellowstone*. Worland, Wyo.: High Plains Publishing Company.

Scott, Jonathan. 1995. *Kingdom of Lions*. London: Kyle Cathie.

Seidensticker, John, and Charles McDougal. 1993. "Tiger Predatory Behaviour, Ecology and Conservation." In Dunstone and Gorman (1993).

Seidensticker, John, Sarah Christie, and Peter Jackson, eds. 1999. *Riding the*

Tiger: Tiger Conservation in Human-Dominated Landscapes. Cambridge: Cambridge University Press.

Servheen, Christopher, Stephen Herrero, and Bernard Peyton. 1999. *Bears: Status Survey and Conservation Action Plan*. Gland, Switz.: IUCN.

Sharma, Arpan. 1998. "Shifting Home: New Horizons on the Anvil for the Asiatic Lion." *Sanctuary Asia*, vol. 18, no. 5.

Sharma, Diwakar, and A. J. T. Johnsingh. 1996. "Impacts of Management Practices on Lion and Ungulate Habitats in Gir Protected Area." Dehra Dun, India: Wildlife Institute of India.

Shaw, Harley. 1989. *Soul Among Lions: The Cougar as Peaceful Adversary*. Boulder, Colo.: Johnson Books.

Shepard, Paul. 1996. *The Others: How Animals Made Us Human*. Washington: Island Press/Shearwater Books.

———. 1998. *The Tender Carnivore and the Sacred Game*. Foreword by George Sessions. Athens, Ga.: University of Georgia Press.

Shepard, Paul and Barry Sanders. 1985. *The Sacred Paw: The Bear in Nature, Myth, and Literature*. New York: Viking.

Simberloff, Daniel. 1997. "Flagships, Umbrellas, and Keystones: Is Single-Species Management Passé in the Landscape Era?" *Biological Conservation*, vol. 83, no. 3.

Sinha, S. K. 1975. "Distribution of the Indian Lion over Centuries." *Cheetal*, vol. 17, no. 1.

Singh, H. S., and R. D. Kamboj. 1996. *Biodiversity Conservation Plan for Gir*, vol. 1. Junagadh, India: Gujarat Forest Department.

Soulé, Michael E., James A. Estes, Joel Berger, and Carlos Martinez del Rio. In press. "Recovery Goals for Ecologically Effective Numbers of Endangered Keystone Species."

Soulé, Michael E., and John Terborgh, eds. 1999. *Continental Conservation: Scientific Foundations of Regional Reserve Networks*. Washington, D.C.: Island Press.

Soulé, Michael E., and Bruce A. Wilcox, eds. 1980. *Conservation Biology: An Evolutionary-Ecological Perspective*. Sunderland, Mass.: Sinauer Associates.

Spear, Percival. 1990. *A History of India*. Vol. 2. New York: Penguin Books.

Srivastav, Asheem, and Suvira Srivastav. 1999. *Asiatic Lion: On the Brink*. Dehra Dun, India: Bishen Singh Mahendra Pal Singh.

Stephan, John J. 1994. *The Russian Far East: A History*. Stanford, Calif.: Stanford University Press.

Stevens, John D., ed. 1999. *Sharks*. New York: Facts on File.

Stern, Horst. 1993. *The Last Hunt*, trans. Deborah Lucas Schneider. New York: Random House.

Stirrat, Simon, David Lawson, and W. J. Freeland. 2000. "Detecting and Responding to Changes in Numbers: The Future of Monitoring *Crocodylus porosus* in the Northern Territory of Australia." Draft paper presented to a meeting of the Crocodile Specialist Group of the IUCN, Cuba, January 2000.

Stokes, J. Lort. 1846. *Discoveries in Australia; with an Account of the Coasts and Rivers Explored and Surveyed during the Voyage of H.M S. Beagle, in the Years 1837–38–39–40–41–42–43*. London: T. and W. Boone.

Stroganov, S. U. 1969. *Carnivorous Mammals of Siberia*, trans. A. Birron. Jerusalem: Israel Program for Scientific Translations.

Subramaniam, Kamala. N.d. *Srimad Bhagavatam*. Bombay: Bharatiya Vidya Bhavan.

Tambs-Lyche, Harald. 1997. *Power, Profit and Poetry: Traditional Society in Kathiawar, Western India*. New Delhi: Manohar.

Tammita-Delgoda, Sinharaja. 1995. *A Traveller's History of India*. New York: Interlink Books.

Taylor, Robert J. 1984. *Predation*. New York: Chapman and Hall.

Terborgh, John. 1974. "Preservation of Natural Diversity: The Problem of Extinction Prone Species." *BioScience*, vol. 24, no. 12.

———. 1988. "The Big Things That Run the World—A Sequel to E. O. Wilson." *Conservation Biology*, vol. 2, no. 4.

Terborgh, John, James A. Estes, Paul Paquet, Katherine Ralls, Diane Boyd-Heger, Brian J. Miller, and Reed F. Noss. 1999. "The Role of Top Carnivores in Regulating Terrestrial Ecosystems." In Soulé and Terborgh (1999).

Terborgh, John, Lawrence Lopez, Percy Nuñez V., Madhu Rao, Ghazala Shahabuddin, Gabriela Orihuela, Mailen Riveros, Rafael Ascanio, Greg H. Adler, Thomas D. Lambert, and Luis Balbas. 2001. "Ecological Meltdown in Predator-Free Forest Fragments." *Science*, vol. 294. November 30, 2001.

Terborgh, John, Lawrence Lopez, José Tello, Douglas Yu, and Ana Rita Bruni. 1997. "Transitory States in Relaxing Ecosystems of Land Bridge Islands." In Laurance and Bierregaard (1997).

Terborgh, John, and Blair Winter. 1980. "Some Causes of Extinction." In Soulé and Wilcox (1980).

Thapar, Valmik. 1997. *Land of the Tiger: A Natural History of the Indian Subcontinent*. Berkeley: University of California Press.

Tharoor, Shashi. 1997. *India: From Midnight to the Millennium*. New York: Arcade Publishing.

Thomas, E. Donnall, Jr. 2001. "Predators: Montanans Confront the Politics of Fang and Claw." *Big Sky Journal*, vol. 8, no. 6.

Thompson, C. J. S. 1968. *The Mystery and Lore of Monsters*. New York: Bell Publishing Company.

Thomson, Donald F. 1949. *Economic Structure and the Ceremonial Exchange Cycle in Arnhem Land*. Melbourne: Macmillan.

Thorbjarnanarson, John. 1992. *Crocodiles: An Action Plan for Their Conservation*. ed. Harry Messel, F. Wayne King, and James Perran Ross. Gland, Switz.: IUCN.

———. 1999. "Crocodile Tears and Skins: International Trade, Economic Constraints, and Limits to the Sustainable Use of Crocodilians." *Conservation Biology*, vol. 13, no. 3.

Tilson, Ronald L., and Ulysses S. Seal, eds. 1987. *Tigers of the World: The Biology, Biopolitics, Management, and Conservation of an Endangered Species*. Park Ridge, N.J.: Noyes Publications.

Todd, Neil B. 1965. "Metrical and Non-metrical Variation in the Skulls of Gir Lions." *Journal of the Bombay Natural History Society*, vol. 62.

Tolkien, J. R. R. 1936. *Beowulf: The Monsters and the Critics*. Sir Israel Gollancz Memorial Lecture, from the Proceedings of the British Academy, Vol. 22. London: Oxford University Press.

Turner, Alan. 1997. *The Big Cats and Their Fossil Relatives*, illus. Mauricio Antón. New York: Columbia University Press.

Underwood, Lamar, ed. 2000. *Man Eaters: True Tales of Animals Stalking, Mauling, Killing, and Eating Human Prey*. New York: Lyons Press.

Van Valen, Leigh. 1969. "Evolution of Dental Growth and Adaptation in Mammalian Carnivores." *Evolution*, vol. 23.

Van Valkenburgh, Blaire. 1988. "Incidence of Tooth Breakage Among Large, Predatory Mammals." *American Naturalist*, vol. 131, no. 2.

———. 1989. "Carnivore Dental Adaptations and Diet: A Study of Trophic Diversity within Guilds." In Gittleman (1989).

Van Valkenburgh, Blaire, and Fritz Hertel. 1993. "Tough Times at La Brea: Tooth Breakage in Large Carnivores of the Late Pleistocene." *Science*, vol. 261, July 23, 1993.

Van Valkenburgh, B., and C. B. Ruff. 1987. "Canine Tooth Strength and Killing Behaviour in Large Carnivores." *Journal of Zoology* (London), vol. 212.

Vereshchagin, N. K., and G. F. Baryshnikov. 1984. "Quaternary Mammalian Extinctions in Northern Eurasia." In Martin and Klein (1984).

Warner, W. Lloyd. 1958. *A Black Civilization: A Social Study of an Australian Tribe*. New York: Harper and Brothers.

Watson, Helen, with the Yolngu community at Yirrkala, and David Wade Chambers. 1989. *Singing the Land, Signing the Land: A Portfolio of Exhibits*. Geelong, Victoria: Deakin University.

Webb, Grahame. 1982. "A Look at the Freshwater Crocodile." *Australian Natural History*, vol. 20, no. 2.

————. 1991a. "'Wise Use' of Wildlife." *Journal of Natural History*, vol. 25.

————. 1991b. "The Influence of Season on Australian Crocodiles." In *Monsoonal Australia: Landscape, Ecology and Man in the Northern Lowlands*, ed. C. D. Haynes, M. G. Ridpath, and M. A. J. Williams. Rotterdam: A. A. Balkema.

————. 1998. *Numunwari*. Chipping Norton, New South Wales, Aus.: Surrey Beatty & Sons.

Webb, Grahame J. W., and S. Charlie Manolis. 1991. "Monitoring Saltwater Crocodiles (*Crocodylus porosus*) in the Northern Territory of Australia." In McCulloch and Barrett (1991).

Webb, Grahame, and Charlie Manolis. 1989. *Crocodiles of Australia*. Frenchs Forest, New South Wales, Aus.: Reed Books.

Webb, Grahame J. W., S. Charlie Manolis, and Peter J. Whitehead, eds. 1987. *Wildlife Management: Crocodiles and Alligators*. Chipping Norton, New South Wales, Aus.: Surrey Beatty & Sons.

Webb, Grahame J. W., S. Charlie Manolis, and Brett Ottley. 1994. "Crocodile Management and Research in the Northern Territory: 1992–94." Paper presented at the 12th Working Meeting of the IUCN-SSC Crocodile Specialist Group, Pattaya, Thailand, May 2–6, 1994.

Webb, Grahame, Charles Missi, and Miriam Cleary. 1996. "Sustainable Use of Crocodiles by Aboriginal People in the Northern Territory." In Bomford and Caughley (1996).

Webb, Grahame J. W., Brett Ottley, Adam R. C. Britton, and S. Charlie Manolis. 2000. "Recovery of Saltwater Crocodiles (*Crocodylus porosus*) in the Northern Territory: 1971–1998." A report to the Parks and Wildlife Commission of the Northern Territory. Sanderson, Northern Territory, Aus.: Wildlife Management International.

Wells, Ann E. 1971. *This Their Dreaming: Legends of the Panels of Aboriginal*

Art in the Yirrkala Church. St. Lucia, Queensland, Aus.: University of Queensland Press.

Wells, Edgar. 1982. *Reward and Punishment in Arnhem Land* 1962–1963. Canberra: Australian Institute of Aboriginal Studies.

Wessing, Robert. 1986. *The Soul of Ambiguity: The Tiger in Southeast Asia.* Special Report, no. 24, Monograph Series on Southeast Asia. Center for Southeast Asian Studies, Northern Illinois University.

Westphal-Hollbusch, S. 1974. *Hinduistische Viehzüchter im Nordwestlichen Indien.* Vol. 1: *Die Rabari.* Berlin: Duncker und Humboldt.

Whitaker, Romulus. 1987. "The Management of Crocodilians in India." In Webb, Manolis, and Whitehead (1987).

Wilberforce-Bell, H. 1980. *The History of Kathiawad from the Earliest Times.* (Reprint of the 1916 edition, London: Heinemann.) New Delhi: Ajay Book Service.

Wildt, D. E., M. Bush, K. L. Goodrowe, C. Packer, A. E. Pusey, J. L. Brown, P. Joslin, and S. J. O'Brien. 1987. "Reproductive and Genetic Consequences of Founding Isolated Lion Populations." *Nature,* vol. 329, September 24, 1987.

Williams, Nancy. 1986. *The Yolngu and Their Land: A System of Land Tenure and the Fight for Its Recognition.* Stanford, Calif.: Stanford University Press.

————. 1998. *Intellectual Property and Aboriginal Environmental Knowledge.* Darwin, Northern Territory, Aus.: Centre for Indigenous Natural and Cultural Resource Management.

Willis, Roy, ed. 1994. *Signifying Animals: Human Meaning in the Natural World.* London: Routledge.

Woodroffe, Rosie. 2000. "Predators and People: Using Human Densities to Interpret Declines of Large Carnivores." *Animal Conservation,* vol. 3.

————. 2001. "Strategies for Carnivore Conservation: Lessons from Contemporary Extinctions." In Gittleman et al. (2001).

Woodroffe, Rosie, and Joshua R. Ginsburg. 1998. "Edge Effects and the Extinction of Populations Inside Protected Areas." *Science,* vol. 280, June 26, 1998.

Wright, David, ed. and trans. 1957. *Beowulf: A Prose Translation.* New York: Penguin Books.

Wynter-Blyth, M. A. 1949. "The Gir Forest and Its Lions, Part I." *Journal of the Bombay Natural History Society,* vol. 48.

————. 1950. "The History of the Lion in Junagadh State 1880 to 1936." *Journal of the Bombay Natural History Society*, vol. 49, no. 3.

————. 1956. "The Lion Census of 1955." *Journal of the Bombay Natural History Society*, vol. 53.

Wynter-Blyth, M. A., and Kumar Shree Dharmakumarsinhji. 1950. "The Gir Forest and Its Lions, Part II." *Journal of the Bombay Natural History Society*, vol. 49, no. 3.

Yirrkala Artists. 1999. *Saltwater: Yirrkala Bark Paintings of the Sea Country*. Neutral Bay, New South Wales, Aus.: Buku-Larrngay Mulka Centre, in association with Jennifer Isaacs Publishing.

Yunupingu, Galarrwuy, ed. 1997. *Our Land Is Our Life: Land Rights—Past, Present and Future*. St. Lucia, Queensland, Aus.: University of Queensland Press.

ACKNOWLEDGMENTS

Before mentioning individuals, I'll note here the depth of my gratitude toward several categories of people whose voices I've listened to, whose lands I've intruded upon, and whose cultures, beliefs, and living histories I've tried to draw from: the Maldharis of Gir; the Yolngu of eastern Arnhem Land; the Barrada, Kunwinku, and Rembaranga of northern Arnhem Land; the villagers of Dangmal and Khamari Sahi in Orissa, India; the shepherds of the Romanian Carpathians; and the Udege of the Bikin River valley in the Russian Far East. These folk, with their collective traditions and all their person-by-person individuality, are the primary human protagonists of my book. I thank them for what they've given, and I honor what they know and have endured. Also, though not mentioned in the book, the Nez Perce of the northern Rocky Mountains (mainly in that portion of their ancestral lands lying within what we now call Idaho) played an important part in my exploration of this subject.

Several institutions indulged, welcomed, aided, or at least tolerated my inquiries and visits. Warm thanks to them all, and notably to this partial list: the Wildlife Institute of India, the Bombay Natural History Society, the United States Information Service, the Taj Group of Hotels, the Gujarat Forest Department, the Orissa Forest Department, the Dhimurru Land Management Aboriginal Corporation, the Bawin-anga Aboriginal Corporation, the Northern Land Council, the Yothu Yindi Foundation, the Carpathian Large Carnivore Project, the Romanian Forest Department, the Hornocker Wildlife Institute, the Institute for Sustainable Use of Natural Resources (of the Russian Far East), the Sikhote-Alin Zapovednik, the Wildlife Conservation Society, and the Mpala Research Centre.

My large debt to Ravi Chellam has already been noted. In addition to him (and his wife Bhooma, his daughter Roshni), a number of scientists and other individuals were especially generous to me with their time, their trust, their influence, and their hospitality. These people assisted me variously—with information, by accepting my presence in field situations while trying to do their own work, by arranging access to situations or people; also, in some cases, by opening their minds, their notebooks, even their homes to me. I'll list them geographically. In India: Ismail Bapu and his family, Divyabhanusinh Chavda, and Bivash Pandav. In Australia: Kelvin Leitch and his family (Jacky, Rosanna, and Tully), Ray Hall, Andrew Cappo, Michael Christie, Nancy Fitzsimmons, Peter Cooke, and David Bowman. In Romania: Annette Mertens, Christoph Promberger, Barbara Promberger-Fuerpass, Andrei Blumer, Ovidiu Ionescu and his family, Viciu Buceloiu and his family, Avram Sandor, Valeria Salvatori, Yves Gasser, Peter Sürth, Gigi Popa and his family, Justine Evans and Nick Turner and Dominic Partridge of the BBC, Ion Micu and his family, Parinte Ignatiu and the other monks of the Slănic Monastery, and Petra Buda and his wife Iolanda. In Russia: Dale Miquelle and his wife Marina, John Goodrich, Dima Pikunov and his wife Olga, Evgeny Smirnov and his family, and Kolya Reebin and his family. In Africa: Laurence Frank, David Western, and Mike Fay. In the United States: Steve Berwick, B. J. Champion, Howard Quigley, Doug Peacock, Andrea Peacock, Steve Primm, Toni Ruth, and E. Donnall Thomas, Jr., and his wife Laurie.

Another form of crucial help came from the good people who

served me as interpreters, field companions and fixers, translators of printed sources: Misha Jones, Ciprian Pavel, Andrei Blumer again, Christi Raab (and his wife Laura), Eduard Érsek, Gigi Bîcleseanu, Nanikiya Munungurritj, and Mandaka Marika. Michael Christie and his colleague Waymamba Gaykamangu supplied translations of some traditional Yolngu stories about Bäru, the Ancestral Crocodile. Jerry Coffey helped me with certain passages of the Old English of *Beowulf*. My faithful transcriber Gloria Thiede performed invaluable labors, again, transcribing interview tapes that in many cases came complicated by unfamiliar accents, and sometimes also by the din of screeching parakeets or the wheezy groan of bad hotel air-conditioners in the background. Chuck West kept me square with the higher powers of the IRS.

Many others gave me assistance and cooperation of various sorts. Some of those helpful roles (such as Jackie Adjarral's) are described in my narrative. Without mentioning every living character again, I thank them all, and in addition the following people, whose contributions went well beyond what I've narrated, or who don't appear by name in the text. Again I'll list them by country. In India: Bittu Sahgal, Asad R. Rahmani, A. J. T. Johnsingh, J. C. Daniel, Mohammad Juma, Nagendra Singh, Dhanna Lakshman, Mahesh Singh, Manoj Dholakia, Ross Mallick, Rom Whitaker, Manoj Mahapatra, Ashley Wills, K. K. Anand, Anne O'Leary, Miriam Caravella, R. K. Vishwanathan, Deepak Mehta, Steve Schermerhorn, D. Subbaram, Terry White, Karuna Singh, Nishith Dharaiya, Ramesh Sabapara, Bhabani Pati, an intelligent young kitchen manager named Saurabh (last name, regrettably, unkown to me) who helped with key information at a bizarre moment, and a tracker named Dumal (again, other name lost) who had survived a cobra bite on the nose to become a great guide in the Bhitarkanika ecosystem. In Australia: Nancy Williams, Grahame Webb, Djalalingba Yunupingu, Galarrwuy Yunupingu, Mandawuy Yunupingu, Charlie Godjuwa, Simon Stirrat, Adam

Britton, Ian Munro, Jimmy Njimanjima, Willie Djorbar, Ellen Jordan, Andrew Blake, Howard and Frances Morphy, Marcia Langton, Ann Palmer, Tom Nichols, Neville Haskins, Rod Kennett, Peter Whitehead, Anna Johnson, Leon Morris, Ursula Zaar, Rae Flannigan, Nik Robinson, and Brett Ottley. In Romania: George Predoiu, Mosorel Surdu and his family, Aurica and Gheorghe Surdu, Vaso Boromin, Marius Scurtu, Victor Mureşan and his family, Alistair Bath, Laszlo Kedves, Mircea Verghelet, Tamás Sandy, Ioan Jantea, Tiberiu Serban, Nicu Buceloiu, Ruxanda Sandu, Liviu Fulga, Marioara and Traian Trif, Radu Sonea, and Ion Petrica. In Russia: Vladimir Aramilev, Vladimir Vasileov, Anatoly Astafiev, Anatoly Khobatnov and his wife Luba, Slava Reebin and his wife Nina, and again a generous British television crew led by Nigel Marvin. In the United States: Mark Bryant, Larry Burke, Hal Espen, Wolf Schroder (on a visit, from the Munich Wildlife Society), Liz Claiborne, Art Ortenburg, Jim Murtaugh, Adrienne Mayor, Josh Ober, Michael Soulé, Chris Servheen, Mike Gilpin, Yvonne Baskin, Bob Rydell, Robert Vare, Marianne Berwick, Vasant Saberwal, Michael Llewellyn, Gordon Wiltsie, Nick Nichols, Scott Bowen, Louisa Wilcox, Lance Craighead, Rosie Woodroffe, Alan Rabinowitz, Josh Ginsberg, Maurice Hornocker, Barrie Gilbert, Jim Halfpenny, Stephen Kellert, Richard Taber, Dave Mattson, Chuck Schwartz, Barrett Golding, Keith Lawrence, Roy Sampsel, Aaron Miles, Silas Whitman, Elmer Crow, Kurt Mack, Glen Contreras, and Brian Horesji. In Kenya: Heather Wallington, Mordecai Ogada, Steven Ekwanga, Aaron Wagner, Ken Wreford-Smith, Tom and Peter Sylvester, and Colin and Rocky Francombe. In addition to these, many other people eased my way with helpful information, good company, and hospitality. I thank them all and apologize for names omitted.

Some of my expert sources and cultural intermediaries performed the additional service of reading sections of the book in draft and offering me corrections, clarifications, further insights. Though blameless for the final results, of course, they are: Ravi Chellam (still again), Bivash Pandav, Steve Berwick, Howard Quigley, Grahame Webb, Andrew Cappo, Ray Hall, Kelvin Leitch, Michael Christie, Peter Cooke, Nancy Williams, Steve Roeger, Annette Mertens, Ovidiu Ionescu, Dale Miquelle, John Goodrich, Adrienne Mayor, Catherine Badgley, Blaire Van Valkenburgh, Michael Soulé, John McCosker, Jane Lubchenco, and Jean Clottes.

The fierce and loving Renée Wayne Golden has been a devoted advocate for this book, as for all my others, since its conception. Maria Guarnaschelli, again, has been my crucial editorial partner during the long process of turning an idea into a book; her patient support and her percipient sense of flow and structure have been invaluable. Katya Rice, again, gave me a smart, keen-eyed copyedit. Erik Johnson provided valuable help with permissions and many other tasks. Justin Morrill and his atelier, The M Factory, deftly met my specifications for the maps. W. W. Norton is a new home to me, but at all levels and in all departments I've found genial and expert colleagues. Susan Moldow, as publisher of Scribner, originally welcomed the project to that house; the excellent treatment, by Scribner, of my three preceding books helped put my exploration of predators on course. Kris Ellingsen endured and aided the early stages of research, and my debt to her can't be plumbed here. Will and Mary Quammen, Kay Hopper, Sallie Quammen, Charlie Fazio, and the rest of my extended family represent the human matrix that becomes only more important with time and distant travel. Betsy Gaines, with her wit and golden smile, has offered vital support through the final stages of work and, even more vitally, given me the opportunity to regrow a life. I thank her here, and I will be thanking her daily amid the raucous, joyous household of creatures (Buddy, Wiley, Skipper) whose custodianship we share.

INDEX